QuickSmart

INTRODUCTORY MATHEMATICS

C. Coady & J. Gosling

**PASCAL
PRESS**

Contents

Preface

Since this text is intended to 'bridge any gap' that exists between a prospective student's current level of mathematical skills and knowledge and that which is assumed by the discipline about to be entered, the writing style is, hopefully, 'easy-to-read' and 'easy-to-understand', without sacrificing any credibility. While the approach is necessarily mathematical it is not rigorous, but rather is designed to instil both an appreciation as well as a knowledge of a number of different mathematical topics.

Included are many worked examples designed to guide the reader to a better understanding of mathematical concepts, skills and applications.

To the reader

We believe that you will find this text a readable and fairly easy to understand treatment of a wide range of mathematical topics. Our intention is to explain carefully and logically the various mathematical concepts and techniques without becoming 'bogged down' in unnecessary rigour and proofs.

We have included many examples within each chapter to allow you to see how each procedure or technique may be applied. Also, since the statement 'practice makes perfect' is very true when attempting to master mathematics, you will find a large number of exercises and problems (with solutions) to help you along the road to success.

Finally, as calculators are now a very necessary tool when solving mathematical problems, those readers without calculator expertise are strongly advised to refer to Appendix E, **Can you use your calculator?**, before beginning chapter 1.

A special request

Despite careful scrutiny during the proof reading of this book, it is probable that some errors will remain both in the text and in the appendices. You are encouraged to notify the authors or the publishers of any errors you detect or of any comments or suggestions you may have as to how the text may be improved.

Acknowledgements

There are a number of people to whom we must extend our thanks. First and foremost are all our colleagues at U.W.S. Nepean (too numerous to name), who responded to our initial plea for help as to the relevant material that should be included in a text of this nature. Without their advice and encouragement, this book would never have progressed beyond the drawing board.

Secondly, to Lorraine Starr and Sharon Bolton a very big thanks for the many hours spent typing up the first draft of much of the manuscript. Without their assistance and hard work, the overall preparation process would have been greatly extended.

To Elien Visscher we extend our very special thanks for all her work and assistance in the preparation of this text, especially from its early hand-written stages. Her patience and dedication are very much appreciated.

To Eunice Roberts, our reader of the 'core' chapters, we say many thanks for her thoroughness of approach, as well as for her resulting comments and suggestions. We are extremely grateful.

To our families, both the two-legged and the four-legged members, MANY, MANY THANKS! Your constant encouragement, support and understanding during the many months of preparation have been truly appreciated - we could not have done it without you!

Finally, our sincere thanks and appreciation to Aidan and his team, Rhonda at Computer Applications and Training (C.A.T.) who have spent several months preparing this text and associated illustrations for publication. Without their hard work, experience and knowledge, this book would not appear as it does today, as a truly professional manuscript both in layout and in content.

Carmel Coady and *Jenny Gosling*

Introduction

As this book is entitled Introductory Mathematics, an appropriate place to begin is with the term 'mathematics'.

What is mathematics?

In a very broad sense, mathematics involves the study of patterns and their relationships, and hence provides precise and efficient methods of representing and describing those patterns and relationships.

These mathematical representations can then be manipulated according to specific rules of logic either to achieve a particular result or to explore further other possible relationships.

In fact, mathematical investigation may be thought of as a creative activity as it involves not only exploration but also invention and intuition. Sometimes, the problems being investigated are actually generated from within the field of mathematics and hence relate directly to mathematical abstractions. At other times, the problems are generated from the 'real' world and hence are embedded in their own particular context. A solution to these types of problems lies in the ability to represent mathematically the main relationships within the context and to re-state the problem in mathematical terms so that the appropriate mathematical techniques may be applied. The final step would then involve the interpretation of the mathematical solution obtained, within the context of the original 'real' world problem.

One further point to note is that solutions to the two different types of problems may actually influence each other. For example, the search for a solution to a real-life problem will often lead to the development of some new mathematical theory.

The preceding discussion gives a very broad and rather abstract response to *'what is mathematics?'*, pointing out that the solution to most practical problems requires either tl e use of readily available mathematical techniques or the development of new theories.

Now, perhaps a question even more relevant to students on the threshold of tertiary studies in some pre-determined discipline is

'Why is mathematics important?'

Whether we like it or not, mathematics at some level is an integral part of our daily lives. Apart from enhancing our understanding of our world and the quality of our participation within our own society, it is extremely important to people, both individually and as a group, as it provides valuable tools that can be used at all levels - personal, social and vocational.

In our everyday lives, the care of our home and family requires the use of mathematics involving counting, measuring and performing simple calculations.

As far as social and civic life is concerned, even making sense of the evening news often requires some mathematical understanding of data collection and handling. In fact, making any sort of informed choice of political preference, banking preference, which paint to buy to paint the house, which clothes drier to purchase etc., requires a general understanding of the underlying mathematics.

In the workplace, mathematics is widely used by everyone, checking daily profits, budgeting, organising payrolls, inventory counting, ordering stock, banking etc.

Although the physical sciences, engineering and computer science have traditionally been regarded as requiring a high level of mathematics, fields such as economics, geography, management, biology, fashion design and even art increasingly use mathematical techniques. In fact, mathematics underlies most social and economic planning as well as all facets of industry, commerce and trade. On a more individual level, it is becoming increasingly clear that fully informed participation in all areas of our lives requires at least some sound knowledge of mathematics.

Having now discussed the 'what' and 'why' of mathematics from a very generalised perspective, we now turn to

The 'what, why and for whom' of this text

Today there is considerable evidence to suggest that many people either dislike or feel intimidated by mathematics. Hence the effect of having to deal with either a real or perceived lack of assumed mathematical knowledge often has a profound bearing on

(i) whether a person decides to undertake tertiary studies in the first place, or

(ii) their overall performance in some chosen field of study at a tertiary institution.

The two basic aims of this text are to provide prospective tertiary students with a means of

(i) 'brushing-up' on their mathematical skills in order to proceed successfully with their studies, and

(ii) improving their self-confidence in mathematics through understanding and practice.

Why did we undertake to write this text?

Again there are two reasons - first, we felt that a text of this nature would be a very valuable addition to the 'mathematics' shelf of any bookshop as it should be most helpful to a number of prospective tertiary students. Secondly, from our own years of experience with tertiary students, there are many who undertake studies in a particular field when they really do not have an appropriate knowledge of the required mathematics and hence often have great difficulty completing some subjects. For these students a text such as this should prove to be extremely beneficial.

Having decided to write the text, our next step was to determine its content. To this end we consulted colleagues in each of our university's faculties, as well as some of our own mathematics and commerce students, for advice as to those mathematical topics that are 'assumed' knowledge for incoming students entering various disciplines. In fact, it is their responses that have formed the basis of our final decision as to content and layout.

Since the text is designed to target students undertaking a very broad range of subjects, clearly the assumed mathematical knowledge will vary across the different disciplines with, for example, physics students being expected to have more mathematical skills at a higher level than, say, humanities students.

To overcome this problem, we have divided the book into two parts:

Part 1: The 'core' chapters, covering those mathematical skills and techniques that are assumed knowledge for **all** students.

Part 2: The 'extra' chapters, covering those mathematical topics that are assumed knowledge only in certain specialised areas – for example, basic financial mathematics for future accountants or matrix algebra and calculus for future scientists.

As you will find, **Chapters 1** to **8** (inclusive) form the 'core' chapters and cover such topics as basic arithmetic and algebra, index numbers, equations, inequalities, an introduction to graphs and functions and a brief look at basic descriptive statistics and probability. **Chapters 9** to **14** form the 'extras', and include the more specialised chapters covering such topics as geometry, trigonometry, matrix algebra, calculus and elementary financial mathematics.

Regardless of the topic to be covered in a particular chapter, the approach taken throughout is one of mathematical logic and practical application rather than theorems and rigorous proofs. Each of the fourteen chapters begins with a brief introduction and finishes with a 'summary of events'. Between the introduction and the summary are several short sections, each dealing with a single concept or technique.

A particular section usually consists of a brief discussion, several worked examples and a number of 'Your turn' exercises, each of which has a solution in **Appendix B**.

At the very end of a chapter, after the summary, you will find a series of additional practice problems all of which, once again, have a solution located in **Appendix C**. Any exercise or problem that we feel is particularly difficult will generally be marked with an asterisk, *.

One further point to note is that in some of the 'core' chapters we have included some additional material under the heading, 'extensions'. As a general rule, these do not involve any new topics, but rather expand upon or delve more deeply into one or more of the topics discussed earlier in the chapter. While this earlier material in the chapter is assumed knowledge for **all** disciplines, the 'extensions' will be relevant only to those disciplines with a more mathematical orientation.

Finally, apart from the fourteen chapters and appendices B and C containing the solutions to all 'Your turn' exercises and practice problems, there is an additional very important 'aid' to keep in mind. It is found in **Appendix E**, entitled '**Can you use your calculator?**, and includes a comprehensive discussion of efficient calculator usage relevant to this text, that should be particularly useful to those readers with little or no calculator expertise.

Keep in mind that, because the majority of readers of this text may not be particularly mathematically oriented, all discussions are presented in as 'everyday-a-language' as possible, although some jargon and technical terminology is clearly unavoidable. Wherever possible, all points discussed are illustrated with carefully worded practical examples, and for better understanding, the 'Your turn' exercises at the end of each section should be attempted before reading further.

How to study mathematics effectively

Recently a great deal of research has been undertaken on how people learn mathematics at different stages of their lives and the overall effects of teachers, attitudes, social expectations etc.

It should be reasonably obvious that the way you *feel* about mathematics can greatly affect your ability to learn, with positive feelings enhancing learning and negative feelings providing almost certain interference.

Hence, one of the important objectives of this text is to encourage positive feelings and personal confidence so that the maximum benefit may be obtained.

The following list provides some suggestions and guidelines, already found to be useful in encouraging a positive attitude, in building confidence and in realising a sense of achievement as each section and each chapter is **completed**.

- Have paper, pencil and a scientific calculator on hand.

- Be sure that you feel comfortable with your calculator, knowing what it can and cannot do. If in doubt, read the discussion given in Appendix E, designed to introduce the uninitiated to the modern scientific calculator. Keep in mind that, in each chapter, all calculator steps required to solve a given problem are listed, where appropriate.

- Although calculators should be used wherever possible, they should be thought of as a 'tool' and used with care. Always write down the operation that is to be performed, and avoid any 'rounding' until the final answer is obtained. (That is, do **not** round any value obtained from an intermediate calculation.)

- Do **not** expect to understand any chapter at first reading. Knowledge will be gained slowly with further reading and by working a number of exercises.

  One recommended procedure for studying a particular chapter is to first,

(i) read the chapter 'straight through' in order to obtain an overview of the content (*not* necessarily an understanding), and secondly,

(ii) re-read one section at a time, concentrating on the worked examples and then attempt the relevant 'Your turn' exercises immediately. Hopefully, when all the 'Your turn' exercises in a section have been completed, you will have a reasonable understanding of the material discussed. Then proceed to the next section, and so on.

- When attempting any exercise or problem, read the question carefully to be sure that you understand what is required. Always take just a little time to 'ponder', especially when a real-life problem is involved.

- Take special care when copying the question onto your work paper or reading from the display of your calculator. A single copying error can create many difficulties.

- Where appropriate, you should draw a diagram, since a simple sketch will generally make the question clearer and often suggest an appropriate method of solution. This is particularly true in geometry and trigonometry, where your diagrams should be large enough for you to mark in all the given information.

- Be careful to set out your answers clearly and logically. Where a series of steps is involved, align the '=' signs vertically, checking that each step follows on immediately from the previous one.

- **Never** falsify your work in an attempt to obtain the correct answer. It is far better to have an incorrect result because of a simple error than to present a correct answer for the *wrong* reasons.

- **Always** check that your answer is 'reasonable'. A probability of 1.62 or −0.38 is impossible just as a *new* car costing $340.00 or a teacher's salary of $8 500 000 is highly unlikely!

- Wherever possible, verify that your answer is correct. *For example*, the solution of an equation must satisfy the original equation; differentiation of an integrated function should give the original function; etc.

  However, although a simple check (mentally, where appropriate) is strongly advised, you should not spend much time and effort on this – checking all working several times over is a sign that you lack confidence in your own ability to solve a given problem.

Note: If you keep in mind the above suggestions as you work your way through the text, you may surprise yourself and find that your learning experience is far more pleasurable and effective than you initially expected.

A final word from the authors

It is hoped that, with the aid of this text, you will successfully 'bridge any gap' that may exist between your current mathematical knowledge and that assumed by the discipline of your choice. It is also hoped that, in the process of gaining the required knowledge, you will also find the material both interesting and enjoyable.

For those of you who are feeling a little apprehensive initially, if you follow the advice given earlier, and approach each new topic with an open mind and a desire to learn, then you surely will.

However, even though some of you will still find the material difficult, some of you will most certainly find it boring and some of you will even find it easy, rest assured that *most* of you will manage to complete the text successfully and should be proud of your achievement.

Carmel Coady   *Jenny Gosling*

1 Basic arithmetic

Introduction

In this chapter two terms will frequently appear and should be mentioned here at the outset.

> An **arithmetic operation** is a process performed on two numbers that will result in a single solution. The specific operation is indicated by an **operating symbol** that is placed between the two numbers.

For example: 6 + 2 involves the arithmetic operation of addition, with the '+' being the operating symbol.

> An **arithmetic expression** is one or more numbers, separated by operating symbols, that may be simplified by performing the arithmetic operations indicated.

In the sections that follow we will examine the four basic operations of addition, subtraction, multiplication and division as they are applied to arithmetic expressions, fractions, decimals, percentages and ratios.

Each section will include definitions (if required), relevant terminology and notation, operating rules, calculator usage, worked examples and activities for you to undertake to check your understanding.

Because calculators are now a part of everyday life, it is often assumed that everyone is familiar with their usage in solving both simple and more complex problems. Therefore, all the examples presented here will be solved using a calculator, with a stepwise indication of the appropriate buttons to press. In some cases, the steps required to solve a problem manually will be discussed in the **extensions** part of this chapter.

Finally, you should remember two points:

(i)   The **extensions** part of this and future chapters covers additional material that may be useful for those of you entering more specialised disciplinary areas or perhaps solely for interest.

(ii)   All the practice exercises you attempt have at least brief answers, with those considered more difficult (marked '*') having a full solution.

1.1   Arithmetic operations

1.1.1   Basic operations

Operation	Symbol	Synonyms
Addition	+	plus, add, sum
Subtraction	–	minus, take away, difference between
Multiplication	x (or · or *)	times, product of
Division	÷ or / or —	divided by

EXAMPLE 1.1
Simplify each of the following arithmetic expressions by carrying out the operations indicated:

(i)   642 + 20 – 83

(ii)   1123 – 21 + 243 + 24

(iii)   23 x 42 ÷ 2

(iv)   169 ÷ 13 x 5

Solution
Although the above expressions can be simplified manually with little difficulty, the calculator (a valuable tool used by mathematicians of *all* abilities) is definitely the 'way to go'!!

(i)   642 $\boxed{+}$ 20 $\boxed{-}$ 83 $\boxed{=}$ $\boxed{579.}$

(ii)   1123 $\boxed{-}$ 21 $\boxed{+}$ 243 $\boxed{+}$ 24 $\boxed{=}$ $\boxed{1369.}$

(iii)   23 $\boxed{x}$ 42 $\boxed{÷}$ 2 $\boxed{=}$ $\boxed{483.}$

(iv)   169 $\boxed{÷}$ 13 $\boxed{x}$ 5 $\boxed{=}$ $\boxed{65.}$

1.1.2 Order of operations

By convention, arithmetic expressions involving mixed operations are performed in the following order, working from left to right.

Step 1: multiplication and/or division

Step 2: addition and/or subtraction.

EXAMPLE 1.2
Simplify, by carrying out the operations indicated in the appropriate order,

$$124 \div 2 + 3 \times 5 - 18$$

Solution

$$124 \div 2 + 3 \times 5 - 18 =$$

Step 1: $\quad 62 \;+\; 15 \;-\; 18 =$

Step 2: $\qquad\quad 77 \;-\; 18 = 59$

Note: If a given arithmetic expression involves brackets (e.g. (...) and/or [...]), then the operations within the brackets **must** be performed first, followed by Steps 1 and 2 above. Where an expression involves brackets within brackets, the innermost bracket must be simplified first.

EXAMPLE 1.3
Simplify, by carrying out the operations in the correct order,

(i) $\quad 38 + (24 - 8) \times 3 - 15$

(ii) $\quad 33 \times 3 + [(20 - 10) \times 3] \div (8 + 2)$

Solution

(i) $\qquad 38 + (24 - 8) \times 3 - 15 =$

$\qquad\quad 38 + \quad 16 \times 3 - \quad 15 =$

$\qquad\quad 38 + \qquad 48 - \quad 15 = 71$

(ii) $\qquad 33 \times 3 + [(20 - 10) \times 3] \div (8 + 2) \;=$

$\qquad\quad 33 \times 3 + \quad [10 \times 3] + \qquad 10 \;=$

$\qquad\quad 33 \times 3 \qquad + 30 \;+ \qquad 10 \;=$

$\qquad\quad\quad 99 \quad + \qquad 3 \qquad\quad = 102$

A *scientific* calculator certainly makes life much easier. To simplify any arithmetic expression, press the appropriate data and operator buttons, **always** working from left to right.

EXAMPLE 1.4
Now let us simplify the two expressions given in Example 1.3 using the calculator.

Solution

(i) $\quad 38 + (24 - 8) \times 3 - 15 =$

38 $\boxed{+}$ $\boxed{(}$ 24 $\boxed{-}$ 8 $\boxed{)}$ $\boxed{\times}$ 3 $\boxed{-}$ 15 $\boxed{=}$ $\boxed{71.}$

(ii) $\quad 33 \times 3 + [(20 - 10) \times 3] \div (8 + 2)$

33 $\boxed{\times}$ 3 $\boxed{+}$ $\boxed{(}$ $\boxed{(}$ 20 $\boxed{-}$ 10 $\boxed{)}$ $\boxed{\times}$ 3 $\boxed{)}$ $\boxed{\div}$ $\boxed{(}$ 8 $\boxed{+}$ 2 $\boxed{)}$ $\boxed{=}$ $\boxed{102.}$

Simplify each of the expressions in exercises 1.1 to 1.7 below.

1.1 $\quad 18 \div 2 + 3 - 5 \times 2 + 7$

1.2 $\quad 8 + 12 \div 6 + 4$

1.3 $\quad (8 + 12) \div (6 + 4)$

1.4 $\quad 15 \times (5 - 2) \div (8 - 3) - 4$

1.5 $\quad 175 - 5 \times (6 + 9 \div (1 + 2) - 5)$

1.6 $\quad 175 - 5 \times [(6 + 9) \div (1 + 2) - 5]$

1.7 $\quad (175 - 5) \times [(6 + 9) \div (1 + 2) - 5]$

1.8 What change should you receive from $50 if you purchase 3 blank video tapes at $5 each and a compact disc costing $30?

*1.9 John regularly works 35 hours per week for $525. If all overtime is paid at *double* time, how many full hours of overtime must he work in one week to earn a total of $765?

*1.10 A wristwatch loses 2 seconds every hour. How many minutes slow will it be after 5 days? (*Note*: There are 24 hours in a day, 60 minutes in an hour and 60 seconds in a minute.)

1.1.3 Further symbols (to be noted here for later use)

Symbol	Meaning	Example
>	greater than	$10 > 6$; $200 > 169$
<	less than	$15 < 20$; $108 < 115$
$\geq$	greater than or equal to	$5 \geq 2$; $5 \geq 5$
$\leq$	less than or equal to	$12 \leq 18$; $12 \leq 12$
$\neq$	not equal to	$100 \neq 62$; $25 \neq 28$
$\approx$	approximately equal to	$1000 \approx 1001$
$\therefore$	therefore	

1.2 Fractions

An example of a fraction is $\frac{3}{7}$, where

 3 is called the *numerator,* and

 7 is called the *denominator.*

1.2.1 Types of fractions

A **proper** fraction is one in which the numerator is less than the denominator,

$$\text{e.g. } \frac{3}{7}$$

An **improper** fraction is one in which the numerator is greater than the denominator,

$$\text{e.g. } \frac{11}{8}$$

A **mixed** fraction consists of a whole number with a proper fraction,

$$\text{e.g. } 1\frac{3}{8}$$

Note: An improper fraction may be written as a mixed fraction by dividing the numerator by the denominator,

$$\text{e.g. } \frac{11}{8} = 1\frac{3}{8}$$

An **equivalent** fraction is formed by multiplying or dividing both the numerator and the denominator by the same number.
For example:

(i) $\quad \frac{1}{2} = \frac{1\,(\times 5)}{2\,(\times 5)} = \frac{5}{10}$

(ii) $\quad \frac{10}{20} = \frac{10(\div 5)}{20(\div 5)} = \frac{2}{4} = \frac{2(\div 2)}{4(\div 2)} = \frac{1}{2}$

1.2.2 Operations with fractions

In this section, the operations of addition, subtraction, multiplication and division are most easily performed on a calculator. The 'fraction' button on your calculator should be similar to $\boxed{a\frac{b}{c}}$ and is used to separate each of the numbers forming the fraction by means of the symbol '$\rfloor$'.

Pressing the 'equals' button, $\boxed{=}$, after entering the fraction will result in the simplest form of the fraction appearing on your calculator display.

For example:

Fraction	Calculator input	Calculator display
$\frac{3}{7}$	3 $\boxed{a\frac{b}{c}}$ 7	3 $\rfloor$ 7.
$\frac{11}{8}$	11 $\boxed{a\frac{b}{c}}$ 8	11 $\rfloor$ 8.
$2\frac{4}{5}$	2 $\boxed{a\frac{b}{c}}$ 4 $\boxed{a\frac{b}{c}}$ 5	2 $\rfloor$ 4 $\rfloor$ 5.
$\frac{11}{8}$ (simplify)	11 $\boxed{a\frac{b}{c}}$ 8 $\boxed{=}$	1 $\rfloor$ 3 $\rfloor$ 8.
$\frac{16}{20}$ (simplify)	16 $\boxed{a\frac{b}{c}}$ 20 $\boxed{=}$	4 $\rfloor$ 5.

Note:

(i) For those of you who may be interested, the procedures for carrying out these operations manually are presented in **extension A** of this chapter.

(ii) The fraction button $\boxed{a\frac{b}{c}}$ can only be used if there are *at most* 3 numerals in either the numerator or the denominator or both.

EXAMPLE 1.5
Simplify each of the following expressions, using both the fraction and operator buttons on your calculator.

(i) $\quad \frac{2}{3} + \frac{1}{4} - \frac{3}{7}$

(ii) $\quad 7 - \frac{3}{5} \times 1\frac{1}{2}$

(iii) $\quad 4\frac{4}{5} \times \left(\frac{3}{4} - \frac{1}{3}\right) + \frac{1}{2}$

Solution

(i) 2 $\boxed{a\frac{b}{c}}$ 3 $\boxed{+}$ 1 $\boxed{a\frac{b}{c}}$ 4 $\boxed{-}$ 3 $\boxed{a\frac{b}{c}}$ 7 $\boxed{=}$ $\boxed{41\rfloor 84.}$

$\therefore \frac{2}{3} + \frac{1}{4} - \frac{3}{7} = \frac{41}{84}$

(ii) 7 $\boxed{-}$ 3 $\boxed{a\frac{b}{c}}$ 5 $\boxed{\times}$ 1 $\boxed{a\frac{b}{c}}$ 1 $\boxed{a\frac{b}{c}}$ 2 $\boxed{=}$ $\boxed{6\rfloor 1\rfloor 10.}$

$\therefore 7 - \frac{3}{5} \times 1\frac{1}{2} = 6\frac{1}{10}$

(iii) 4 $\boxed{a\frac{b}{c}}$ 4 $\boxed{a\frac{b}{c}}$ 5 $\boxed{\times}$ $\boxed{(}$ 3 $\boxed{a\frac{b}{c}}$ 4 $\boxed{-}$ 1 $\boxed{a\frac{b}{c}}$ 3 $\boxed{)}$

$\boxed{+}$ 1 $\boxed{a\frac{b}{c}}$ 2 $\boxed{=}$ $\boxed{2\rfloor 1\rfloor 2.}$

$\therefore 4\frac{4}{5} \times \left(\frac{3}{4} - \frac{1}{3}\right) + \frac{1}{2} = 2\frac{1}{2}$

Simplify the expressions in exercises 1.11 to 1.13, using the fraction button on your calculator.

1.11 $\dfrac{1}{2} + \dfrac{2}{3} - \dfrac{5}{8}$

1.12 $2\dfrac{3}{8} - \dfrac{3}{4} \times \dfrac{1}{2} + 1\dfrac{1}{3}$

1.13 $1\dfrac{3}{4} \div \dfrac{1}{2} \times \left(\dfrac{2}{3} - \dfrac{1}{6}\right) + 3\dfrac{1}{4}$

1.14 A certain alloy contains (by mass) $\dfrac{3}{8}$ copper, $\dfrac{1}{4}$ tin and the remainder zinc. In a sample of this alloy, what fraction is zinc?

1.15 Jill purchased $10\dfrac{1}{2}$ metres of wire mesh to make a rabbit hutch with measurements shown in the following diagram.

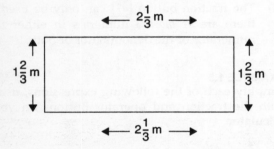

If she puts the mesh around each of the 4 sides with no overlap, how much of the original $10\dfrac{1}{2}$ metres will remain?

1.3 Decimals and percentages

The following is an example of a decimal number to three decimal places.

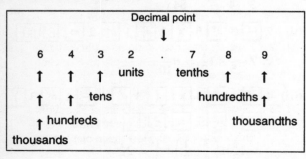

1.3.1 Types of decimals

A **terminating** decimal has a **fixed** number of places after the decimal point, e.g. 0.752.

A **repeating** (or **recurring**) decimal has one or more of the numbers after the decimal point repeating continuously.

e.g.	0.333.....	is	denoted $0.\dot{3}$
	1.245245245.....	is	denoted $1.\dot{2}\,4\,\dot{5}$
	0.34555.....	is	denoted $0.34\dot{5}$

1.3.2 Percentages

Percentages may be interpreted as **hundredths**.

e.g.	25%	=	25 hundredths	=	0.25
	253%	=	253 hundredths	=	2.53
	5%	=	5 hundredths	=	0.05

1.3.3 Operations with decimals and percentages

Once again, the operations discussed in this section are most easily performed on a calculator. However, those readers interested in manual calculations involving decimals and percentages should refer to **extension A** at the end of this chapter.

EXAMPLE 1.6

(a) Simplify each of the following expressions by performing the required operations.

    (i) $1.4 \times 0.03 + 10.864 \div 0.2$

    (ii) $10.025 + [(2.45 + 0.05) \times 3.27]$

(b) Find

    (i) 25% of 150

    (ii) 157% of 40.

Solution

(a) **(i)** 1.4 [x] .03 [+] 10.864 [÷] .2
    [=] [54.362]

    (ii) 10.025 [+] [(] [(] 2.45 [+] .05 [)] [x]
    3.27 [)] [=] [18.2]

(b) **(i)** .25 [x] 150 [=] [37.5]

    (ii) 1.57 [x] 40 [=] [62.8]

1.3.4 Conversions

	Conversion	Method	Example
a.	decimal → percentage	Multiply by 100; attach % sign.	$.34 \rightarrow .34 \times 100 = 34\%$
b.	percentage → decimal	Drop % sign; divide by 100.	$34\% \rightarrow 34 \div 100 = 0.34$
c.	fraction → percentage	Multiply by 100; attach % sign.	$\frac{3}{4} \rightarrow \frac{3}{4} \times 100 = 75\%$
d.	percentage → fraction	Drop % sign; the percent number becomes the numerator with 100 forming the denominator. Simplify to an equivalent fraction if possible.	$75\% \rightarrow \frac{75}{100} = \frac{75\,(\div 25)}{100\,(\div 25)} = \frac{3}{4}$
e.	fraction → decimal	Divide the numerator by the denominator.	$\frac{5}{8} \rightarrow 5 \div 8 = 0.625$
f.	decimal → fraction	The numbers following the decimal point form the numerator while the position of the final decimal number determines the denominator. Simplify to an equivalent fraction if possible.	$0.625 \rightarrow \frac{625}{1000} = \frac{625\,(\div 125)}{1000\,(\div 125)} = \frac{5}{8}$

EXAMPLE 1.7

Express:

(i) 0.875 as a percentage

(ii) 0.875 as a fraction

(iii) 68% as a decimal

(iv) 68% as a fraction

(v) $\frac{17}{20}$ as a decimal

(vi) $\frac{17}{20}$ as a percentage.

Solution

(i) $.875$ ⊠ $\boxed{\times}$ 100 $\boxed{=}$ $\boxed{87.5}$

∴ $0.875 = 87.5\%$

(ii) $\frac{875}{1000} = \frac{875 \div 5}{1000 \div 5} = \frac{175 \div 5}{200 \div 5} = \frac{35 \div 5}{40 \div 5} = \frac{7}{8}$

or $\frac{875}{1000}$ is 875 $\boxed{a\frac{b}{c}}$ 1000 $\boxed{=}$ $\boxed{7\,\rfloor 8.}$

∴ $0.875 = \frac{7}{8}$

(iii) 68 $\boxed{\div}$ 100 $\boxed{=}$ $\boxed{0.68}$

∴ $68\% = 0.68$

(iv) $\frac{68}{100} = \frac{68 \div 2}{100 \div 2} = \frac{34 \div 2}{50 \div 2} = \frac{17}{25}$

or $\frac{68}{100}$ is 68 $\boxed{a\frac{b}{c}}$ 100 $\boxed{=}$ $\boxed{17\,\rfloor 25.}$

∴ $68\% = \frac{17}{25}$

(v) 17 $\boxed{\div}$ 20 $\boxed{=}$ $\boxed{0.85}$

∴ $\frac{17}{20} = 0.85$

(vi) 17 $\boxed{a\frac{b}{c}}$ 20 $\boxed{\times}$ 100 $\boxed{=}$ $\boxed{85.}$

or 17 $\boxed{\div}$ 20 $\boxed{\times}$ 100 $\boxed{=}$ $\boxed{85.}$

∴ $\frac{17}{20} = 85\%$

YOUR TURN

1.16 Evaluate $0.068 + 29.9 \times 3.02 - 1.56 \div 2$

1.17 Evaluate $(0.068 + 29.9) \times (3.02 - 1.56) \div 2$

1.18 Add $10.32, $15.76, $4.25, $0.30 and $22.18.

What is my change from $60.00?

1.19 Find 12% of 75.

1.20 Express the percentage as a fraction and evaluate
(i) 50% of 62   **(ii)** 18% of 250.

1.21 Express the percentage as a decimal and evaluate
(i) 15% of 23.6   **(ii)** 25% of 1.6.

1.22 Express the percentage as a fraction and then evaluate the result both as a fraction and as a decimal.
(i) 30% of $\frac{2}{3}$ **(ii)** 8% of $6\frac{3}{4}$

1.23 I wish to purchase 5 packets of lifesavers at 45 cents (i.e. $0.45) each. I hand the shopkeeper $2.50. How much change should I receive?

1.24 One tin of Spot dog food costs $1.09. A carton containing 12 tins costs $12.50. Which is the cheaper way to buy Spot dog food — individually or by the carton? How much do I save?

1.25 Convert
  (i) 35.26% to a decimal
  (ii) 0.045 to a percentage
  (iii) $\frac{39}{50}$ to a percentage
  (iv) $52\frac{1}{2}$% to a fraction and simplify.

1.26 Round off each of the following correct to the nearest 100.
  (i) 9064   **(ii)** 7830
  (iii) 1099   **(iv)** 60828

1.27 Round off each of the following correct to 2 decimal places.
  (i) 26.708   **(ii)** 0.0007
  (iii) 32.824   **(iv)** 0.1064

1.28 Round off each of the following correct to the nearest whole number.
  (ii) 13.84   **(ii)** 0.005
  (iii) 9.635   **(iv)** 16.498

1.29 Truncate each of the following correct to 2 decimal places.
  (i) 26.708   **(ii)** 0.0007
  (iii) 32.825   **(iv)** 0.1064

1.30 Calculate each of the following, rounding off your answer to 1 decimal place.
  (i) 36% of 124   **(ii)** 25.87 x 1.5
  (iii) 8.452 ÷ 0.75

1.3.5 Approximations with decimals

A. Rounding off

Rounding off involves writing a numeral to a specified number of decimal places.

For example:

(i) $2.58 \approx 2.6$ (rounded off to 1 decimal place)

(ii) $0.934 \approx 0.93$ (rounded off to 2 decimal places)

(iii) $7.343 \approx 7.3$ (rounded off to 1 decimal place)

(iv) $11.0545 \approx 11.055$ (rounded off to 3 decimal places)

(v) $102.1397 \approx 102.14$ (rounded off to 2 decimal places)

(vi) $0.3965 \approx 0.40$ (rounded off to 2 decimal places)

Note: Whenever the rounding decision involves the numeral 5, (in this text) the preceding number is rounded up to the next highest number, (as in examples (iv) and (vi) above).

B. Truncating

Truncating involves deleting all the numerals occurring after the specified number of decimal places.

For example:

(i) $2.58 \approx 2.5$ (truncated to 1 decimal place)

(ii) $0.934 \approx 0.93$ (truncated to 2 decimal places)

(iii) $7.343 \approx 7.3$ (truncated to 1 decimal place)

(iv) $11.0545 \approx 11.054$ (truncated to 3 decimal places)

(v) $102.1397 \approx 102.13$ (truncated to 2 decimal places)

(vi) $0.3965 \approx 0.39$ (truncated to 2 decimal places)

1.4 Ratios

A **ratio** is a comparison of two quantities.

For example: In an experiment involving the breeding of flowers with a green stigma, Mendelian theory predicts that if we have 16 such flowers, 12 of them should be magenta and 4 should be red. This comparison (or ratio) may be written in the form

$$12 : 4$$

where the symbol ':'is read as 'is to'.

An **equivalent** ratio is formed by multiplying or dividing **both** numbers of the ratio by the same numeral.

For example:

Ratio	Operation	Equivalent ratio
12 : 4	÷ 2	6 : 2
6 : 2	÷ 2	3 : 1
12 : 4	÷ 4	3 : 1
3 : 1	x 3	9 : 3
3 : 1	x 4	12 : 4

Note: The concept of a ratio may be quite readily extended in order to compare more than two quantities, e.g. the 4 main blood groups (O, A, B, AB) may be found to occur in the ratio 12 : 5 : 2 : 1.

1.4.1 Operations with ratios

Important: When operating with ratios, be sure that the quantities refer to the **same** units.
For example:

(i) 15 seconds is to 20 seconds(same units)
→ 15 : 20

(ii) 15 seconds is to 2 minutes(unlike units)
→ 15 seconds is to 120 seconds(same units)
→ 15 : 120

EXAMPLE 1.8
Divide $20 between 2 people in the ratio 1 : 4.

Solution
The ratio 1 : 4 can be interpreted as 1 part is to 4 parts, implying a *total* of 5 parts.
Hence,

Person A	Person B
receives: $\frac{1}{5}$ of $20	receives: $\frac{4}{5}$ of $20
$= \frac{1}{5} \times 20$	$= \frac{4}{5} \times 20$
= $4	= $16

EXAMPLE 1.9
A metal rod 30 metres in length is cut into 3 sections in the ratio 3 : 2 : 5. Find the length of each section.

Solution
The ratio 3 : 2 : 5 implies 10 parts altogether.
Therefore,

(i) length of section 1: $\frac{3}{10} \times 30 = 9$ metres

(ii) length of section 2: $\frac{2}{10} \times 30 = 6$ metres

(iii) length of section 3: $\frac{5}{10} \times 30 = 15$ metres.

EXAMPLE 1.10
The fuel for a 2-stroke lawn mower combines the petrol and oil in the ratio 19:1. How much petrol and oil must I buy if I wish to completely fill my mower's 1 litre tank?

Solution

Petrol	Oil
$\frac{19}{20} \times 1$ litre	$\frac{1}{20} \times 1$ litre
$= \frac{19}{20} \times 1000$ (Since 1 litre = 1000 mls)	$= \frac{1}{20} \times 1000$ mls
= 950 mls	= 50 mls

Therefore, my 1 litre tank would require 950 mls of petrol and 50 mls of oil.

EXAMPLE 1.11
In order to make a light flaky pastry, the flour and butter must be mixed in the ratio 5 : 3 respectively. If 125 grams of flour is used, determine the amount of butter that is required.

Solution
```
          flour    :    butter
in the ratio 5     :      3
```
Since 125 = 5 x 25, we can answer the question using equivalent ratios.

```
      5 (x 25)    :    3 (x 25)
→      125        :      75
```

Therefore, if 125 grams of flour is used, 75 grams of butter will be required.

YOUR TURN

For exercises 1.31 to 1.34 use the given examples as a guide in order to simplify each ratio.

1.31

e.g. 3 : 9, divide by 3 = 1 : 3

(i) 7 : 49 (ii) 16 : 12 (iii) 24 : 9
(iv) 9 : 30 (v) 8 : 72

1.32

e.g. 2.0 : 2.5, multiply by 10 = 20 : 25, divide by 5 = 4 : 5

(i) 0.5 : 0.3 (ii) 0.9 : 2.1 (iii) 0.02 : 0.08
(iv) 1 : 0.025 (v) 24 : 3.6

1.33

e.g. $\frac{1}{4} : \frac{3}{8}$, multiply by 8 $= \frac{8}{4} : \frac{24}{8} = 2 : 3$

(i) $\frac{2}{9} : \frac{4}{9}$ (ii) $\frac{4}{5} : \frac{7}{10}$ (iii) $\frac{7}{12} : \frac{2}{3}$

(iv) $\frac{1}{2} : 1\frac{1}{2}$ (v) $2\frac{1}{2} : \frac{3}{4}$

1.34

e.g. 75 cm is to 2 m = 75 cm is to 200 cm = 75 : 200 = 3 : 8

(i) 250 g is to 1 kg
(ii) 40 secs is to 2 mins
(iii) $2.40 is to 80 cents
(iv) 40 mm is to 8 cm
(v) $2\frac{1}{2}$ litres is to 750 mls.

1.35 Divide a 250 g block of chocolate between Jan and Kate in the ratio 6 : 4.

1.36 If $500 is to be divided among John, Mark and Andrew in the ratio 8 : 7 : 5, how much does each person receive?

1.37 Four business partners, Joy, Pam, Ted and Peter share their annual profits in the ratio 7 : 6 : 4 : 3. How much does each person receive if the profit is $250 000?

1.38 Sarah and Alex started a small business by investing $10 000 and $15 000 respectively. If the profit of $12 500 is to be divided between them in the ratio of their investments, how much will each one receive?

1.39 To make jam out of most stoned fruits, we must mix together 5 parts of fruit with 3 parts of white sugar. How much white sugar should be added to 4 kg of plums in order to make plum jam?

1.40 A particular concrete mixture is made up of 2 parts of cement, 3 parts of coarse sand and 5 parts of blue metal. If I have 60 kg of blue metal, what quantities of cement and sand do I need in order to obtain the required concrete mixture?

1.5 The real number system

So far, our discussion has concentrated on **positive** numbers only. However, these numbers form only a part of what we call the **real number system**. The system may be best represented using the following diagram.

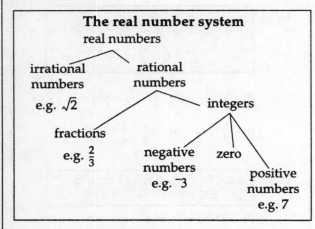

The real number system

real numbers

irrational numbers
e.g. $\sqrt{2}$

rational numbers

fractions
e.g. $\frac{2}{3}$

integers

negative numbers
e.g. $^-3$

zero

positive numbers
e.g. 7

1.5.1 Operations involving negatives

Important: In this section we will indicate that a number is negative by placing the symbol '$^-$' at the top left of the number, e.g. negative 2 is written $^-2$.

A. Addition and subtraction
Rules by example:

(i) $6 + {}^-2 = 6 - 2 = 4$

(ii) $6 - {}^-2 = 6 + 2 = 8$

(iii) $^-6 + 2 = {}^-4$

(iv) $^-6 - 2 = {}^-8$

(v) $^-6 + {}^-2 = {}^-6 - 2 = {}^-8$

B. Multiplication and division
Rules by example:

(vi) $6 \times {}^-2 = {}^-12$

(vii) $^-6 \times 2 = {}^-12$

(viii) $^-6 \times {}^-2 = 12$

(ix) $6 \div {}^-2 = {}^-3$

(x) $^-6 \div 2 = {}^-3$

(xi) $^-6 \div {}^-2 = 3$

Comment: Many people find that operations involving negative quantities are not as easily understood as their positive counterparts. Hence, for those readers that may be interested, the answers obtained to the examples given in A and B above may also be explained in terms of the *laws* associated with the real number system (to be discussed in Chapter 2).

Remember: Most expressions may be simplified or questions answered using a calculator, as we shall see when we repeat the '*rules by example*' given above, indicating the appropriate calculator buttons to press.

Stop: When operating with negatives, you must be sure that you can distinguish between the '$-$' sign implying the process of subtraction and the '$^-$' sign denoting a negative number.

For example: $6 - {}^-2$ reads 6 minus negative 2.

To evaluate this using your calculator, you would press the following buttons:

$$6 \;\boxed{-}\; 2 \;\boxed{+/_-}\; \boxed{=} \quad \boxed{8.}$$

You can see that $\boxed{-}$ indicates the operations of subtraction whereas $\boxed{+/_-}$ implies that the numeral preceding this 'change-sign' button is negative.

EXAMPLE 1.12
Using your calculator, evaluate all the '*rules by example*' given in A and B above .

Solution

(i) $6 + {}^-2 : 6 \;\boxed{+}\; 2 \;\boxed{+/_-}\; \boxed{=} \quad \boxed{4.}$

(ii) $6 - {}^-2 : 6 \;\boxed{-}\; 2 \;\boxed{+/_-}\; \boxed{=} \quad \boxed{8.}$

(iii) ⁻6 + 2 : 6 [+/−] [+] 2 [=] $\boxed{-4.}$

(iv) ⁻6 − 2 : 6 [+/−] [−] 2 [=] $\boxed{-8.}$

(v) ⁻6 + ⁻2 : 6 [+/−] [+] 2 [+/−] [=] $\boxed{-8.}$

(vi) 6 x ⁻2 : 6 [x] 2 [+/−] [=] $\boxed{-12.}$

(vii) ⁻6 x 2 : 6 [+/−] [x] 2 [=] $\boxed{-12.}$

(viii) ⁻6 x ⁻2 : 6 [+/−] [x] 2 [+/−] [=] $\boxed{12.}$

(ix) 6 ÷ ⁻2 : 6 [÷] 2 [+/−] [=] $\boxed{-3.}$

(x) ⁻6 ÷ 2 : 6 [+/−] [÷] 2 [=] $\boxed{-3.}$

(xi) ⁻6 ÷ ⁻2 : 6 [+/−] [÷] 2 [+/−] [=] $\boxed{3.}$

EXAMPLE 1.13

Evaluate each of the following expressions.

(i) 2 x (⁻5 + 4)

(ii) 10 − 5 x ⁻3

(iii) (10 − 5) x ⁻3

(iv) $\dfrac{1}{2} - \dfrac{1}{3} + \dfrac{-2}{5}$

(v) 0.421 + 1.84 ÷ ⁻2

(vi) (0.421 + 1.84) ÷ ⁻2

(vii) (⁻8 − 3 − ⁻4 − 4 + 3) x ⁻1

(viii) ⁻8 − 3 − ⁻4 − 4 + 3 x ⁻1

(ix) $3 + {}^-1\dfrac{1}{2} \times 4 - {}^-\dfrac{1}{2}$

(x) $(3 + {}^-1\dfrac{1}{2}) \times (4 - {}^-\dfrac{1}{2})$

Solution (using the calculator)

(i) 2 [x] [(] 5 [+/−] [+] 4 [)] [=] $\boxed{-2.}$
Solution = ⁻2

(ii) 10 [−] 5 [x] 3 [+/−] [=] $\boxed{25.}$
Solution = 25

(iii) [(] 10 [−] 5 [)] [x] 3 [+/−] [=] $\boxed{-15.}$
Solution = ⁻15

(iv) 1 [aᵇ𝒸] 2 [−] 1 [aᵇ𝒸] 3 [+] 2 [aᵇ𝒸] 5 [+/−]
[=] $\boxed{-7\rfloor30.}$ Solution = $\dfrac{-7}{30}$

(v) .421 [+] 1.84 [÷] 2 [+/−] [=] $\boxed{-0.499}$
Solution = ⁻0.499

(vi) [(] .421 [+] 1.84 [)] [÷] 2 [+/−]
[=] $\boxed{-1.1305}$ Solution = ⁻1.1305

(vii) [(] 8 [+/−] [−] 3 [−] 4 [+/−] [−] 4 [+] 3 [)]
[x] 1 [+/−] [=] $\boxed{8.}$ Solution = 8

(viii) 8 [+/−] [−] 3 [−] 4 [+/−] [−] 4 [+] 3 [x] 1 [+/−]
[=] $\boxed{-14.}$ Solution = ⁻14

(ix) 3 [+] 1 [aᵇ𝒸] 1 [aᵇ𝒸] 2 [+/−] [x] 4 [−] 1 [aᵇ𝒸] 2
[+/−] [=] $\boxed{-2\rfloor1\rfloor2.}$ Solution = $-2\dfrac{1}{2}$

(x) [(] 3 [+] 1 [aᵇ𝒸] 1 [aᵇ𝒸] 2 [+/−] [)] [x] [(] 4
[−] 1 [aᵇ𝒸] 2 [+/−] [)] [=] $\boxed{6\rfloor3\rfloor4.}$
Solution = $6\dfrac{3}{4}$

Evaluate each of the expressions in exercises 1.41 to 1.46.

1.41 10 − ⁻8 x ⁻2 + 4

1.42 (10 − ⁻8) x ⁻2 + 4

1.43 (10 − ⁻8) x (⁻2 + 4)

1.44 $\dfrac{3}{4} + \dfrac{5}{8} - {}^-1\dfrac{1}{4}$

1.45 $^-8 \times \dfrac{1}{4} \times \dfrac{-3}{4} \div \dfrac{-1}{2} \times \dfrac{1}{4}$

1.46 $24 \div (5 + {}^-2\dfrac{3}{5})$

1.47 Show that 2 ÷ (⁻8 ÷ ⁻4) is *not* equal to (2 ÷ ⁻8) ÷ ⁻4 by calculating the difference between the two expressions.

1.48 Most deep-sea divers descend below the surface in stages. If Judy descends in 6 stages, each of 8 metres, write a mathematical statement that indicates her final position (distance) below the surface.

1.49 The temperature at Perisher at 6 am one winter's morning was ⁻15°C. By 10 am it had risen by 10°C and by 3 pm it had risen a further 8°C. What was the temperature at 3 pm?

At midnight the temperature was down to ⁻20°C. What was the drop in temperature from the 3 pm reading?

Extensions to chapter 1

Extension A: Operating with fractions and decimals manually

A-1 Fractions

a. Addition and/or subtraction

> Only 'like' fractions can be added or subtracted – that is, they must have the same denominator. Whenever this is not the case, the lowest common denominator is chosen by forming equivalent fractions.

EXAMPLE E1.1
Evaluate the following without using your calculator:

(i) $\dfrac{3}{7} + \dfrac{5}{7}$

(ii) $\dfrac{1}{3} + \dfrac{1}{2}$

(iii) $\dfrac{2}{5} - \dfrac{3}{10}$

(iv) $1\dfrac{1}{4} + \dfrac{3}{8} - \dfrac{2}{3}$

Solution

(i) Since both fractions have the same denominator, all that is required is to add together the numerators.

$$\frac{3}{7} + \frac{5}{7} = \frac{8}{7}$$

$$= 1\frac{1}{7}$$

(ii) In this example the lowest common denominator is 6.

$$\frac{1}{3} + \frac{1}{2} = \frac{1 \times 2}{3 \times 2} + \frac{1 \times 3}{2 \times 3} \quad \text{(form equivalent fractions)}$$

$$= \frac{2}{6} + \frac{3}{6} \quad \text{(add 'like' fractions)}$$

$$= \frac{5}{6}$$

(iii)

$$\frac{2}{5} - \frac{3}{10} = \frac{2 \times 2}{5 \times 2} - \frac{3}{10} \quad \text{(form equivalent fractions)}$$

$$= \frac{4}{10} - \frac{3}{10} \quad \text{(subtract 'like' fractions)}$$

$$= \frac{1}{10}$$

(iv)

$$1\frac{1}{4} + \frac{3}{8} - \frac{2}{3} = \frac{5}{4} + \frac{3}{8} - \frac{2}{3} \quad \text{(convert to an improper fraction)}$$

$$= \frac{5 \times 6}{4 \times 6} + \frac{3 \times 3}{8 \times 3} - \frac{2 \times 8}{3 \times 8} \quad \text{(form equivalent fractions)}$$

$$= \frac{30}{24} + \frac{9}{24} - \frac{16}{24} \quad \text{(add and subtract 'like' fractions)}$$

$$= \frac{23}{24}$$

b. Multiplication

> In order to multiply fractions, first simplify if possible and then multiply together the numerators and the denominators respectively.

EXAMPLE E1.2
Evaluate the following without using your calculator:

(i) $\dfrac{2}{3} \times \dfrac{1}{5}$

(ii) $\dfrac{4}{5} \times 1\dfrac{1}{2}$

Solution

(i)

$$\frac{2}{3} \times \frac{1}{5} = \frac{2 \times 1}{3 \times 5} \quad \text{(no simplification possible)}$$

$$= \frac{2}{15}$$

(ii)

$$\frac{4}{5} \times 1\frac{1}{2} = \frac{4}{5} \times \frac{3}{2} \quad \text{(convert to an improper fraction)}$$

$$= \frac{\overset{2}{\cancel{4}}}{5} \times \frac{3}{\underset{1}{\cancel{2}}} \quad \text{(simplify)}$$

$$= \frac{2 \times 3}{5 \times 1} \quad \text{(multiply numerators and denominators)}$$

$$= \frac{6}{5} = 1\frac{1}{5} \quad \text{(convert to a mixed fraction)}$$

c. Division

> In order to divide fractions, multiply the first fraction by the **reciprocal** of the second. The simplest way to form a reciprocal is to 'invert' the fraction.

Note: Multiplying any rational number by its reciprocal always results in 1.
For example:

$$\frac{2}{5} \times \frac{5}{2} = 1 \ ; \ 7 \times \frac{1}{7} = 1$$

EXAMPLE E1.3
Evaluate the following without using your calculator:

(i) $\quad \dfrac{1}{3} \div \dfrac{2}{5}$

(ii) $\quad 1\dfrac{1}{3} \div \dfrac{4}{9}$

Solution

(i) $\quad \dfrac{1}{3} \div \dfrac{2}{5} = \dfrac{1}{3} \times \dfrac{5}{2}$ $\qquad$ (multiply by reciprocal)

$\qquad = \dfrac{1 \times 5}{3 \times 2}$ $\qquad$ (multiply numerators and denominators)

$\qquad = \dfrac{5}{6}$

(ii) $\quad 1\dfrac{1}{3} \div \dfrac{4}{9} = \dfrac{4}{3} \div \dfrac{4}{9}$ $\qquad$ (convert to an improper fraction)

$\qquad = \dfrac{4}{3} \times \dfrac{9}{4}$ $\qquad$ (multiply by reciprocal)

$\qquad = \dfrac{\overset{1}{4}}{\underset{1}{3}} \times \dfrac{\overset{3}{9}}{\underset{1}{4}}$ $\qquad$ (simplify)

$\qquad = \dfrac{1 \times 3}{1 \times 1}$ $\qquad$ (multiply numerators and denominators)

$\qquad = \dfrac{3}{1} = 3$

A-2 Decimals

a. Addition and/or subtraction

> In order to add or subtract decimals, first align the decimal points and then perform the calculation.

EXAMPLE E1.4
Evaluate 1.32 + 0.156

Solution

```
  1.320 +
  0.156
  ─────
  1.476
```

Note: 1.32 was written as 1.320 so that both numerals contain the same number of places after the decimal point.

b. Multiplication and/or division

> The simplest way to multiply or divide decimals is to convert each to a fraction first and then follow the operating rules for fractions stated above.

EXAMPLE E1.5
Evaluate 2.4 ÷ 0.03

Solution

$2.4 \div 0.03 = 2\dfrac{4}{10} \div \dfrac{3}{100}$ $\quad$ (convert each decimal to a fraction)

$\qquad = \dfrac{24}{10} \div \dfrac{3}{100}$ $\quad$ (convert to an improper fraction)

$\qquad = \dfrac{24}{10} \times \dfrac{100}{3}$ $\quad$ (multiply by the reciprocal)

$\qquad = \dfrac{\overset{8}{24}}{\underset{1}{10}} \times \dfrac{\overset{10}{100}}{\underset{1}{3}}$ $\quad$ (simplify)

$\qquad = \dfrac{8 \times 10}{1 \times 1}$ $\quad$ (multiply numerators and denominators)

$\qquad = \dfrac{80}{1} = 80$

Extension B: Metrics

Metrics (sometimes called **Standard International or SI Units**) refers to a decimal system of length, mass and capacity.
The following table lists the common units and their corresponding relationships for each measure.

Measure	Common units	Abbrev-iation	Relation-ships
Length	kilometre metre centimetre millimetre	km m cm mm	1 km = 1000 m 1 m = 100 cm 1 cm = 10 mm
Mass	tonne kilogram gram milligram	t kg g mg	1 t = 1000 kg 1 kg = 1000 g 1 g = 1000 mg
Capacity	kilolitre litre millilitre	kL L ml	1 kL = 1000 L 1 L = 1000 ml

Note: Further units of measure are given in Appendix D (e.g. megametre, hectogram, decalitre).

Conversions

> To convert from a *smaller* unit to a *larger* unit, **divide** by 10, 100, 1000 etc., as appropriate.
> To convert from a *larger* unit to a *smaller* unit, **multiply** by 10, 100, 1000 etc., as appropriate.

EXAMPLE E1.6

(i)   1.25 m = 125 cm (multiply by 100)

(ii)   2200 mg = 2.2 g (divide by 1000)

(iii) 7.6 L = 7600 ml (multiply by 1000)

(iv)   350 mm = 35 cm (divide by 10)

(v)   125 g = 0.125 kg (divide by 1000)

(vi)   67.25 cm = 672.5 mm (multiply by 10)

(vii) 50 kg = 0.05 t (divide by 1000)

(viii) 0.6 kL = 600 mls (multiply by 1000)

Extension C: Absolute value

The concept of **absolute value** relies (at this stage) on an understanding of the number line often used to represent the real number system (*Section 5*). A useful representation of the number line is

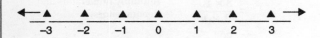

> The **absolute value** of a number may now be defined to be its *distance* from zero. It is denoted by the symbol '| |'.

EXAMPLE E1.7

(i) $|3| = 3$

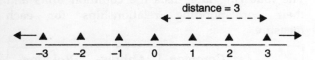

(ii) $|-2| = 2$

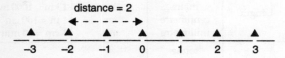

(iii) $\left|\dfrac{-3}{4}\right| = \dfrac{3}{4}$

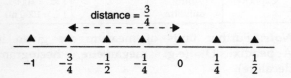

(iv) $|0.5| = 0.5$

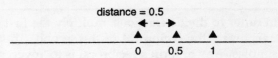

(v) $\left|-1\dfrac{1}{4}\right| = 1\dfrac{1}{4}$

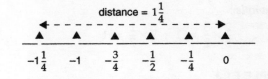

(vi) $|6 - 2| = |4| = 4$

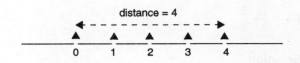

(vii) $|4 - 7| = |-3| = 3$

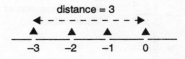

(viii) $\left|\dfrac{1}{2} - \dfrac{3}{4}\right| = \left|-\dfrac{1}{4}\right| = \dfrac{1}{4}$

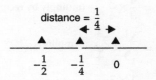

(ix) $\left|-8 + 4\dfrac{1}{2}\right| = \left|-3\dfrac{1}{2}\right| = 3\dfrac{1}{2}$

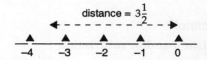

(x) $|-2 + 3| = |1| = 1$

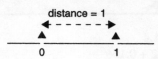

EXAMPLE E1.8

Evaluate the following expressions requiring a mixture of basic operations on two or more absolute values.

(i) $|-2| + |3| = 2 + 3 = 5$

(ii) $|2| + |-3| = 2 + 3 = 5$

(iii) $|-2| - |-3| = 2 - 3 = -1$

(iv) $|-5| \times |-2| - |-3| = 5 \times 2 - 3 = 7$

(v) $|-5| \times -2 + \left|\dfrac{-1}{2}\right| = 5 \times (-2) + \dfrac{1}{2} = -20$

Extension D: Significant figures

'**Significant figures**' is a term used to indicate a required level of accuracy.

The concept is best explained in terms of zeros and non-zero numerals, as follows:

(a) All non-zero numerals are **significant**. e.g. 345.6 is correct to 4 significant figures.

(b) One or more zeros occurring between two non-zero numerals are **significant**. e.g. 60.23, 1002, .2406 are all correct to 4 significant figures.

(c) With decimal numbers (less than one), all zeros **to the left** of the first non-zero numeral are **non-significant**. e.g. 0.008765 is correct to 4 significant figures since the zeros to the left of 8 are non-significant.

(d) With whole numbers, all zeros **to the right** of the last non-zero numeral are generally **non-significant**. e.g. 709200 is correct to 4 significant figures since the zeros to the right of 2 are non-significant.

> **In summary**
> - All non-zero figures and any zeros between non-zero figures are **significant**.
> - All zeros at the beginning of a decimal or at the end of a whole number are **non-significant**.

EXAMPLE E1.9

Given number	Number correct to 3 significant figures
(i) 0.04672 (4 significant figures)	0.0467 (the zeros are non-significant)
(ii) 48.008 (5 significant figures)	48.0 (the zeros are significant)
(iii) 0.056000 (5 significant figures)	0.0560
(iv) 700.735 (6 significant figures)	701 (the zeros are significant)
(v) 0.0007048 (4 significant figures)	0.000705 (all zeros in front of 7 are non-significant)
(vi) 430800 (4 significant figures)	431000 (the zeros after the 8 are non-significant)

Summary

This chapter has presented several topics that may be included under the broad heading of **basic arithmetic**, namely,

1. *Arithmetic Operations*, covering

(i) the four basic operations of $+$, $-$, $\times$ and $\div$

(ii) the order of operations when simplifying an expression involving mixed operations. (*Note:* most calculators do this automatically)

(iii) further symbols for later use, e.g. $>$, $<$, $\geq$, $\leq$, $\neq$, $\approx$, $\therefore$

2. *Fractions*, e.g. $\dfrac{2 \leftarrow \text{numerator}}{5 \leftarrow \text{denominator}}$, covering

(i) types of fractions

- proper e.g. $\dfrac{2}{5}$

- improper e.g. $\dfrac{8}{3}$

- mixed e.g. $2\dfrac{2}{3}$

- equivalent e.g. $\dfrac{1}{2} = \dfrac{2}{4} = \dfrac{5}{10} = \dfrac{6}{12}$

(ii) operations $(+, -, \times, \div)$ applied to fractions. These are most easily carried out using the fraction button $\boxed{a^b_c}$ on your calculator.

e.g. $\dfrac{2}{5}$ using calculator, 2 $\boxed{a^b_c}$ 5 gives result on display $\boxed{2 \rfloor 5.}$

3. *Decimals and percentages*, covering

(i) types of decimals

- terminating e.g. 0.752, 1.4

- repeating (recurring) e.g. $0.\dot{3} = 0.333...$, $1.\dot{2}\dot{7} = 1.272727...$

(ii) operations $(+, -, \times, \div)$ with decimals and percentages. These are most easily performed on a calculator remembering to convert all percentages to decimal form first.

(iii) conversions

- decimal $\rightarrow$ percentage e.g. $0.34 = 34\%$

- percentage $\rightarrow$ decimal e.g. $8\% = 0.08$

- fraction $\rightarrow$ percentage e.g. $\dfrac{3}{4} = 75\%$

- percentage $\rightarrow$ fraction e.g. $25\% = \dfrac{1}{4}$

- fraction $\rightarrow$ decimal e.g. $\dfrac{7}{8} = 0.875$

- decimal $\rightarrow$ fraction e.g. $0.35 = \dfrac{35}{100} = \dfrac{7}{20}$

4. *Approximations,* **covering**

(i) **rounding off** e.g. 2.58 ≈ 2.6, rounded off to 1 decimal place.

(ii) **truncating** e.g. 2.58 ≈ 2.5, truncated to 1 decimal place.

5. *Ratios.* These compare two or more quantities using the symbol ':'
For example: 6 : 4 is read as '6 is to 4'.

(i) 5 : 2 and 10 : 4 are called **equivalent** ratios, as are 8 : 6 and 4 : 3.

(ii) **Operations** (+, −, ×, ÷) applied to ratios. First check that all numerals refer to the same units and then rewrite each in fraction form before performing any calculations.

6. The *real number system.* This is best represented using the diagram on page 8.

(i) When operating with negatives, e.g. ⁻2, *note* the following:

- multiplying or dividing two numbers with **unlike** signs (i.e. one positive and one negative) results in a **negative** answer.

- multiplying or dividing two numbers with **like** signs (i.e. both positive or both negative) results in a **positive** answer.

- do **not** confuse the '−' sign used to denote the operation of subtraction and the '⁻' sign used to denote a negative. The change-sign $\boxed{+/-}$ button on a calculator is used to indicate a negative.

  e.g. ⁻2: Using calculator, 2 $\boxed{+/-}$, gives on display $\boxed{-2.}$

7. **Extensions to chapter 1, covering**

A. *Manual operations* applied to fractions and decimals.
For example:

(i) $\frac{1}{3} + \frac{1}{2} = \frac{2}{6} + \frac{3}{6} = \frac{5}{6}$ (common denominator)

(ii) $\frac{2}{3} \times \frac{1}{5} = \frac{2}{15}$ (multiply 'top' and 'bottom')

(iii) $\frac{4}{7} \div \frac{2}{5} = \frac{4}{7} \times \frac{5}{2} = \frac{10}{7} = 1\frac{3}{7}$ (multiply by reciprocal)

(iv) $2.3 + 6.45 = \begin{array}{r} 2.30 + \\ 6.45 \\ \hline 8.75 \\ \hline \end{array}$ (align decimal points and add zeros if necessary)

(v) $1.2 + 0.5 = 1\frac{2}{10} + \frac{5}{10}$ (convert to fractions)

$= \frac{12}{10} \times \frac{10^1}{5}$ (multiply by reciprocal)

$= \frac{12}{5} = 2\frac{2}{5} = 2.4$

B. *Metrics* **(S.I. units).** This refers to a decimal system of length, mass and capacity. Refer to the table on page 11 for a list of common units of measure and their corresponding relationships.

When converting

(i) a smaller unit to a larger, **divide** by 10, 100, 1000 etc. as appropriate, e.g. 500 g = 0.5 kg.

(ii) a larger unit to a smaller, **multiply** by 10, 100, 1000 etc. as appropriate, e.g. 2.5 m = 250 cm.

C. The *absolute value* of a number, denoted '| |', is defined to be its **distance** from zero.

e.g. $\left| {}^-4 + 2 \right| = \left| {}^-2 \right| = 2$

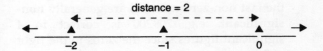

distance = 2

−2   −1   0

D. *Significant figures* indicate a required level of accuracy.
As a rule:

(i) All non-zero figures and any zeros between non-zero figures are **significant**.
e.g. 23.804 and 240.08 both have five significant figures.

(ii) all zeros at the beginning of a decimal or at the end of a whole number are *non-significant*.
e.g. 63700 and 0.00239 both have three significant figures.

1.1 $-6 + 3 + \dfrac{5}{2} - 3(1 + 6) = ?$

1.2 $-\dfrac{9}{2} - \dfrac{2}{3} + \dfrac{17}{-2} = ?$

1.3 $2 \times 3 + 3 \times 4 = ?$

1.4 $-2 \times -3 \times -5 = ?$

1.5 $9 - 3[6 - 2(4 + 2) - 3(1 - 2)] = ?$

1.6 $\dfrac{20}{4} - \dfrac{25}{5} = ?$

1.7 $\dfrac{1}{6} + \dfrac{4}{3} \times \dfrac{1}{2} = ?$

1.8 $\dfrac{1}{2} \times 1\dfrac{4}{5} \div \dfrac{2}{3} = ?$

1.9 $6 + 5 \div 2 - \dfrac{104}{52} = ?$

1.10 A carpenter is paid at the rate of $24.50 per hour. What is his wage for
(a) an 8-hour day?
(b) a 38-hour week?

1.11 A fruit picker receives $5.80 per basket. How much does Sam earn in a week where the number of baskets picked were: Monday 17, Tuesday 10, Wednesday 23, Thursday 18 and Friday 30.

*1.12 A metal worker is paid at the rate of $10.35 per hour for working 36 hours per week. Any overtime is paid at one-and-a-half times the hourly rate for each hour worked. Find the wage in a week where a worker works 50.2 hours.

1.13 A cosmetics sales person works part-time, receiving 23% commission on all sales. If sales totalled $625 in a week, how much is earned in commission?

1.14 Which is cheaper and by how much per week:
(i) a shop let at $205 a week, or
(ii) a shop leased at $10 556 per year?

1.15 The Fast Print Shop has a production capacity of 30 000 units per week. The daily production last week was: Monday 3 275; Tuesday 10 321; Wednesday 1 245; Thursday 3 158 and Friday 8 213. What is the difference between the actual production last week and the production capacity?

1.16 Fluffy socks cost $2.534 per pair to make and they sell for $4.55 per pair. Determine the profit on 175 pairs. (Round your answer to the nearest cent.)

1.17 If three teaspoons of honey are required to make a particular pie, how many pies can be made with 108 teaspoons of honey?

*1.18 To make beer, a recipe says that 3.5 cups of sugar and 3 tablespoons of malt are needed as well as other ingredients. If this recipe will serve 5, how many cups of sugar and tablespoons of malt would be necessary
(a) to serve 10?
(b) to serve 12?

1.19 Statistics show that $73\dfrac{1}{2}$ out of every 100 people in Australia earn in excess of $10 100 annually. If we gathered 554 people at random from the population, how many would you expect to earn less than $10 100 annually?

1.20 Each sheet of steel is 0.35 cm thick. If 355 sheets are stacked together, find the total thickness of the stack.

*1.21 A sum of money is invested at 8.5% per annum. The interest is $825. How much money is invested?

1.22 A student has 78 points out of a possible 125 on a test. Find the percentage to the nearest whole number.

*1.23 Given 60 ml of a 12% solution, what volume of 0.5% solution can be prepared?

1.24 If 45% of a senior citizen's income is spent on drugs, doctors and hospitals, how much of an income of $10 500 per year is spent in this way?

1.25 The doctor asks you to give your patient 8 tablets each day: one-half of them after breakfast, one-quarter of them after lunch and the rest after dinner. How many tablets will you administer
(a) after breakfast?
(b) after lunch?
(c) after dinner?

* Indicates problems are more difficult.

Elementary algebra

Introduction

Algebra is often thought of as the symbolic language of mathematics just as shorthand is the symbolic language of the written or spoken word. You will see as you progress through this and later chapters that it allows us to describe mathematical relationships clearly and concisely.

As with the previous chapter, there are again several terms that will appear frequently and hence require some attention before we proceed further.

> A **variable**, in an algebraic context, represents a quantity that may assume any of a set of values and is usually denoted by a letter or a symbol, sometimes referred to as a **pronumeral**.

> An **algebraic expression** is a mathematical statement that contains one or more terms involving either symbols, letters or a combination of symbols, letters or numerals, separated by basic operating symbols.

For example: $x + y - 3$ is an algebraic expression with three terms.

This chapter is the first of two algebra chapters and should be thought of as an 'entree' only, with the 'main course' to follow in Chapter 4.

Here there are three sections to be considered. The *first* discusses operations as they are applied to the simplification of and the substitution into algebraic expressions, while the *second* and *third* present the algebraic laws and properties. All sections include a number of worked examples together with activities for you to undertake to check your understanding of the material presented.

In this chapter, there are only two topics to be presented under the heading of **extensions**. The first is due to the reference made in Chapter 1 to the use of the algebraic laws in Chapter 2 to explain some rather difficult concepts associated with negatives. The second involves a brief

discussion of simple algebraic fractions. This topic is presented here for both interest and information as it will appear in later chapters.

Notation

When two or more variables and numbers are combined using the operation of multiplication, the corresponding sign is omitted.

$a \times b$	is written as ab
$2 \times x \times y$	$\rightarrow 2xy$
$c \times 3 \times d$	usually $\rightarrow 3cd$
$5 \times z \times 2 \times x$	usually $\rightarrow 10xz$
$1 \times c$	$\rightarrow c$ (since $1c = c$)
$r \times 1$	$\rightarrow r$

2.1 Operations

2.1.1 Simplification

a. Addition and subtraction

> In any algebraic expression, only **'like'** terms can be added or subtracted.

Note: The word 'like' in this context refers to terms that are of the same type.

For example: a and $2a$ are 'like' terms, whereas a and $3b$ are not.

EXAMPLE 2.1

Simplify each of the following algebraic expressions by collecting the 'like' terms.

(i) $a + b - 2a + 3b + 5a$

(ii) $4r - 3s - 5t + r - 2t + 4s$

(iii) $6xy - 4x + 2y - 3xy - y$

(iv) Add $2x - 3y + 5$ to $5x + y - 4$

(v) Find the sum of $4m + 2$, $2n - 5$, $n - 2m - 1$

Solution

(i) $\quad a + b - 2a + 3b + 5a$

$\quad = a - 2a + 5a + b + 3b$

$\quad = 6a - 2a + 4b$

$\quad = 4a + 4b$

(ii) $4r - 3s - 5t + r - 2t + 4s$

$\quad = 4r + r - 3s + 4s - 5t - 2t$

$\quad = 5r + s - 7t$

(iii) $6xy - 4x + 2y - 3xy - y$

$\quad = 6xy - 3xy - 4x + 2y - y$

$\quad = 3xy - 4x + y$

(iv) $5x + y - 4 + 2x - 3y + 5$

$\quad = 5x + 2x + y - 3y - 4 + 5$

$\quad = 7x - 2y + 1$

(v) $4m + 2 + 2n - 5 + n - 2m - 1$

$\quad = 4m - 2m + 2n + n + 2 - 5 - 1$

$\quad = 2m + 3n - 4$

b. Multiplication and/or division

> ***Multiplication:*** Multiply together the components of each term and omit the × sign.
> e.g. $\quad 2x \times 3y = 6xy$
>
> ***Division:*** Write the two terms in fraction form with the first as the numerator, then simplify as for fractions.
>
> e.g. $\quad 6x \div 2x = \dfrac{\cancel{6}^3 \cancel{x}^1}{\cancel{2}_1 \cancel{x}_1} = 3$

EXAMPLE 2.2

Simplify each of the following algebraic expressions by performing the operations indicated.

(i) $9ab \times 6$

(ii) $9ab \div 6a$

(iii) $5r \times 2s \times 3t$

(iv) $15ab \div 3ac$

(v) $2y \div 20x \times 3xz$

(vi) $4mn \div (2p \times 5m)$

Solution

(i) $9ab \times 6 = 54ab$

(ii) $9ab \div 6a = \dfrac{9ab}{6a}$

$\quad = \dfrac{\cancel{9}^3 \cancel{a}^1 b}{\cancel{2}\cancel{6}_1 \cancel{a}}$

$\quad = \dfrac{3b}{2}$

(iii) $5r \times 2s \times 3t = 10rs \times 3t$

$\quad = 30rst$

(iv) $15ab \div 3ac = \dfrac{15ab}{3ac}$

$\quad = \dfrac{\cancel{15}^5 \cancel{a}^1 b}{\cancel{3}_1 \cancel{a} c}$

$\quad = \dfrac{5b}{c}$

(v) $2y \div 20x \times 3xz = \dfrac{2y}{20x} \times 3xz$

$\quad = \dfrac{\cancel{2}^1 y}{\cancel{20}_{10} \cancel{x}} \times 3x^1 z$

$\quad = \dfrac{3yz}{10}$

(vi) $4mn \div (2p \times 5m) = \dfrac{4mn}{2p \times 5m}$

$\quad = \dfrac{4mn}{10pm}$

$\quad = \dfrac{\cancel{4}^2 \cancel{m}^1 n}{\cancel{5}\cancel{10} p_1 \cancel{m}}$

$\quad = \dfrac{2n}{5p}$

c. Mixed operations

When a mixture of basic operations is involved, simplify the algebraic expression by following the 'order of operations' rule given in Chapter 1.

EXAMPLE 2.3

Simplify each of the following expressions.

(i) $(3a + 6a) \div a$

(ii) $2a \times b + 4c \div 2$

(iii) $12st \div 3s - 4t + 5s$

Solution

(i) $(3a + 6a) \div a = \dfrac{9a}{a}$

$\quad = \dfrac{9\cancel{a}^1}{1\cancel{a}}$

$\quad = 9$

(ii) $2a \times b + 4c \div 2 = 2ab + \dfrac{4c}{2}$

$\quad = 2ab + \dfrac{\cancel{4}^2 c}{\cancel{2}_1}$

$\quad = 2ab + 2c$

(iii)

$$12st + 3s - 4t + 5s = \frac{12st}{3s} - 4t + 5s$$

$$= \frac{\overset{4}{\cancel{12}}\,\overset{1}{\cancel{s}}t}{\cancel{3}_1\,\cancel{s}_1} - 4t + 5s$$

$$= 4t - 4t + 5s$$

$$= 5s$$

2.1.2 Numerical substitution

Numerical substitution involves replacing one or more variables in an algebraic expression with a numeral.

EXAMPLE 2.4

Solve each of the following:

(i) If $p = 3$ and $q = 2$, find the value of $5pq$.

(ii) Given $x = a + 4b$, find the value of x if $a = -2$ and $b = \frac{1}{2}$.

(iii) If $r = -3$, $s = 5$ and $t = -\frac{1}{2}$, evaluate the expression $(r + s + 2t) \times \frac{r}{t} + 2$.

(iv) If $F = \frac{9c}{5} + 32$, find the value of F when $c = 10$ and when $c = 35$.

(v) If $A = \frac{1}{2}bh$, find the value of h given $A = 50$ and $b = 10$.

Solution

(i) If $p = 3$ and $q = 2$, then

$$5pq = 5 \times 3 \times 2$$

$$= 30$$

(ii) If $a = -2$ and $b = \frac{1}{2}$, then

$$x = a + 4b$$

$$= -2 + \overset{2}{\cancel{4}} \times \frac{1}{\cancel{2}_1}$$

$$= -2 + 2$$

$$= 0$$

(iii) If $r = -3$, $s = 5$ and $t = -\frac{1}{2}$, then

$$(r + s + 2t) \times \frac{r}{t} + 2$$

$$= \left(-3 + 5 + \overset{1}{\cancel{2}}\,z \times \left(-\frac{1}{\cancel{2}_1}\right)\right) \times \frac{-3}{\left(-\frac{1}{2}\right)} + 2$$

$$= (-3 + 5 - 1) \times -3 \times \left(-\frac{2}{1}\right) + 2$$

$$= 1 \times 6 + 2$$

$$= 8$$

(iv) When $c = 10$, $F = \frac{9c}{5} + 32$

$$= \frac{9 \times \overset{2}{\cancel{10}}}{\cancel{5}_1} + 32$$

$$= 18 + 32$$

$$= 50$$

When $c = 35$, $F = \frac{9c}{5} + 32$

$$= \frac{9 \times \overset{7}{\cancel{35}}}{\cancel{5}_1} + 32$$

$$= 63 + 32$$

$$= 95$$

(v) Let $A = \frac{1}{2}bh$ $\therefore$ if $A = 50$ and $b = 10$,

$$50 = \frac{1}{2} \times 10 \times h$$

$$= 5 \times h$$

Hence, $10 = h$ or $h = 10$ (as $5 \times 10 = 50$).

YOUR TURN

2.1 Express the following algebraic expressions in their simplest form by collecting the 'like' terms.
  (i) $10 - 2a - 7b - 12 + 10b + 2a$
  (ii) $3x + 5z - 9x + 4y - 2x + 2y$
  (iii) $1\frac{1}{2}t + 2s - \frac{1}{2}t - \frac{3}{4}r + s - \frac{1}{4}r$
  (iv) $5n - 3m - 2m + 7n - 4m$
  (v) $3xy - zx + 2yz - 2zx - xy + 4zx$

2.2 Simplify each of the following by carrying out the operations indicated.
  (i) $6r \times 3st$
  (ii) $-2a \times -4b$
  (iii) $x \times 2yz \times 4$
  (iv) $-m \times -2n \times -3p \times 5q$
  (v) $4xyz \times -\frac{1}{2}$

2.3 Express each of the following in its simplest form.
  (i) $2a \times 4b \div 4a$

2.3 *Cont.*

   (ii) $3abc \div \frac{1}{2}ac$

   (iii) $3(x + y) \div 3$
   (iv) $3(x + y) \div (x + y)$
   (v) $12r(2s - t) \div 4(2s - t)$

2.4 If $a = -1$, $b = 4$ and $c = \frac{1}{2}$, evaluate each of the following algebraic expressions.
   (i) $a + b + 2c$
   (ii) $3abc$

   (iii) $\dfrac{6a}{2b + 4c}$

   (iv) $\dfrac{2(a + c)}{4c - a}$

   (v) $\dfrac{ab - 6ac}{a}$

2.5 Answer each of the following.
   (i) If $a = 6.4$, $b = 3.2$ and $c = 0.8$ find, rounded off to 1 decimal place, the value of $\frac{1}{2}abc$.

   (ii) Calculate, rounded off to 2 decimal places, the value of $\frac{xy}{z}$ when $x = 2.86$, $y = 4.23$ and $z = 1.93$.

   (iii) If $s = ut + \frac{1}{2}at$, find s when $u = 20$, $t = 4$ and $a = -3$.

   (iv) The formula $C = \frac{5}{9}(F - 32)$ can be used to convert a Fahrenheit temperature reading to a Celsius reading. Find the equivalent temperature in degrees centigrade for a reading of 100°F.

   (v) If a vehicle is travelling at k kilometres per hour, then its speed m (in metres per second) is given by the formula
$$m = \frac{5k}{18}.$$
If Susan is driving along the motorway at 97.2 km/h, calculate her speed in metres per second.

2.2   Algebraic laws

a. The commutative law

> **(i)** for **addition:** $a + b = b + a$
> **(ii)** for **multiplication:** $ab = ba$

For example:

(i) $3 + 5 = 5 + 3 = 8$

(ii) $a + 2a = 2a + a = 3a$

(iii) $3 \times 5 = 5 \times 3 = 15$

(iv) $xy = yx$

(v) $2a \times 3b = 3b \times 2a$

> The order in which two real numbers are added or multiplied does not affect the result.

b. The associative law

> **(i)** for **addition** : $(a + b) + c = a + (b + c)$
> **(ii)** for **multiplication:** $(ab)c = a(bc)$
>   i.e. $(a \times b) \times c = a \times (b \times c)$

For example:

(i) $(4 + 6) + 3 = 10 + 3 = 13$;
      $4 + (6 + 3) = 4 + 9 = 13$
      $\therefore (4 + 6) + 3 = 4 + (6 + 3)$

(ii) $(3t + 2t) + t = 5t + t = 6t$;
      $3t + (2t + t) = 3t + 3t = 6t$
      $\therefore (3t + 2t) + t = 3t + (2t + t)$

(iii) $(2 \times 4) \times 3 = 8 \times 3 = 24$;
      $2 \times (4 \times 3) = 2 \times 12 = 24$
      $\therefore (2 \times 4) \times 3 = 2 \times (4 \times 3)$

(iv) $(3x \times y) \times 2z = 3xy \times 2z = 6xyz$;
      $3x \times (y \times 2z) = 3x \times (2yz) = 6xyz$
      $\therefore (3x \times y) \times 2z = (3xy)2z = 3x(2yz)$

c. The distributive law

> **(i)** $a(b + c) = ab + ac$
> **(ii)** $(b + c)a = ba + ca$

For example:

(i) $3(2 + 5) = 3 \times 7 = 21$;
      $3 \times 2 + 3 \times 5 = 6 + 15 = 21$
      $\therefore 3(2 + 5) = 3(2) + 3(5)$

(ii) $6m(3n + n) = 6m \times 4n = 24mn$
      $6m \times 3n + 6m \times n = 18mn + 6mn = 24mn$
      $\therefore 6m(3n + n) = 18mn + 6mn$

d. The identity law

> **(i)** for **addition:** $a + 0 = 0 + a = a$
>   Here the number 0 is called the *additive identity*.
> **(ii)** for **multiplication:** $a \times 1 = 1 \times a = a$
>   Here the number 1 is called the *multiplicative identity*.

For example:

(i) $6 + 0 = 0 + 6 = 6$

(ii) $d + 0 = 0 + d = d$

(iii) $7 \times 1 = 1 \times 7 = 7$

(iv) $rs \times 1 = 1 \times rs = rs$

e. The inverse law

(i) for **addition:** $a + (\overline{}a) = (\overline{}a) + a = 0$
  Here $\overline{}a$ is called the *additive inverse*.

(ii) for **multiplication:** $a \times \dfrac{1}{a} = \dfrac{1}{a} \times a = 1$
  Here $\dfrac{1}{a}$ is called the *multiplicative inverse* (provided $a \neq 0$, since $\dfrac{1}{0}$ is undefined).

For example:

(i) $2 + \overline{}2 = 0$

(ii) $\overline{}p + p = 0$

(iii) $6 \times \dfrac{1}{6} = 1$

(iv) $\dfrac{1}{x} \times x = 1$ (for $x \neq 0$)

EXAMPLE 2.5

Using one or more of the algebraic laws, change the form of the following expressions if possible.

(i) $r(3 + 2s)$

(ii) $12mn + 6m$

(iii) $-2x + 2$

(iv) $xy + x$

(v) $-1(c - d)$

Solution

(i) $r(3 + 2s) = 3r + 2rs$ *(distributive law)*

(ii) $12mn + 6m = m(12n + 6)$
$= 6m(2n + 1)$

(iii) $-2x + 2 = 2 - 2x$
$= 2(1 - x)$

(iv) $xy + x = x(y + 1)$

(v) $-1(c - d) = -c - 1(-d)$
$= -c + d \text{ or } d - c$

EXAMPLE 2.6

Show that

(i) $-(b - c) + a = a - b + c$

(ii) $-2(x + 3) - 4x = -6(x + 1)$

Solution

(i) $-(b - c) + a = -b - (-c) + a$
$= -b + c + a \text{ or } a - b + c$

(ii) $-2(x + 3) - 4x = -2x - 6 - 4x$
$= -6x - 6$
$= -6(x + 1)$

2.6 Change the form of the following expressions using the algebraic laws.
  (i) $3c(4d + 2)$
  (ii) $14u + 7uv$
  (iii) $ab + abc$
  (iv) $7(m - 2n)$
  (v) $\overline{}4 + 2a$

2.7 Show that each of the following is true.
  (i) $2x + 5 - (2x + 6) = \overline{}1$
  (ii) $7 - 5(3s - 2) = 17 - 15s$
  (iii) $2(3p - 5) - 3(p - 6) = 3p + 8$
  (iv) $m - n - 3(m - n) = m - n + 3(n - m)$
  (v) $4(3 - \dfrac{1}{2}) = 12 - 2 = 10$

2.8 Which of the algebraic laws is used in each of the following?
  (i) $4 + 3 = 3 + 4$
  (ii) $6 \times 1 = 6$
  (iii) $\overline{}q \times \dfrac{-1}{q} = 1$
  (iv) $4(5 \times 6) = (5 \times 6) \times 4$
  (v) $c + \overline{}c = 0$

2.9 Which of the algebraic laws is used in each of the following?
  (i) $(8 \times 5) \times 2 = 8 \times (5 \times 2)$
  (ii) $4 \times 3 = 3 \times 4$
  (iii) $9 \times 4 - 9 \times 2 = 9(4 - 2)$
  (iv) $xy(a + 2b) = xya + 2xyb$
  (v) $2cd - 6c = 2c(d - 3)$

2.10 Find both the additive inverse and the multiplicative inverse of each of the following.
  (i) 6 **(ii)** $\dfrac{1}{5}$ **(iii)** 2.4
  (iv) $\overline{}4$ **(v)** $2\dfrac{1}{4}$

2.3 Some useful properties of real numbers

a. The cancellation property

(i) for **addition:** If $a + c = b + c$, then $a = b$.

(ii) for **multiplication:** If $ac = bc$, then $a = b$
  (provided $c \neq 0$).

For example:

(i) If $5 + 3 = b + 3$,
   then $b = 5$.

(ii) If $7 \times 2 = b \times 2$,
   then $b = 7$.

b. Multiplication by zero

$$a \times 0 = 0 \times a = 0$$

For example:

(i) $x \times 0 = 0$

(ii) $5 \times 0 = 0$

(iii) $0 \times {}^-6 = 0$

(iv) $0 \times \dfrac{3}{4} = 0$

c. An extension of the inverse law

(i) for **addition**: If $a + b = 0$, then $a = {}^-b$.

(ii) for **multiplication**: If $ac = 1$, then $a = \dfrac{1}{c}$.

For example:

(i) $x + 2y = 0$, so $x = {}^-2y$.

(ii) $x + \dfrac{{}^-1}{2} = 0$, so $x = \dfrac{1}{2}$.

(iii) $s - 3t = 0$, so $s = 3t$.

(iv) $2xy = 1$, so $x = \dfrac{1}{2y}$.

(v) $5d = 1$, so $d = \dfrac{1}{5}$.

d. Subtraction

The difference between a and b may be written
as $\qquad a - b = a + {}^-b$
Hence, $a(b - c) = a(b + {}^-c)$
$\qquad\qquad\quad = ab + {}^-ac$ *(by the distributive law)*
$\qquad\qquad\quad = ab - ac$

For example:

(i) $3(x - y) = 3x - 3y$

(ii) $6m - 9n = 3(2m - 3n)$

(iii) $5a + {}^-3a = 5a - 3a = 2a$

Extensions to chapter 2

Extension A: Rules when operating with negatives

Although these rules were presented in *Section 5* of Chapter 1, they are now re-interpreted here, for any interested readers, using the algebraic laws.

1. $\quad {}^-a + {}^-b = {}^-(a + b)$ **Why?**

$\begin{aligned}
{}^-a + {}^-b &= 1({}^-a) + 1({}^-b) && \textit{multiplicative identity}\\
&= {}^-1(a) + {}^-1(b) && \textit{commutative law}\\
&= {}^-1(a + b) && \textit{distributive law}\\
&= {}^-(a + b)
\end{aligned}$

2. $\quad a \times {}^-b = {}^-(ab)$ **Why?**

$\begin{aligned}
a \times {}^-b &= a \times 1 \times {}^-b && \textit{multiplicative identity}\\
&= {}^-1 \times a \times b && \textit{commutative law}\\
&= {}^-(ab)
\end{aligned}$

3. $\quad {}^-({}^-a) = a$ **Why?**

$\begin{aligned}
{}^-({}^-a) + {}^-a &= 0 && \textit{additive inverse}\\
{}^-({}^-a) + {}^-a &= a + {}^-a && \textit{additive inverse}\\
{}^-({}^-a) &= a && \textit{cancellation property}
\end{aligned}$

4. $\quad {}^-a \times {}^-b = ab$ **Why?**

$\begin{aligned}
{}^-a \times {}^-b &= 1 \times {}^-a \times 1 \times {}^-b && \textit{multiplicative identity}\\
&= {}^-1 \times a \times {}^-1 \times b && \textit{commutative law}\\
&= {}^-1 \times {}^-1 \times a \times b && \textit{commutative law}\\
&= {}^-({}^-1 \times 1) \times a \times b && \textit{property 2 above}\\
&= {}^-({}^-1)(ab) && \textit{multiplicative identity}\\
&= 1(ab) && \textit{property 3 above}\\
&= ab
\end{aligned}$

EXAMPLE E2.1

	Examples	Rule applied
(i)	${}^-10 + {}^-4\frac{1}{2} = {}^-(10 + 4\frac{1}{2})$ $= {}^-14\frac{1}{2}$	Rule 1
(ii)	${}^-4x + {}^-2x = {}^-(4x + 2x)$ $= {}^-6x$	Rule 1
(iii)	$5 \times \dfrac{{}^-2}{5} = {}^-(\cancel{5}^1 \times \dfrac{2}{\cancel{5}_1})$ $= {}^-2$	Rule 2
(iv)	$x \times {}^-3y = {}^-(x \times 3y)$ $= {}^-(3xy)$	Rule 2
(v)	${}^-8 \times {}^-2 = 8 \times 2$ $= 16$	Rule 4
(vi)	${}^-4m \times {}^-5n = 4m \times 5n$ $= 20mn$	Rule 4

Extension B: Algebraic fractions

An algebraic fraction is any fraction whose component parts are letters or letters and numerals rather than just numerals.

Note: Algebraic fractions follow the same operating rules with respect to addition, subtraction, multiplication and division as arithmetic fractions.

Operating rules for algebraic fractions

a. Cancellation

$$\frac{am}{bm} = \frac{a}{b}$$

Note: In $\frac{a+m}{b+m}$ the '*m*' *cannot* be cancelled.

b. Addition and subtraction

(i) $\dfrac{a}{c} + \dfrac{b}{c} = \dfrac{a+b}{c}$ (for $c \neq 0$)

$\dfrac{a}{c} - \dfrac{b}{c} = \dfrac{a-b}{c}$ (for $c \neq 0$)

(ii) $\dfrac{a}{b} + \dfrac{c}{d} = \dfrac{ad+cb}{bd}$ (for $b, d \neq 0$)

$\dfrac{a}{b} - \dfrac{c}{d} = \dfrac{ad-cb}{bd}$ (for $b, d \neq 0$)

c. Multiplication and division

$$\frac{a}{b} \times \frac{c}{d} = \frac{ac}{bd} \qquad \text{(for } b, d \neq 0)$$

$$\frac{a}{b} \div \frac{c}{d} = \frac{a}{b} \times \frac{d}{c} = \frac{ad}{bc} \text{ (for } b, c, d \neq 0)$$

EXAMPLE E2.2

Simplify each of the following algebraic fractions.

(i) $\dfrac{3 \times 9x}{3}$

(ii) $\dfrac{3 + 9x}{3}$

(iii) $\dfrac{8xy - 2x}{2x}$

(iv) $\dfrac{a}{s} - \dfrac{2}{s}$

(v) $\dfrac{m}{m+n} + \dfrac{n}{m+n}$

Solution

(i) $\dfrac{3 \times 9x}{3} = \dfrac{\cancel{3}^1 \times 9x}{\cancel{3}_1}$

$= 9x$

(ii) $\dfrac{3 + 9x}{3} = \dfrac{3}{3} + \dfrac{9x}{3}$

$= 1 + \dfrac{\cancel{9}^3 x}{\cancel{3}_1}$

$= 1 + 3x$

(iii) $\dfrac{8xy - 2x}{2x} = \dfrac{8xy}{2x} - \dfrac{\cancel{2}^1 x}{\cancel{2x}_1}$

$= \dfrac{\cancel{8}^4 \cancel{x}^1 xy}{\cancel{2}_1 \cancel{x}_1} - 1$

$= 4y - 1$

(iv) $\dfrac{9}{s} - \dfrac{2}{s} = \dfrac{9-2}{s} = \dfrac{7}{s}$

(v) $\dfrac{m}{m+n} + \dfrac{n}{m+n} = \dfrac{\cancel{(m+n)}^1}{\cancel{(m+n)}_1} = 1$

EXAMPLE E2.3

Express each of the following as a single algebraic fraction.

(i) $\dfrac{b}{4} - \dfrac{d}{3}$

(ii) $\dfrac{1-q}{pq} + \dfrac{p+1}{p}$

(iii) $\dfrac{3st}{r} \times \dfrac{2r}{3s}$

(iv) $\dfrac{14xy}{t} \div \dfrac{7x}{2t}$

(v) $1 + \dfrac{m+n}{m-n}$

Solution

(i) $\dfrac{b}{4} - \dfrac{d}{3} = \dfrac{b \times 3}{4 \times 3} - \dfrac{d \times 4}{3 \times 4}$

$= \dfrac{3b}{12} - \dfrac{4d}{12}$

$= \dfrac{3b - 4d}{12}$

(ii) $\dfrac{1-q}{pq} + \dfrac{p+1}{p} = \dfrac{1-q}{pq} + \dfrac{q(p+1)}{qp}$

$= \dfrac{1-q}{pq} + \dfrac{qp+q}{qp}$

$= \dfrac{1-q+pq+q}{pq}$

$= \dfrac{1+pq}{pq}$

(iii) $\dfrac{3st}{r} \times \dfrac{2r}{3s} = \dfrac{3st \times 2r}{r \times 3s}$

$= \dfrac{6rst}{3rs}$

$= \dfrac{\overset{2}{\cancel{6}}\,\overset{1}{\cancel{r}}\,\overset{1}{\cancel{s}}\,t}{\underset{1}{\cancel{3}}\,\underset{1}{\cancel{r}}\,\underset{1}{\cancel{s}}}$

$= 2t$

(iv) $\dfrac{14xy}{t} \div \dfrac{7x}{2t} = \dfrac{14xy}{t} \times \dfrac{2t}{7x}$

$= \dfrac{\overset{2}{\cancel{14}}\,\overset{1}{\cancel{x}}y}{\underset{1}{\cancel{t}}} \times \dfrac{2\,\overset{1}{\cancel{t}}}{\underset{1}{\cancel{7}}\,\underset{1}{\cancel{x}}}$

$= 4y$

(v) $1 + \dfrac{m+n}{m-n} = \dfrac{m-n}{m-n} + \dfrac{m+n}{m-n}$

$= \dfrac{m-n+m+n}{m-n}$

$= \dfrac{2m}{m-n}$

Summary

This chapter is the first of two chapters that together discuss the more important and relevant topics under the broad heading of **algebra**. Here the notion of a variable and an algebraic expression are presented at the outset together with some notational conventions, followed by several sections dealing with the more basic algebraic topics.

1. Algebraic operations

A. Simplification

- addition and subtraction:
  e.g. $2a + 3a = 5a$, $6a - 2a = 4a$;
  but $6a - 2b$ cannot be simplified.

- multiplication:
  e.g. $4x \times 2y = 8xy$

- division:

  e.g. **(i)** $8x \div 2x = \dfrac{\overset{4}{\cancel{8}}x}{2\underset{1}{\cancel{x}}} = 4$

(ii) $8xy \div 2y = \dfrac{\overset{4}{\cancel{8}}x\,\overset{1}{\cancel{y}}}{\underset{}{\cancel{2}}\underset{1}{\cancel{y}}} = 4x$

Note: When simplifying an algebraic expression that involves a mixture of basic operations, always follow the **'order of operations'** rule given in Chapter 1.

B. Substitution

This involves replacing one or more variables in an algebraic expression by a number or a letter as directed, and then simplifying the result where possible.

e.g. If $x = 4$ and $y = 3$, then $2xy = 2 \times 4 \times 3 = 24$.

2. Algebraic laws

a. The *commutative* law for addition and multiplication

   (i) $a + b = b + a$

   (ii) $ab = ba$

b. The *associative* law for addition and multiplication

   (i) $(a + b) + c = a + (b + c)$

   (ii) $(ab)c = a(bc)$

c. The *distributive* law
   $a(b + c) = ab + ac$

d. The *identity* law for addition and multiplication

   (i) $a + 0 = 0 + a = 0$; '0' is the additive identity

   (ii) $a \times 1 = 1 \times a = a$; '1' is the multiplicative identity

e. The *inverse* law for addition and multiplication

   (i) $a + (^-a) = (^-a) + a = 0$; $'^-a'$ is the additive inverse

   (ii) $a \times \dfrac{1}{a} = \dfrac{1}{a} \times a = 1$; $'\dfrac{1}{a}'$ is the multiplicative inverse (for $a \neq 0$)

3. Useful properties

a. The *cancellation* property for addition and multiplication

   (i) $a + c = b + c$ implies that $a = b$

   (ii) $ac = bc$ implies that $a = b$ (for $c \neq 0$)

b. Multiplication *by zero*
   $a \times 0 = 0 \times a = 0$

c. An *extension* of *the inverse law* for addition and multiplication

   (i) $a + b = 0$ implies that $a = {}^-b$

   (ii) $ac = 1$ implies that $a = \dfrac{1}{c}$

d. The *subtraction* property

  (i) $a - b = a + {}^-b$

  (ii) $a(b - c) = ab - ac$

4. Extensions to chapter 2, covering

A. **Rules** when operating with negatives

  (i) ${}^-a + {}^-b = {}^-(a + b)$

  (ii) $a \times {}^-b = {}^-(ab)$

  (iii) ${}^-({}^-a) = a$

  (iv) ${}^-a \times {}^-b = ab$

B. **Algebraic fractions – operating rules**

  (i) **Cancellation**

  $$\frac{am}{bm} = \frac{a}{b}$$

  (ii) **Addition and subtraction**

  $$\frac{a}{c} + \frac{b}{c} = \frac{a + b}{c} \qquad (\text{for } c \neq 0)$$

  $$\frac{a}{c} - \frac{b}{c} = \frac{a - b}{c} \qquad (\text{for } c \neq 0)$$

  $$\frac{a}{b} + \frac{c}{d} = \frac{ad + cb}{bd} \qquad (\text{for } b, d \neq 0)$$

  $$\frac{a}{b} - \frac{c}{d} = \frac{ad - cb}{bd} \qquad (\text{for } b, d \neq 0)$$

  (iii) **Multiplication and division**

  $$\frac{a}{b} \times \frac{c}{d} = \frac{ac}{bd} \qquad (\text{for } b, d \neq 0)$$

  $$\frac{a}{b} \div \frac{c}{d} = \frac{a}{b} \times \frac{d}{c} = \frac{ad}{bc} \qquad (\text{for } b, c, d \neq 0)$$

2.1 Write an expression for the number that is
  (a) 5 greater than y
  (b) 3 less than x
  (c) three times as large as b
  (d) the sum of c and 10
  (e) 4 less than 5 times z.

2.2 Simplify
  (a) $2a + 4a$
  (b) $5x + 2x - 3x$
  (c) $8ac + ac$
  (d) $b + 4 + b - 5 + b + 1$
  (e) $5 \times 2x$
  (f) $2b \times 3$

2.3 Expand
  (a) $5(x + 2)$
  (b) $4(1 - 3b)$
  (c) $b(3 + x)$
  (d) $-4(x - 5)$
  (e) $-x(2 - c)$

2.4 If $a = 7$, $b = 3$, $c = -2$ and $d = 1$, find the value of:
  (a) $b - c$
  (b) ac
  (c) $a + c - d$
  (d) $b + bc$
  (e) $abcd$

2.5 Given $V = lbh$, find V when $l = 5$, $b = 2$, $h = 4$.

2.6 Given $M = \dfrac{r}{r - R}$, find M when $r = 13$ and $R = 3$.

2.7 Complete the following table, given that $y = \dfrac{12 - 3x}{4}$.

x	−2	0	5
y			

2.8 Given $3a + 2b = 12$, find a when $b = 3$.

2.9 Given $q = 6 + p$, find p when $q = -3$.

2.10 Write down a formula for each of the following:
  (a) the number c of cents in e dollars
  (b) the number s of seconds in n minutes
  (c) the number m of metres in y centimetres.

2.11 If $x = -2$ and $y = 4$, find $3xy - x$.

2.12 Use the formula $s = \dfrac{n}{2}(a + l)$ to find s, if $n = 60$, $a = 2$ and $l = 4$.

2.13 Find x if $\dfrac{x}{8} = \dfrac{65}{32}$.

2.14 Simplify $4x - 3y - 4(2x + 5y)$

2.15 If $A = 3(c - d)$, find c when $A = 60$ and $d = 4$.

3 Indices (plural of index)

Introduction

Index notation is used in a wide variety of mathematical contexts and certainly makes the life of a mathematician, a statistician, a number cruncher, a student and many others much easier. It is a relatively simple shorthand method of writing down and operating with either very large or very small numbers. In particular, the use of index numbers allows us to solve problems dealing with the metric system in a far more efficient and effective manner.

Before we proceed further, it is appropriate to define exactly what we mean by index notation.

> When any non-zero number is written in the form
> $$a^n = a.a.a. \dots a$$
> (the term 'a' is multiplied together n times), it is said to be written in **index** form (also referred to as **exponential** or **power** form).

Note:

(i) a^n consists of two parts: a, called the base and n, called the index, power or exponent.

That is,

$a^n \longleftarrow$ index

base

(ii) a^n is read 'a to the power of n' and is called 'the nth power of a'.

Now that everyone is familiar with the terminology and notation associated with indices, we proceed to look at what you can expect in the chapter ahead.

There are three sections in the core part of this chapter. The *first* presents examples of index numbers and index notation and introduces the special 'power' key on the calculator making computations far more efficient. The *second* section lists all the very important **index laws** together with examples of each, while the *third* section discusses the notion of **surds** and their associated operating rules.

As usual, all sections include a number of worked examples, followed by a set of activities for you to complete to check your understanding of the material presented.

This chapter will include two separate topics under the heading of **extensions**. The first discusses *scientific notation* - what it means as well as the interpretation of the calculator output, while the second examines both *logarithms* and *exponentials* – what they mean and how to operate with them using a calculator.

3.1 Index notation

In the introduction above, we defined
$$a^n = a \times a \times a \times \dots \times a$$
('a' multiplied together n times)
where a^n is called the nth power of a.

Now let us work through some examples and also see how a calculator may be used to simplify matters.

EXAMPLE 3.1
Expand each of the following and simplify where possible.

(i) 10^2

(ii) $(^-3)^4$

(iii) x^5

(iv) $(2a)^3$

(v) $2a^3$

(vi) $^-(\dfrac{-1}{2})^2$

(vii) $(1.3)^3$

(viii) $(^-4t)^3$

(ix) $(2.4m)^4$

Solution

(i) $10^2 = 10 \times 10 = 100$

(ii) $(^-3)^4 = {}^-3 \times {}^-3 \times {}^-3 \times {}^-3 = 9 \times 9 = 81$

Note: Do *not* confuse $(^-3)^4$ with $^-(3)^4$

$$(^-3)^4 = ^-3 \times ^-3 \times ^-3 \times ^-3 = 81$$

$$^-(3)^4 = ^-(3 \times 3 \times 3 \times 3) = ^-81$$

(iii) $x^5 = x \times x \times x \times x \times x$ or $x.x.x.x.x$

(iv) $(2a)^3 = 2a \times 2a \times 2a = 8a^3$

(v) $2a^3 = 2 \times a \times a \times a$

(vi) $^-(\frac{-1}{2})^2 = ^-(\frac{-1}{2} \times \frac{-1}{2}) = \frac{-1}{4}$

(vii) $(1.3)^3 = 1.3 \times 1.3 \times 1.3 = 2.197$

(viii) $(^-4t)^3 = ^-4t \times ^-4t \times ^-4t = ^-64t^3$

(ix) $2.4m \times 2.4m \times 2.4m \times 2.4m = 33.1776m^4$

Calculator usage

Indices involving only numeric values may be simplified more easily using your calculator. Look for the symbol x^y located either *on* a button or (in brackets) directly *above* a button. If your x^y is *above* a key then you must first press the 'shift' key or similar followed by the appropriate button (see Appendix E for further information on the 'shift' button).

To sum up, when simplifying an index number using your calculator,

(i) enter the base value

(ii) press the $\boxed{x^y}$ (raise to the power) key, preceded by 'shift' if necessary

(iii) enter the index number

(iv) press $\boxed{=}$ to obtain the answer.

EXAMPLE 3.2

Evaluate each of the following using your calculator.

(i) $(1.3)^3$

(ii) $(\frac{-1}{2})^2$

(iii) 6^4

(iv) $(^-3)^5$

(v) $(1\frac{3}{4})^3$

(vi) $(5.82)^4$

Solution

(i) $(1.3)^3 : 1.3\ \boxed{x^y}\ 3\ \boxed{=}\ \boxed{2.197}$

(ii) $(\frac{-1}{2})^2 : \boxed{(}\ 1\ \boxed{a_c^b}\ 2\ \boxed{^+/_-}\ \boxed{)}\ \boxed{x^y}\ 2\ \boxed{=}\ \boxed{0.25}$

Note:

a) With examples of this type, even though the data is fed into the calculator in fraction form, the answer on the display is in *decimal* form.

b) In examples involving squares (i.e. numbers raised to the power 2, e.g. 3^2), the $\boxed{x^2}$ button may be used instead of the $\boxed{x^y}$ key.

So, $(\frac{-1}{2})^2 : \boxed{(}\ 1\ \boxed{a_c^b}\ 2\ \boxed{^+/_-}\ \boxed{)}\ \boxed{x^2}\ \boxed{0.25}$

(iii) $6^4 : 6\ \boxed{x^y}\ 4\ \boxed{=}\ \boxed{1296.}$

(iv) $(^-3)^5 : \boxed{(}\ 3\ \boxed{^+/_-}\ \boxed{)}\ \boxed{x^y}\ 5\ \boxed{=}\ \boxed{^-243.}$

(v) $(1\frac{3}{4})^3 : \boxed{(}\ 1\ \boxed{a_c^b}\ 3\ \boxed{a_c^b}\ 4\ \boxed{)}\ \boxed{x^y}$
$3\ \boxed{=}\ \boxed{5.359375}$

(vi) $5.82^4 : 5.82\ \boxed{x^y}\ 4\ \boxed{=}\ \boxed{1147.3395}$

3.1 Write each of the following algebraic expressions in index form:

(i) $a \times a \times a \times a$

(ii) $5 \times m \times m \times m$

(iii) $x \times y \times x \times y$

(iv) $st \times st \times st \times st \times st$

(v) $4 \times p \times p$

(vi) $^-c \times ^-c \times ^-c$

(vii) $16 \times y \times y \times y \times y$

(viii) $7 \times a \times a \times b \times b \times b$

(ix) $3 \times x \times x \times y \times z \times z \times z$

(x) $27 \times d \times f \times e \times f \times d \times e \times e \times d \times f$

3.2 Evaluate each of the following indices using your calculator:

(i) 8^4

(ii) $(^-2)^6$

(iii) $(^-2)^7$

(iv) $^-2^6$

(v) $(6.72)^5$

(vi) $(4\frac{2}{3})^3$

(vii) $(41)^2$

(viii) $(\frac{-5}{8})^3$

(ix) $^-2.7^4$

(x) 0.38^2

3.2 Index laws

The table below states each of the index laws together with examples given in both numeric and algebraic form. It is assumed here that a and b are **positive rational numbers**, p and q are **non–zero real numbers** and m and n are **positive integers**.

Law	Numeric example	Algebraic example
(i) $a^p \times a^q = a^{p+q}$	$3^4 \times 3^5 = 3^{4+5} = 3^9$	$t^2 + t^3 = t^{2+3} = t^5$
(ii) $a^p \div a^q = a^{p-q}$ or $\dfrac{a^p}{a^q} = a^{p-q}$	$5^3 \div 5^2 = 5^{3-2} = 5^1 = 5$ or $\dfrac{5^3}{5^2} = 5^{3-2} = 5$	$d^5 \div d^3 = d^{5-3} = d^2$ or $\dfrac{d^5}{d^3} = d^{5-3} = d^2$
(iii) $(a^p)^q = a^{pq}$	$(2^2)^3 = 2^{2\times3} = 2^6 = 64$	$(k^2)^5 = k^{2\times5} = k^{10}$
(iv) $a^0 = 1$ (for $a \neq 0$)	$2 \times 6^0 = 2 \times 1 = 2$	$3w^0 = 3 \times 1 = 3$
(v) $a^{-p} = \dfrac{1}{a^p}$	$4^{-3} = \dfrac{1}{4^3} = \dfrac{1}{64}$	$2x^{-4} = 2 \times \dfrac{1}{x^4} = \dfrac{2}{x^4}$
(vi) $(ab)^p = a^p b^p$	$(4 \times 3)^2 = 4^2 \times 3^2 = 16 \times 9 = 144$	$(xy)^4 = x^4 . y^4$
(vii) $\left(\dfrac{a}{b}\right)^p = \dfrac{a^p}{b^p}$	$\left(\dfrac{8}{5}\right)^3 = \dfrac{8^3}{5^3} = \dfrac{512}{125} = 4.096$	$\left(\dfrac{e}{f}\right)^2 = \dfrac{e^2}{f^2}$
(viii) $\sqrt[n]{a} = a^{1/n}$ *see below for *note* on $\sqrt{\ }$ symbol	$\sqrt[3]{8} = 8^{1/3} = 2$	$\sqrt[4]{3y} = (3y)^{1/4}$
(ix) $\sqrt[n]{a^m} = a^{m/n}$	$\sqrt[3]{4^6} = 4^{6/3} = 4^2 = 16$	$\sqrt[9]{x^3} = x^{3/9} = x^{1/3} = \sqrt[3]{x}$
(x) $\sqrt[n]{\dfrac{a}{b}} = \dfrac{\sqrt[n]{a}}{\sqrt[n]{b}} = \dfrac{a^{\frac{1}{n}}}{b^{\frac{1}{n}}}$	$\sqrt[4]{\dfrac{8}{5}} = \dfrac{\sqrt[4]{8}}{\sqrt[4]{5}} = \dfrac{8^{\frac{1}{4}}}{5^{\frac{1}{4}}}$	$\sqrt[5]{\dfrac{r}{t}} = \dfrac{\sqrt[5]{r}}{\sqrt[5]{t}} = \dfrac{r^{\frac{1}{5}}}{t^{\frac{1}{5}}}$

* Note:

1. The symbol '$\sqrt[n]{a}$' is used to denote the nth root of a. However, when $n = 2$, mathematical convention says that the 2 is omitted and the '$\sqrt{\ }$' symbol is called the **square root**.

2. $\sqrt{a^2} = a$ and $(\sqrt{a})^2 = a$, implying $\sqrt{a^2} = (\sqrt{a})^2$.

EXAMPLE 3.3
Simplify each of the following:

(i) $\dfrac{3xy^2 \times 4x^3 \times y}{2xy \times 4y^2}$

(ii) $8^{4/3}$

(iii) $(^-2m^2t^3r^4)^2$

(iv) $8a^{-1/2} \div 2a^{-2/3}$

(v) $\dfrac{2x^3 \times 3x^2 \times 2x^0}{4x \times 6x^4}$

Solution

(i)
$$\dfrac{3xy^2 \times 4x^3 \times y}{2xy \times 4y^2} = \dfrac{12x^{(1+3)}y^{(2+1)}}{8xy^{(1+2)}}$$
$$= \dfrac{12x^4 y^3}{8xy^3}$$
$$= \dfrac{^3\cancel{12}x^{\cancel{4}^3}\cancel{y^3}}{\cancel{8}_2 \cancel{x}\cancel{y^3}}$$
$$= \dfrac{3}{2}x^3 \text{ or } 1\dfrac{1}{2}x^3$$

(ii) $8^{\frac{4}{3}} = \sqrt[3]{8^4}$

$\qquad = \sqrt[3]{\left(2^3\right)^4}$

$\qquad = \sqrt[3]{\left(2^4\right)^3}$

$\qquad = \sqrt[3]{16^3}$

$\qquad = 16^{\frac{3}{3}}$

$\qquad = 16^1$

$\qquad = 16$

(iii) $\left(^-2m^2t^3r^4\right)^2 = 4m^4t^6r^8$ (multiplying each index inside the bracket by 2)

(iv) $8a^{-\frac{1}{2}} \div 2a^{-\frac{2}{3}} = 8 \times \dfrac{1}{a^{\frac{1}{2}}} \div 2 \times \dfrac{1}{a^{\frac{2}{3}}}$

$\qquad\qquad = \dfrac{8}{a^{\frac{1}{2}}} \div \dfrac{2}{a^{\frac{2}{3}}}$

$\qquad\qquad = \dfrac{8}{a^{\frac{1}{2}}} \times \dfrac{a^{\frac{2}{3}}}{2}$

$\qquad\qquad = \dfrac{4a^{\frac{2}{3}}}{a^{\frac{1}{2}}}$

$\qquad\qquad = 4a^{\frac{2}{3}-\frac{1}{2}}$

$\qquad\qquad = 4a^{\frac{1}{6}}$ (since $\frac{2}{3} - \frac{1}{2} = \frac{4}{6} - \frac{3}{6} = \frac{1}{6}$)

(v) $\dfrac{2x^3 \times 3x^2 \times 2x^0}{4x \times 6x^4} = \dfrac{6x^{3+2} \times 2\,(1)}{24x^{(1+4)}}$

$\qquad\qquad = \dfrac{12x^5}{24x^5}$

$\qquad\qquad = \dfrac{1}{2}$

3.3 Simplify each of the following:

(i) $xy \times y \times x^2y^2$

(ii) $\dfrac{6q^8}{24q^6}$

(iii) $\dfrac{15y^7z^5}{20y^6z^4}$

(iv) $\dfrac{a^2b^3 \times a^2b^2 \times b}{b^4 \times ab}$

(v) $(5n^3)^3$

3.4 Simplify each of the following and evaluate if possible:

(i) $(4^2m^3)^0$

(ii) $\dfrac{a^2bc \times a^3c}{a^2 \times ab}$

(iii) $12^0 \div 12^{-2}$

(iv) $\dfrac{3h^2 \times 2h^{-5}}{6h^{-4}}$

(v) $\dfrac{6^{\frac{1}{3}} \times 18^{\frac{1}{2}}}{12^{\frac{3}{4}}}$

3.5 (i) The area of a rectangle of length, l, and breadth, b, is given by $A = lb$. Determine the area if $l = 10^3$ metres and $b = 10^2$ metres.

(ii) If an object travels a distance, s, during a time interval, t, its average velocity is given by $v = s \div t$. If $s = 6 \times 10^3$ metres and $t = 2 \times 10^5$ seconds, calculate the average velocity, v.

3.3 Surds

A **surd** (latin for 'deaf') is the name given to an irrational number than can be expressed in the form $\sqrt[n]{a}$.

That is, a **surd** is a number that involves the n-th *root* of a positive rational number where only an *approximate* value can be obtained.
For example:

$\sqrt{7} \begin{cases} \text{rounded to 3 decimal places is } 2.646 \\ \text{rounded to 4 decimal places is } 2.6458 \\ \text{rounded to 5 decimal places is } 2.64575. \end{cases}$

However, no matter how many decimal places are involved, the resulting value is always an approximation since the decimal never terminates and never recurs.

Similarly, irrational numbers such as $\sqrt{2}$, $5\sqrt{3}$, $\sqrt[3]{7}$ ... are *surds*, whereas $\sqrt{4}$, $\sqrt[3]{27}$, $3\sqrt{9}$ ... are *not*, since they can be evaluated *exactly*.

3.3.1 Operations with surds

The following properties may prove useful when we are required to operate with surds.

A. Addition

> If b is a positive rational number, then
> $$a\sqrt{b} + c\sqrt{b} = (a+c)\sqrt{b}$$

For example:

1. $3\sqrt{2} + 4\sqrt{2} = (3+4)\sqrt{2} = 7\sqrt{2}$

2. $^-5\sqrt{7} + 3\sqrt{7} = (^-5+3)\sqrt{7} = {}^-2\sqrt{7}$

B. Multiplication

> If a, b, c and d are positive rationals, then
>
> (i) $\sqrt{a} \times \sqrt{b} = \sqrt{ab}$, implying that
> $$\sqrt[n]{a} \times \sqrt[n]{b} = \sqrt[n]{ab}$$
>
> (ii) $\sqrt{ab} = \sqrt{a} \times \sqrt{b}$, implying that
> $$\sqrt[n]{ab} = \sqrt[n]{a} \times \sqrt[n]{b}$$
>
> (iii) $a\sqrt{b} \times c\sqrt{d} = ac\sqrt{bd}$
>
> (iv) $\sqrt{a^2 b} = a\sqrt{b}$

For example:

1. $\sqrt{2} \times \sqrt{3} = \sqrt{6}$

2. $\sqrt[3]{5 \times 8} = \sqrt[3]{5} \times \sqrt[3]{8} = \sqrt[3]{5} \times 2 = 2\sqrt[3]{5}$

3. $3\sqrt{2} \times 5\sqrt{3} = 15\sqrt{2 \times 3} = 15\sqrt{6}$

4. $\sqrt{48} = \sqrt{16 \times 3} = \sqrt{4^2 \times 3} = 4\sqrt{3}$

C. Division

> If a and b are positive rationals, then
>
> (i) $\dfrac{\sqrt{a}}{\sqrt{b}} = \sqrt{\dfrac{a}{b}}$, implying that $\dfrac{\sqrt[n]{a}}{\sqrt[n]{b}} = \sqrt[n]{\dfrac{a}{b}}$
>
> (ii) $\sqrt{\dfrac{a}{b}} = \dfrac{\sqrt{a}}{\sqrt{b}}$, implying that $\sqrt[n]{\dfrac{a}{b}} = \dfrac{\sqrt[n]{a}}{\sqrt[n]{b}}$

For example:

1. $\dfrac{\sqrt{27}}{\sqrt{3}} = \sqrt{\dfrac{27}{3}} = \sqrt{9} = 3$

2. $\sqrt{\dfrac{3}{4}} = \dfrac{\sqrt{3}}{\sqrt{4}} = \dfrac{\sqrt{3}}{2} = \dfrac{1}{2}\sqrt{3}$

EXAMPLE 3.4
Using one or more of the surdic properties, simplify each of the following:

(i) $3\sqrt{12}$ (ii) $\sqrt{10} \times \sqrt{5}$

(iii) $\sqrt{\dfrac{27}{12}}$ (iv) $^-5\sqrt{3} + \sqrt{12}$

(v) $\dfrac{1}{3}\sqrt{27}$

Solution

(i) $3\sqrt{12} = 3\sqrt{4 \times 3}$
$$= 3 \times \sqrt{4} \times \sqrt{3}$$
$$= 3 \times 2 \times \sqrt{3}$$
$$= 6\sqrt{3}$$

(ii) $\sqrt{10} \times \sqrt{5} = \sqrt{10 \times 5}$
$$= \sqrt{50}$$
$$= \sqrt{25 \times 2}$$
$$= \sqrt{25} \times \sqrt{2}$$
$$= 5\sqrt{2}$$

(iii) $\sqrt{\dfrac{27}{12}} = \dfrac{\sqrt{27}}{\sqrt{12}}$
$$= \dfrac{\sqrt{9 \times 3}}{\sqrt{4 \times 3}}$$
$$= \dfrac{\sqrt{9} \times \sqrt{3}}{\sqrt{4} \times \sqrt{3}}$$
$$= \dfrac{3\sqrt{3}}{2\sqrt{3}}$$
$$= \dfrac{3}{2} = 1\dfrac{1}{2}$$

(iv) $^-5\sqrt{3} + \sqrt{12} = {}^-5\sqrt{3} + \sqrt{4 \times 3}$
$$= {}^-5\sqrt{3} + \sqrt{4} \times \sqrt{3}$$
$$= {}^-5\sqrt{3} + 2\sqrt{3}$$
$$= (^-5 + 2)\sqrt{3}$$
$$= {}^-3\sqrt{3}$$

(v) $\frac{1}{3}\sqrt{27} = \frac{\sqrt{27}}{3}$ or $\frac{1}{3}\sqrt{27} = \frac{1}{3}\sqrt{9 \times 3}$

$$= \frac{\sqrt{27}}{\sqrt{9}} \qquad\qquad = \frac{1}{3} \times \sqrt{9} \times \sqrt{3}$$

$$= \sqrt{\frac{27}{9}} \qquad\qquad = \frac{1}{3} \times 3 \times \sqrt{3}$$

$$= \sqrt{3} \qquad\qquad = \sqrt{3}$$

YOUR TURN

3.6 Express the following surds in their simplest form.

   (i) $\sqrt{80}$

   (ii) $\sqrt{450}$

   (iii) $\sqrt{54}$

   (iv) $\sqrt{108}$

   (v) $\sqrt{242}$

3.7 Simplify each of the following expressions:

   (i) $4\sqrt{5} - 2\sqrt{5} + 3\sqrt{5}$

   (ii) $\sqrt{12} + 2\sqrt{48}$

   (iii) $\sqrt{75} - \sqrt{27} + \sqrt{32}$

   (iv) $\sqrt{75} - \sqrt{48}$

   (v) $2\sqrt{a} + \sqrt{a^3}$

3.8 Simplify the following:

   (i) $\sqrt{\dfrac{5}{16}}$

   (ii) $\sqrt{\dfrac{25}{72}}$

   (iii) $\sqrt{\dfrac{27}{8}}$

   (iv) $\sqrt{\dfrac{25}{10}}$

   (v) $\sqrt{\dfrac{36}{18}}$

3.9 Express each of the following in a simpler form:

   (i) $(\sqrt{7})^2$

   (ii) $\sqrt{5} \times \sqrt{5}$

   (iii) $3\sqrt{6} \times \sqrt{2}$

   (iv) $(3\sqrt{2})^2$

   (v) $2\sqrt{7} \times \sqrt{63}$

3.10 Simplify each of the following to form a single surd:

   (i) $\sqrt{32} + \sqrt{8}$

   (ii) $\sqrt{72} - \sqrt{18} + 4\sqrt{2}$

   (iii) $^{-}2\sqrt{5} \times {}^{-}3\sqrt{10}$

   (iv) $\dfrac{4\sqrt{6} \times 5\sqrt{12}}{10\sqrt{2}}$

   (v) $\dfrac{2}{3}\sqrt{18}$

Extensions to chapter 3

Extension A: Scientific notation

> **Scientific notation** provides a useful means of expressing both very large and very small positive numbers. It involves re-writing the given number in the form $R \times 10^n$, where R is a number between 1 and 10 ($1 \le R < 10$).

For example:

$$1.86 \times 10^7 \qquad = 18600000$$
$$1.86 \times 10^{-7} \qquad = 0.000000186$$
$$170000 \qquad = 1.7 \times 10^5$$
$$0.0000247 \qquad = 2.47 \times 10^{-5}$$

A. **A quick method of determining the power of ten** (i.e. n in the formula $R \times 10^n$)

EXAMPLE E3.1

Express 6340000 in scientific notation.

Solution

(i) The place between the first and second significant figures will always be the position of the decimal point in scientific notation.

Here, 6 340000
     ↑

(ii) Count how many numerals are between the arrow and the decimal point, remembering that with integers the decimal point is understood to be after the final numeral. Counting to the *right* results in a **positive** power.

Here, 6 340000
     ↑

6 numerals to the right
Hence, the power of 10 is 6.
$\therefore 6340000 = 6.34 \times 10^6$

EXAMPLE E3.2

Express 0.000375 in scientific notation.

Solution

As before, the decimal point will lie between the first and second significant digits (see arrow) and the number of figures between the answer and the original decimal point will determine the power. Counting to the *left* results in a **negative** power.

Here, 0.0003 75

4 numerals to the left
Hence, the power of 10 is –4
$\therefore$ $0.000375 = 3.75 \times 10^{-4}$.

B. Interpreting scientific notation on your calculator.

Whenever an arithmetic operation involves either very large or very small numbers, the calculator displays the result using scientific notation.

For example:

(i)   256987 $\boxed{\times}$ 7531 $\boxed{=}$ $\boxed{1.9353691^{09}}$

Now, 1.9353691^{09} on your display should be interpreted as $1.9353691 \times 10^9 = 1935369100$.

(ii)   .045 $\boxed{\times}$ .0003 $\boxed{=}$ $\boxed{1.35^{-05}}$

Now, 1.35^{-05} on your display should be interpreted as $1.35 \times 10^{-5} = 0.0000135$.

Extension B: Logarithms.

If $N = a^x$ where $a \neq 0$ or 1, then $x = log_a N$ is called the **logarithm** of N to the **base** a.

Note carefully: A logarithm is simply an index (or exponent).

For example:

(i)   Since $8 = 2^3$, then $3 = \log_2 8$.

(ii) $2 = \log_{10} 100$ because $100 = 10^2$.

(iii)   If $3^x = \dfrac{1}{27}$, then $x = \log_3 \dfrac{1}{27}$.

(iv)   If $\log_{10} k = t$, then $k = 10^t$.

(v)   If $y = \log_a 7$, then $7 = a^y$.

(vi)   If $x = 5^{-2}$, then $-2 = \log_5 x$.

(vii) If $p = \log_e t$, then $t = e^p$.

Special Notes:

1.   Logarithms to the base 10, i.e. $\log_{10} N$, are usually written $\log N$ (omitting the subscript 10). Logarithms of this type are known as *common* logarithms.

2.   Logarithms to the base e, i.e. $\log_e N$ (where, in a mathematical context, $e \approx 2.718$), are usually written $\ln N$. Logarithms of this type are known as *natural* logarithms.

A. Laws of logarithms

Law 1: $\log_a xy = \log_a x + \log_a y$

For example:

(i) $\log_5 12 = \log_5 2 + \log_5 6$
$= \log_5 2 + \log_5 (2 \times 3)$
$= \log_5 2 + \log_5 2 + \log_5 3$
$= 2\log_5 2 + \log_5 3$

or   (ii) $\log_5 12 = \log_5 3 + \log_5 4$
$= \log_5 3 + \log_5 (2 \times 2)$
$= \log_5 3 + \log_5 2 + \log_5 2$
$= \log_5 3 + 2\log_5 2$

Law 2: $\log_a \dfrac{x}{y} = \log_a x - \log_a y$

For example:

$\log_e \dfrac{4}{3} = \log_e 4 - \log_e 3$
$= \log_e (2 \times 2) - \log_e 3$
$= \log_e 2 + \log_e 2 - \log_e 3$
$= 2\log_e 2 - \log_e 3$

Law 3: $\log_a x^m = m \log_a x$

For example:

(i) $\log_7 32 = \log_7 2^5$
$= 5\log_7 2$

or   (ii) $\log_7 32 = \log_7 (2 \times 16)$
$= \log_7 2 + \log_7 16$
$= \log_7 2 + \log_7 4^2$
$= \log_7 2 + 2\log_7 4$
$= \log_7 2 + 2\log_7 2^2$
$= \log_7 2 + 4\log_7 2$
$= 5\log_7 2$

Special note: Two very useful properties of logarithms are:

(i) $\log_a a = 1$ (since $a = a^1$ is true)

(ii) $\log_a 1 = 0$ (since $a^0 = 1$ is true)

B. Change of base of logarithms

The logarithm of a number N to the base a is

related to the logarithm of that same number N to the base b by means of the formula

$$\log_a N = \frac{\log_b N}{\log_b a}$$

For example:

(i) $\log_5 11 = \dfrac{\log_e 11}{\log_e 5} = \dfrac{\ln 11}{\ln 5}$

(ii) $\log_3 5 = \dfrac{\log_2 5}{\log_2 3}$

(iii) $\log_2 100 = \dfrac{\log_{10} 100}{\log_{10} 2}$

$$= \frac{\log 10 \times 10}{\log 2}$$

$$= \frac{\log 10 + \log 10}{\log 2}$$

$$= \frac{1 + 1}{\log 2}$$

$$= \frac{2}{\log 2}$$

C. Calculator usage.

The simplest way to evaluate any logarithmic expression is to use your calculator. However, the calculator will only allow you to determine the value of an expression involving *common* (i.e. base 10) logarithms and/or *natural* (i.e. base e) logarithms. Therefore, all expressions involving bases other than 10 or e must be re-written applying the *'change of base'* formula above before using the calculator.

Note: Common logarithms (base 10) are evaluated using the $\boxed{\text{log}}$ button while natural logarithms (base e) are evaluated using the button $\boxed{\text{In}}$.

For example:

(i) Find $\log_{10} 6.73$: 6.73 $\boxed{\text{log}}$ $\boxed{=}$ $\boxed{0.828015}$

(ii) Find $\log_e 0.859$: .859 $\boxed{\text{In}}$ $\boxed{=}$ $\boxed{^-0.1519863}$

(iii) Find $\log_2 37$:

1. $\log_2 37 = \dfrac{\log 37}{\log 2}$:37 $\boxed{\text{log}}$ $\boxed{\div}$ 2 $\boxed{\text{log}}$

$\boxed{=}$ $\boxed{5.2094534}$

or

2. $\log_2 37 = \dfrac{\ln 37}{\ln 2}$:37 $\boxed{\text{In}}$ $\boxed{\div}$ 2 $\boxed{\text{In}}$

$\boxed{=}$ $\boxed{5.2094534}$

EXAMPLE E3.3

(i) Given that $\log x = 2.5$ and $\log y = 1.8$, evaluate $\log x^2 y^3$.

(ii) Given that $\ln a = 3.6$ and $\ln b = 2.4$, evaluate $\dfrac{\ln a^2}{\ln b^4}$.

Solution

(i) $\log x^2 y^3 = \log x^2 + \log y^3$

$$= 2\log x + 3\log y$$

$$= (2 \times 2.5) + (3 \times 1.8)$$

$$= 5 + 5.4$$

$$= 10.4$$

(ii) $\dfrac{\ln a^2}{\ln b^4} = \ln a^2 - \ln b^4$

$$= 2\ln a - 4\ln b$$

$$= (2 \times 3.6) - (4 \times 2.4)$$

$$= 7.2 - 9.6$$

$$= -2.4$$

EXAMPLE E3.4

Express each of the following as a single logarithm and evaluate.

(i) $\log_2 5 + \log_2 3 - \log_2 30$

(ii) $\log_5 \sqrt{2} - \log_5 \sqrt{10}$

Solution

(i) $\log_2 5 + \log_2 3 - \log_2 30 = \log_2 (5 \times 3) - \log_2 30$

$$= \log_2 15 - \log_2 30$$

$$= \log_2 \frac{15}{30}$$

$$= \log_2 \frac{1}{2}$$

$$= \log_2 1 - \log_2 2$$

$$= 0 - 1$$

$$= -1$$

(ii) $\log_5 \sqrt{2} - \log_5 \sqrt{10} = \log_5 \dfrac{\sqrt{2}}{\sqrt{10}}$

$$= \log_5 \sqrt{\frac{2}{10}}$$

$$= \log_5 \sqrt{\frac{1}{5}}$$

$$= \log_5 \left(\frac{1}{5}\right)^{\frac{1}{2}}$$

$$= \frac{1}{2}\log_5\left(\frac{1}{5}\right)$$

$$= \frac{1}{2}\log_5 1 - \frac{1}{2}\log_5 5$$

$$= 0 - \frac{1}{2}$$

$$= -\frac{1}{2}$$

EXAMPLE E3.5

Evaluate each of the following using a calculator.

(i)　$\log_5 3$, rounded to 3 decimal places.

(ii)　$\log_3 \sqrt{7}$, rounded to 2 decimal places.

Solution

(i)　$\log_5 3 = \dfrac{\log 3}{\log 5}$

By calculator: $3\ \boxed{\log}\ \boxed{\div}\ 5\ \boxed{\log}$

$\boxed{=}\ \boxed{0.6826061}$

$\therefore \log_5 3 = 0.683$

(ii)　$\log_3 \sqrt{7} = \dfrac{\ln \sqrt{7}}{\ln 3}$

By calculator: $7\ \boxed{\sqrt{\ }}\ \boxed{\ln}\ \boxed{\div}\ 3\ \boxed{\ln}$

$\boxed{=}\ \boxed{0.8856218}$

$\therefore \log_3 \sqrt{7} = 0.89$

Summary

In this chapter we have discussed the notion of index numbers and it should now be clear that the use of index notation is a most efficient method of handling very large and very small numbers.

Remember: A number written in index notation must be of the form a^n where 'a' is the base, 'n' is the index, power or exponent and a^n is called the nth power of a.

Section 1 dealt with the use of the $\boxed{x^y}$ button to evaluate indices. In our calculator, where x^y is written in brackets just *above* the key, $\boxed{x}$, press $\boxed{\text{shift}}$ $\boxed{x}$ to access $\boxed{x^y}$. Check your own calculator.

Section 2 states the important *index laws* in tabular form, with examples. These laws include:

(i)　$a^p \times a^q = a^{p+q}$

(ii)　$a^p \div a^q = \dfrac{a^p}{a^q} = a^{p-q}$

(iii)　$(a^p)^q = a^{pq}$

(iv)　$a^0 = 1$ (for $a \neq 0$)

(v)　$a^{-p} = \dfrac{1}{a^p}$

(vi)　$(ab)^p = a^p b^p$

(vii)　$\left(\dfrac{a}{b}\right)^p = \dfrac{a^p}{b^p}$

(viii)　$\sqrt[n]{a} = a^{\frac{1}{n}}$

(ix)　$\sqrt[n]{a^m} = a^{\frac{m}{n}}$

(x)　$\sqrt[n]{\dfrac{a}{b}} = \dfrac{\sqrt[n]{a}}{\sqrt[n]{b}}$

Note: The above laws assume that:

(i)　a and b are positive rational numbers

(ii)　p and q are non-zero real numbers

(iii)　m and n are positive integers

Section 3 introduces the term **surd**, defined to be an irrational number of the form $\sqrt[n]{a}$.

Useful properties when operating with surds

(i)	$a\sqrt{b} + c\sqrt{b} = (a+c)\sqrt{b}$	(b is a positive rational)
(ii)	$\sqrt{a} \times \sqrt{b} = \sqrt{ab}$	
(iii)	$a\sqrt{b} \times c\sqrt{d} = ac\sqrt{bd}$	(a,b,c and d are all positive rationals)
(iv)	$\sqrt{a^2 b} = a\sqrt{b}$	
(v)	$\dfrac{\sqrt{a}}{\sqrt{b}} = \sqrt{\dfrac{a}{b}}$	

The **extensions** to this chapter cover two topics, scientific notation and logarithms.

1.　*Scientific notation*

This involves re-writing very large or very small numbers in the form $R \times 10^n$, where $1 \leq R < 10$.
For example:

(i)　$5890000 = 5.89 \times 10^6$

(ii)　$0.00024 = 2.4 \times 10^{-4}$

2.　*Logarithms*

A logarithm is just an index or exponent. **Why?**

If $N = a^x$ ($a \neq 0$ or 1), then $x = \log_a N$ is the logarithm of N to the base a.

Note:

(i) $\log_{10} N = \log N$ is called a *common* logarithm.

(ii) $\log_e N = \ln N$ is called a *natural* logarithm.

Logarithmic laws

1. $\log_a xy = \log_a x + \log_a y$

2. $\log_a \dfrac{x}{y} = \log_a x - \log_a y$

3. $\log_a x^m = m\log_a x$

Useful properties

(i) $\log_a a = 1$

(ii) $\log_a 1 = 0$

Changing the base of a logarithm

$$\log_a N = \frac{\log_b N}{\log_b a}$$

This is a very useful technique, especially when calculators are involved as they are able to evaluate common (base 10) and natural (base e) logarithms only.

Therefore, using your calculator,

$$\log_a N = \begin{cases} N \boxed{\log} \boxed{\div} a \boxed{\log} \boxed{=} \boxed{\text{ANSWER}} \\ \text{or} \\ N \boxed{\ln} \boxed{\div} a \boxed{\ln} \boxed{=} \boxed{\text{ANSWER}} \end{cases}$$

3.1 Evaluate each of the following:

(a) $10^2 - 8^2$

(b) $2^4 - 3^2$

(c) $5^3 - 10^2$

3.2 Simplify the following expressions:

(a) $3x^2 \times 4x$

(b) $a^{10} \div a^2$

(c) $\sqrt{25x^6}$

(d) $-4y^3 \times -6y$

(e) $(2x^2)^3$

3.3 Write 93000000 in scientific notation.

3.4 Find the value of $\sqrt{\dfrac{1}{0.067}}$, correct to 1 decimal place.

3.5 If $x = 6$ and $y = -2$, find $x^2 - 3y$.

3.6 If $x = -2$ and $y = 4$, find xy^2.

3.7 Evaluate $(7.9)^3$.

3.8 Simplify $a^3b^2 \div ab$.

3.9 Simplify each of the following:

(a) $\dfrac{\sqrt{14}}{\sqrt{2}}$

(b) $\dfrac{10\sqrt{3}}{\sqrt{5}}$

(c) $\dfrac{10}{\sqrt{2}}$

(d) $\dfrac{6ab\sqrt{cd}}{3\sqrt{ad}}$

(e) $\dfrac{4\sqrt{5}}{\sqrt{2}}$

3.10 Simplify

(a) $5\sqrt{3} \times 2\sqrt{5}$

(b) $2\sqrt{2} \times \sqrt{8}$

(c) $2\sqrt{10} \times 4\sqrt{15}$

(d) $\sqrt{7} \times 2\sqrt{7} \times 3\sqrt{7}$

3.11 (a) Express log 20 as the sum of two logarithms in two different ways.

(b) Using your answer in a), simplify log 20 − log 5.

3.12 Given that $\log_{10} a = 1.2$ and $\log_{10} b = 1.8$, evaluate

(a) $\log_{10} \dfrac{\sqrt{b}}{a^2}$

(b) $\dfrac{\log_{10} b}{\log_{10} a}$

3.12 *Cont.*

(c) $\log_{10} a^3 b^3$

*3.13 If $x = \ln 2$ and $y = \ln 3$, express each of the following in terms of x and y:

(a) $\ln 9$

(b) $\ln 3\sqrt{3}$

(c) $\ln 12$

*3.14 Calculate to 3 decimal places,

$$\log_e(1.2) - \left[0.2 - \frac{(0.2)^2}{2} - \frac{(0.2)^3}{3} - \frac{(0.2)^4}{4} \right]$$

*3.15 Determine, to 3 decimal places, the value of:

(a) $\log_3 5 + \log_3 3 - \log_3 30$

(b) $\log_3 10 - \log_3 45 + 2\log_3 3$

(c) $\log_2 \sqrt{3} - \log_3 \sqrt{6}$

(d) $2\log_{10} 20 - \log_4 6 + 2\log_5 2 + \log_4 3$

4 Algebra revisited

Introduction

In Chapter 2, Elementary algebra, we discussed several basic topics including algebraic operations, the laws of algebra and some further useful properties. This groundwork, together with the material on indices in Chapter 3, now allows us to examine more complex topics under the heading of **algebra**.

This chapter consists of four major sections.

Section 1 has another look at the concept of an algebraic expression and extends our previous knowledge to include more complicated expressions involving indices. As you will see, there are a number of new terms being introduced in this section and these are very important as they will continue to be used in future chapters.

Section 2 examines what are called *binomial products,* where the term 'binomial' is defined in *Section 1* and the 'product' of any two quantities is the result of their multiplication. *For example*, the product of 2 x 3 is 6.

Section 3 is headed *factorization* and as the name implies involves determining the 'factors' of an algebraic expression, a process made a little easier with some special factorization methods. The terms 'factor' and 'factorization' will be fully defined at the beginning of the section.

Finally, *Section 4* looks again at numerical substitution, but here we are interested in the substitution of numbers into, and hence the evaluation of, far more complex algebraic expressions than were presented in Chapter 2.

The **extensions** to this chapter include another look at two of the topics introduced in Chapters 2 and 3 – namely, surds and algebraic fractions. Here we add to our previous knowledge by using more complex mathematical expressions of the type dealt with in the core of this chapter.

One further topic covered in the **extensions** follows on from the material presented in *Section 2* (binomial products) and examines what are generally termed **multinomial** products.

Each of the areas discussed in the **extensions**, as well as being included for general interest, is useful for those students intending to pursue the physics, chemistry and mathematics disciplines.

4.1 Another look at algebraic expressions

Terminology

1. A **monomial** is an algebraic expression with only one term.

    e.g. $3x^2$

2. A **multinomial** is an algebraic expression with more than one term.

    e.g. $6x + 2y + 3z + 4y^3 + 8$

3. A **binomial** is a multinomial that has exactly *two* terms.

    e.g. $3x + 4y$

4. A **trinomial** is a multinomial that has exactly *three* terms.

    e.g. $a^2 + 4b - 6$

5. A **coefficient** is the number by which a variable in an algebraic expression is multiplied.

    e.g. In $a^2 + 4b - 6$, the coefficient of a^2 is 1 and the coefficient of b is 4.

6. A **constant** is a term in a mathematical expression that has a *fixed* value.

    e.g. In $a^2 + 4b - 6$, the constant is ‾6.

7. A **polynomial** is a special case of a multinomial. It is an algebraic expression with one or more terms involving **one** variable only, raised to a non–negative integer power.

    e.g. $2x^2 - 3x^3 + x - 5$

8. The **degree** of a polynomial is the value of the highest power of the variable.

    e.g. $2x^2 - 3x^3 + x - 5$ is a polynomial of degree 3.

EXAMPLE 4.1

For each of the following expressions, write down

(i)   the type of expression – multinomial, binomial, trinomial, polynomial

(ii)   the number of terms in the expression

(iii)   the coefficient of each variable

(iv)   the constant term, if it exists

(v)   the degree of the polynomial, if applicable.

(a) $7ab - 7bc + 21b - 4$

(b) $4x + 2y^2 - c$

(c) $a^2 - 9b^3$

(d) $y^3 - 7y^2 + 3$

(e) $0.5s^2 - 3s + \sqrt{7} + s^4$

Solution

(a) $7ab - 7bc + 21b - 4$

(i)   multinomial

(ii)   4 terms

(iii) coeff. of $ab = 7$

coeff. of $bc = {}^-7$

coeff. of $b = 21$

(iv) constant $= {}^-4$

(b) $4x + 2y^2 - c$

(i)   trinomial

(ii)   3 terms

(iii) coeff. of $x = 4$

coeff. of $y^2 = 2$

coeff. of $c = {}^-1$

(iv) no constant

(c) $a^2 - 9b^3$

(i)   binomial

(ii)   2 terms

(iii) coeff. of $a^2 = 1$

coeff. of $b^3 = -9$

(iv) no constant

(d) $y^3 - 7y^2 + 3$

(i)   polynomial in y

(ii)   3 terms

(iii) coeff. of $y^3 = 1$

coeff. of $y^2 = {}^-7$

(iv) constant $= 3$

(v)   degree $= 3$

(e) $0.5s^2 - 3s + \sqrt{7} + s^4$

(i)   polynomial in s

(ii)   4 terms

(iii) coeff. of $s^2 = 0.5$

coeff. of $s = -3$

coeff. of $s^4 = 1$

(iv) constant $= \sqrt{7}$

(v)   degree $= 4$

4.1   For each of the following expressions, write down

(i)   the type of expression – multinomial, binomial, trinomial, polynomial

(ii)   the number of terms in the expression

(iii)   the coefficient of each variable

(iv)   the constant term, if it exists

(v)   the degree of the polynomial, if applicable.

(a) $6ax - 9x + 8ay - 12y^2 + 2$

(b) $5mn - \dfrac{1}{4}m^2$

(c) $6 - 2p^3 + q$

(d) $2y^2 - y^4 + \dfrac{1}{2}y - 5$

(e) $3t^5 - 5t + t^3 + 0.62t^2 - 4\sqrt{2}$

4.2   Binomial products

A **binomial product** is the result obtained when two binomial algebraic expressions are multiplied together.

The simplest way to perform the required multiplication is to use the *distributive law* (see Chapter 2).

EXAMPLE 4.2

(i) $(x - 1)(x + 3) = x(x + 3) - 1(x + 3)$
            (distributive law)
$= x^2 + 3x - x - 3$
            (distributive law)
$= x^2 + 2x - 3$ (simplify)

(ii) $(3y + 4)(y - 5) = 3y(y - 5) + 4(y - 5)$
$= 3y^2 - 15y + 4y - 20$
$= 3y^2 - 11y - 20$

(iii) $(x + a)(y + b) = x(y + b) + a(y + b)$
$= xy + xb + ay + ab$

(iv) $(5 + 2t)(3 - 9y) = 5(3 - 9y) + 2t(3 - 9y)$
$= 15 - 45y + 6t - 18ty$
$= 3(5 - 15y + 2t - 6ty)$

Special binomial products

| 1. | $(a+b)^2 = a^2 + 2ab + b^2$ |

e.g. $(x+3)^2 = (x)^2 + 2(x)(3) + (3)^2$
$$= x^2 + 6x + 9$$

| 2. | $(a-b)^2 = a^2 - 2ab + b^2$ |

e.g. $(x-2)^2 = (x)^2 - 2(x)(2) + (2)^2$
$$= x^2 - 4x + 4$$

| 3. | $(a-b)(a+b) = a^2 - b^2$ |

e.g. $(x-4)(x+4) = (x)^2 - (4)^2$
$$= x^2 - 16$$

Note:
In 1. and 2. above, $(a+b)^2$ and $(a-b)^2$ are called *perfect squares*.

EXAMPLE 4.3
Show that $(a+b)^2 = a^2 + 2ab + b^2$

Solution
$(a+b)^2 = (a+b)(a+b)$
$$= a(a+b) + b(a+b)$$
$$= a^2 + ab + ba + b^2$$
$$= a^2 + 2ab + b^2, \text{ as required.}$$

EXAMPLE 4.4
Expand each of the following:

(i) $(3a-2)(2a+5)$

(ii) $(2x+3y)(x+2y)$

(iii) $(5-x)(5+x)$

(iv) $(ab+1)(ab-1)$

(v) $(6n-5m)^2$

Solution

(i) $(3a-2)(2a+5)$
$$= 3a(2a+5) - 2(2a+5)$$
$$= (3a)(2a) + 15a - 4a - 10$$
$$= 6a^2 + 11a - 10 \text{ (a polynomial of degree 2)}$$

(ii) $(2x+3y)(x+2y)$
$$= 2x(x+2y) + 3y(x+2y)$$
$$= (2x)(x) + (2x)(2y) + 3xy + (3y)(2y)$$
$$= 2x^2 + 4xy + 3xy + 6y^2$$
$$= 2x^2 + 7xy + 6y^2 \text{ (a trinomial)}$$

(iii) $(5-x)(5+x)$
$$= 5(5+x) - x(5+x)$$
$$= 25 + 5x - 5x - x^2$$

$$= 25 - x^2 \text{ (a binomial and a polynomial of degree 2)}$$

(iv) $(ab+1)(ab-1)$
$$= ab(ab-1) + 1(ab-1)$$
$$= (ab)^2 - ab + ab - 1$$
$$= (ab)^2 - 1$$
$$= a^2b^2 - 1 \text{ (a binomial)}$$

(v) $(6n-5m)^2$
$$= (6n)^2 - 2(6n)(5m) + (5m)^2$$
$$= 36n^2 - 60nm + 25m^2 \text{ (a trinomial)}$$

YOUR TURN

4.2 Show that
(i) $(a-b)^2 = a^2 - 2ab + b^2$
(ii) $(a-b)(a+b) = a^2 - b^2$

4.3 Expand and simplify if possible
(i) $(3x+1)(3x-1)$
(ii) $(2t^2+t)(t+4)$
(iii) $(x^2+4y)(x^2-4y)$
(iv) $(1-5y)^2$
(v) $(2m-n)(3m+n)$

4.4 Expand the following and state whether the resulting expression is multinomial, trinomial, binomial or polynomial of degree n.
(i) $(a-7)(3a^2-a)$
(ii) $(3a^2-5b^2)(3a^2+5b^2)$
(iii) $(1-2x)^2$
(iv) $(3xy-z)^2$
(v) $(x-2y)(3x-y)$

4.3 Factorization

If two or more algebraic expressions are multiplied together, then these expressions are termed the **factors** *of the resulting product.*

The process of determining the factors of an algebraic expression is called **factorization**, and involves the distributive law 'in reverse'.

For example:
By the distributive law,
$$a(b+c) = ab + ac.$$
However, we may also write
$$ab + ac = a(b+c)$$
Hence, we see that the expression $ab + ac$ is the product of 2 factors, a and $(b+c)$.

(That is, $ab + ac$ can be *factorized* into two *factors, a* and $(b + c)$, and clearly *factorization* is just the reverse of the process of *expansion*.)

Factorization methods:

Method 1

$$ab + ac = a(b + c)$$

For example:

(i) $2a + 2b = 2(a + b)$

(ii) $5a^2 - 15a = 5(a^2 - 3a)$
$$= 5 \times a(a - 3)$$
$$= 5a(a - 3)$$

Method 2

$$ab + ac + pb + pc = (a + p)(b + c)$$

For example:

(i) $3b + 3c + ab + ac = (3 + a)(b + c)$

(ii) $ab + 3bc + 4a + 12c = ba + b3c + 4a + 4(3c)$
$$= (b + 4)(a + 3c)$$

Method 3

$$x^2 + (a + b)x + ab = (x + a)(x + b)$$

For example:

(i) $m^2 + 5m + 6 = (m + 2)(m + 3)$

(ii) $x^2 - 7x + 10 = x^2 + (-7x) + 10$
$$= (x + (-5))(x + (-2))$$
$$= (x - 5)(x - 2)$$

Method 4

$$a^2 + 2ab + b^2 = (a + b)^2$$

For example:

(i) $x^2 + 6x + 9 = (x + 3)^2$

(ii) $4x^2 + 20x + 25 = (2x + 5)^2$

Method 5

$$a^2 - 2ab + b^2 = (a - b)^2$$

For example:

(i) $y^2 - 6y + 9 = (y - 3)^2$

(ii) $4y^2 - 20y + 25 = (2y - 5)^2$

Method 6

$$a^2 - b^2 = (a - b)(a + b)$$

For example:

(i) $9t^2 - 16 = (3t - 4)(3t + 4)$

(ii) $36 - s^2 = (6 - s)(6 + s)$

or $36 - s^2 = -(s^2 - 36)$
$$= -(s - 6)(s + 6)$$
$$= (6 - s)(6 + s)$$

EXAMPLE 4.5

Factorize the following expressions by extracting the *highest common factor.*

(i) $x^2 - 9x$

(ii) $2by^2 + 4by - 6y$

(iii) $12x - 3y + 4mx - my$

(iv) $p^{\frac{1}{2}} + p^{\frac{3}{2}}$

Solution

(i) $x^2 - 9x = x(x - 9)$ (common factor $= x$)

(ii) $2by^2 + 4by - 6y = 2(by^2 + 2by - 3y)$
$$= 2 \times y(by + 2b - 3)$$
$$= 2y(by + 2b - 3)$$
$$\text{(common factor} = 2y)$$

(iii) $12x - 3y + 4mx - my = (12x - 3y) + (4mx - my)$
$$= 3(4x - y) + m(4x - y)$$
$$= (3 + m)(4x - y)$$

or $12x - 3y + 4mx - my = (12x + 4mx) - (3y + my)$
$$= 4(3x + mx) - y(3 + m)$$
$$= 4x(3 + m) - y(3 + m)$$
$$= (4x - y)(3 + m)$$

(iv) $p^{\frac{1}{2}} + p^{\frac{3}{2}} = p^{\frac{1}{2}} + p^{1\frac{1}{2}}$
$$= p^{\frac{1}{2}} + p^{\left(1 + \frac{1}{2}\right)}$$
$$= p^{\frac{1}{2}} + (p^1)\left(p^{\frac{1}{2}}\right)$$
$$= p^{\frac{1}{2}}(1 + p^1)$$
$$= p^{\frac{1}{2}}(1 + p) \text{ (common factor} = p^{\frac{1}{2}})$$

EXAMPLE 4.6

Use Methods 4, 5 or 6 to factorize each of the following.

(i) $25a^2 - 9b^2$

(ii) $64 - 36m^2$

(iii) $5t^2 + 30t + 45$

(iv) $8x - 16 - x^2$

Solution

(i) $25a^2 - 9b^2 = (5a - 3b)(5a + 3b)$
$\qquad$ (using Method 6)

(ii) $64 - 36m^2 = 4(16 - 9m^2)$
$\qquad = 4(4 - 3m)(4 + 3m)$

or $\quad 64 - 36m^2 = (8 - 6m)(8 + 6m)$
$\qquad = 2(4 - 3m)\ 2(4 + 3m)$
$\qquad = 4(4 - 3m)(4 + 3m)$
$\qquad$ (using Method 6)

(iii) $5t^2 + 30t + 45 = 5(t^2 + 6t + 9)$
$\qquad = 5(t + 3)^2$ (using Method 4)

(iv) $8x - 16 - x^2 = -1(-8x + 16 + x^2)$
$\qquad$ (common factor = –1)
$\qquad = -(x^2 - 8x + 16)$
$\qquad = -(x - 4)^2$ (using Method 5)

EXAMPLE 4.7

Factorize each of the following trinomials.

(i) $\quad 3m^2 + 6m + 3$

(ii) $\quad x^2 + 3x - 4$

(iii) $\quad 2t^2 + 3t - 2$

(iv) $\quad 16y - 3 - 5y^2$

(v) $\quad 4x^2 - 11x + 6$

Solution

(i) $3m^2 + 6m + 3 = 3(m^2 + 2m + 1)$
$\qquad$ (common factor = 3)
$\qquad = 3(m + 1)^2$ (using Method 4)

(ii) $x^2 + 3x - 4 = (x + 4)(x - 1)$ **why??**

From Rule 3 we know that

$\qquad x^2 + (a + b)x + ab = (x + a)(x + b)$

$\qquad \uparrow \qquad \uparrow \qquad \nearrow$

Here we have $x^2 + 3x + {}^-4 = (x + a)(x + b)$.
Find a and b.

Hence, we require a and b such that

$(a + b) = 3$ and $ab = {}^-4$. Thus, $a = 4$ and $b = {}^-1$
is the only possibility.

$\therefore x^2 + 3x - 4 = (x + 4)(x - 1)$

(iii) $2t^2 + 3t - 2 = (2t - 1)(t + 2)$ **why??**

This trinomial is more difficult to factorize
than the previous example because the x^2
term has a coefficient different from one.
Hence, Rule 3 is not applicable.

However, there is a procedure called the
'*cross method*' that has been found to be useful
in determining the factors of trinomials of
this form. This method, involving 'cross

products', is best described diagrammatically
as follows.

Trinomial $\qquad 2t^2 + 3t - 2$

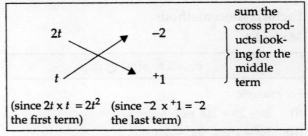

(since $2t \times t = 2t^2$ $\quad$ (since $^-2 \times {}^+1 = {}^-2$
the first term) $\qquad$ the last term)

Other possible combinations of the first and last
terms are

$\qquad \begin{matrix} 2t & {}^+1 \\ & \times \\ t & ^-2 \end{matrix} \qquad \begin{matrix} 2t & {}^+2 \\ & \times \\ t & ^-1 \end{matrix} \qquad \begin{matrix} 2t & ^-1 \\ & \times \\ t & {}^+2 \end{matrix}$

To determine the *correct* combination, sum the two
cross products in each case looking for the answer
$3t$, the middle term of the given trinomial.
Here, the last combination

is the only one where the cross products sum to $3t$,
(that is, $(2t \times 2) + (t \times -1) = 4t + -t = 3t$)
Hence, the factors of the given trinomial must be
$(2t - 1)$ and $(t + 2)$.
$\qquad \therefore 2t^2 + 3t - 2 = (2t - 1)(t + 2)$

Note: After locating the correct combination using
the *cross method*, the factors are formed by reading
across the line, not **down** the diagonal.

(iv) $16x - 3 - 5x^2 = {}^-(5x^2 - 16x + 3)$ (common
$\qquad\qquad\qquad\qquad\qquad\qquad$ factor $^-1$)
$\qquad\qquad\qquad = {}^-(5x - 1)(x - 3)$ **why?**

Possible combinations, using the '*cross
method*', that give a positive x^2 term, a
positive constant and a negative x term:

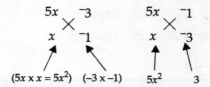

$\qquad (5x \times x = 5x^2)\quad (-3 \times -1)\qquad 5x^2 \qquad 3$

The second combination is the correct one
since

$\qquad (5x \times {}^-3) + (x \times {}^-1) = {}^-15x + {}^-x = {}^-16x$
$\qquad\qquad\qquad\qquad\qquad$ (the middle term)

Hence, the factors are $(5x - 1)$ and $(x - 3)$.

$$\therefore 16x - 3 - 5x^2 = -(5x^2 - 16x + 3)$$
$$= -(5x - 1)(x - 3)$$
$$= (-5x + 1)(x - 3)$$
$$= (1 - 5x)(x - 3)$$
$$= (5x - 1)(^-x + 3)$$
$$= (5x - 1)(3 - x)$$

(v) $4x^2 - 11x + 6$

Using the *'cross method'* the following combinations will give a positive first and last term and a negative middle term, as required.

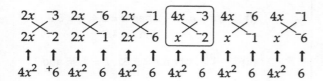

The boxed combination is the correct one since its cross products sum to ^-11x as required,

(that is, $(4x \times ^-2) + (x \times ^-3) = ^-8x + ^-3x = ^-11x$).

Hence, the factors are $(4x - 3)$ and $(x - 2)$.

$$\therefore 4x^2 - 11x + 6 = (4x - 3)(x - 2)$$

YOUR TURN

4.5 Factorize the following expressions by extracting the highest common factor:

(i) $81x^3 - 27x$

(ii) $8m^2n + 6mn^2 + 10mn$

(iii) $8b - 80c - b^2 + 10bc$

(iv) $4 - 4y + cy - c$

(v) $x(a - b) + 5(b - a)$

4.6 Factorize each of the following (using Methods 1, 4, 5 or 6):

(i) $p^2 - 20p + 100$

(ii) $3x^2 + 12x + 12$

(iii) $36t^2 - 4s^2$

(iv) $1 - 25q^2$

(v) $45x^2y^2 - 80$

4.7 Factorize each of the following trinomials:

(i) $2x^2 + 7x + 3$

(ii) $4y^2 - 12y + 9$

(iii) $6m^2 - 7m - 3$

(iv) $12 - 7t - 10t^2$

(v) $6a^2 + 30a - 36$

4.4 Numerical substitution

In Chapter 2 we learned how to substitute numbers into simple algebraic expressions and evaluate the answer. Since this chapter involves more complex algebraic expressions, including multinomials and indices, it is important to review the processes of substitution and evaluation and check that they can still be handled effectively.

EXAMPLE 4.8

If $a = ^-1$, $b = 5$ and $c = ^-2$, evaluate each of the following expressions.

(i) $(a + b)(a - b)$

(ii) $b(3a^2 + 2a)$

(iii) $\dfrac{(a + c)^2}{b^2}$

Solution

(i) $(a + b)(a - b) = (^-1 + 5)(^-1 - 5)$
$$= 4 \times ^-6$$
$$= ^-24$$

(ii) $b(3a^2 + 2a) = 5[(3 \times ^-1 \times ^-1) + (2 \times ^-1)]$
$$= 5[3 \times 1 + ^-2]$$
$$= 5[3 - 2]$$
$$= 5 \times 1$$
$$= 5$$

(iii) $\dfrac{(a + c)^2}{b^2} = \dfrac{(-1 + (-2))^2}{5^2}$
$$= \dfrac{(-3)^2}{5^2}$$
$$= \dfrac{9}{25}$$

EXAMPLE 4.9

If $x = \dfrac{1}{2}$, $y = \dfrac{1}{3}$ and $z = \dfrac{1}{4}$, find the value of

(i) $4x^2 + 2x - 3z$

(ii) $(x + 3y + 2z)^3$

Solution

(i) $4x^2 + 2x - 3z = \left(4 \times \dfrac{1}{2} \times \dfrac{1}{2}\right) + \left(2 \times \dfrac{1}{2}\right)$
$$- \left(3 \times \dfrac{1}{4}\right)$$
$$= 1 + 1 - \dfrac{3}{4}$$
$$= 2 - \dfrac{3}{4} = 1\dfrac{1}{4}$$

or $4x^2 + 2x - 3z = 2x(2x + 1) - 3z$

$$= 2 \times \frac{1}{2}\left(2 \times \frac{1}{2} + 1\right) - 3 \times \frac{1}{4}$$

$$= 1(2) - \frac{3}{4}$$

$$= 2 - \frac{3}{4}$$

$$= 1\frac{1}{4}$$

(ii) $(x + 3y + 2z)^3 = \left(\frac{1}{2} + 3 \times \frac{1}{3} + 2 \times \frac{1}{4}\right)^3$

$$= \left(\frac{1}{2} + 1 + \frac{1}{2}\right)^3$$

$$= 2^3$$

$$= 8$$

EXAMPLE 4.10
Evaluate the following expressions, correct to 2 decimal places:

(i) $yz - x$, when $x = {}^-7.8$, $y = 5.2$ and $z = 6.4$

(ii) $\left(\dfrac{3a}{2b}\right)^2$, when $a = {}^-19.8$ and $b = 7.3$

Solution

(i) $yz - x = 5.2 \times 6.4 - {}^-7.8$

Using the calculator, we have

5.2 $\boxed{\times}$ 6.4 $\boxed{-}$ 7.8 $\boxed{+/-}$ $\boxed{=}$ $\boxed{41.08}$

$\therefore yz - x = 41.08$

(ii) $\left(\dfrac{3a}{2b}\right)^2 = \left(\dfrac{3 \times {}^-19.8}{(2 \times 7.3)}\right)^2$

Using the calculator, we have

$\boxed{(}$ 3 $\boxed{\times}$ 19.8 $\boxed{+/-}$ $\boxed{\div}$ $\boxed{(}$ 2 $\boxed{\times}$ 7.3 $\boxed{)}$ $\boxed{)}$

$\boxed{x^2}$ $\boxed{16.552637}$

$\therefore \left(\dfrac{3a}{2b}\right)^2 = 16.55$, correct to 2 decimal places.

YOUR TURN

4.8 If $r = {}^-2$, $s = 6$ and $t = {}^-1$, evaluate each of the following expressions.

(i) $2r(r - 2s + 3t)$

(ii) $\dfrac{r^2 + r + 2s}{s + t^2}$

4.8 *Cont.*

(iii) $\sqrt[3]{t + 3s - 5r}$

(iv) $(4r^2 - t^2)(3s^2 + t^2)$

4.9 Find the value of each of the following expressions if $a = \dfrac{1}{4}$, $b = \dfrac{1}{3}$ and $c = \dfrac{1}{2}$.

(i) $4c + 6b - 8a$

(ii) $24abc^3$

(iii) $(3b + 2c)(6b - 8a)$

(iv) $\sqrt{\dfrac{2bc}{ab}}$

4.10 Using your calculator, evaluate each of the following expressions correct to one decimal place.

(i) $\sqrt[3]{d + e + f}$, when $d = 8.4$, $e = 6.2$ and $f = 4.3$

(ii) $\sqrt{\dfrac{ab}{c}}$, when $a = 5.63$, $b = 1.96$ and $c = 2.79$

(iii) $\sqrt{pq^2 - t}$, when $p = 12.6$, $q = -3.82$ and $t = 13.45$

(iv) $\sqrt{\dfrac{xy}{z}}$, when $x = 0.923$, $y = 2.841$ and $z = 5.296$

Extensions to chapter 4

Extension A: Multinomial products
As with binomial products, the distributive law is useful in determining multinomial products.

EXAMPLE E4.1
Expand and simplify each of the following algebraic expressions.

(i) $(a + 1)(a^2 + 5a - 2)$

(ii) $(2y^2 - 3y + 1)(y^3 + 2y - 1)$

(iii) $(pq - p^2 - 2q + 3)(2p - q)$

Solution

(i) $(a + 1)(a^2 + 5a - 2)$

$= a(a^2 + 5a - 2) + 1(a^2 + 5a - 2)$
(distributive law)

$= a^3 + 5a^2 - 2a + a^2 + 5a - 2$
(distributive law)

$= a^3 + 5a^2 + a^2 - 2a + 5a - 2$
(commutative law)

$= a^3 + 6a^2 + 3a - 2$
(simplify)

(ii) $(2y^2 - 3y + 1)(y^3 + 2y - 1)$

$= 2y^2(y^3 + 2y - 1) - 3y(y^3 + 2y - 1) + 1(y^3 + 2y - 1)$

$= 2y^5 + 4y^3 - 2y^2 - 3y^4 - 6y^2 + 3y + y^3 + 2y - 1$

$= 2y^5 - 3y^4 + 4y^3 + y^3 - 2y^2 - 6y^2 + 3y + 2y - 1$

$= 2y^5 - 3y^4 + 5y^3 - 8y^2 + 5y - 1$

(iii) $(pq - p^2 - 2q + 3)(2p - q)$

$= (2p - q)(pq - p^2 - 2q + 3)$
    (commutative law)

$= 2p(pq - p^2 - 2q + 3) - q(pq - p^2 - 2q + 3)$

$= 2p^2q - 2p^3 - 4pq + 6p - pq^2 + p^2q + 2q^2 - 3q$

$= 2p^2q + p^2q - 2p^3 - 4pq + 2q^2 - pq^2 + 6p - 3q$

$= 3p^2q - pq^2 - 2p^3 - 4pq + 2q^2 + 6p - 3q$

Extension B: More complicated algebraic fractions

A first step in simplifying more complex algebraic fractions often involves the use of the factorization methods discussed in *Section 3*.

EXAMPLE E4.2
Simplify the following:

(i) $\dfrac{3a^2b - 6ab}{2a^2b^2 - 4ab^2}$

(ii) $\dfrac{y^2 - y - 6}{y^2 - 10y + 21}$

(iii) $\dfrac{(x+y)^2 - z^2}{4x + 4y - 4z}$

(iv) $\dfrac{(a-b)^2}{3} \times \dfrac{1}{(a^2 - b^2)}$

(v) $\dfrac{t^2 - t - 20}{t^2 - 25} \div \dfrac{t+1}{t^2 + 5t}$

(vi) $\dfrac{c+1}{1-c} + \dfrac{c-1}{1+c}$

(vii) $5 - \dfrac{3m - n}{3m + n}$

(viii) $\dfrac{p^2}{p^2 + 3p + 2} - \dfrac{2p}{p+2}$

Solution

(i) $\dfrac{3a^2b - 6ab}{2a^2b^2 - 4ab^2}$

$= \dfrac{3ab(a-2)}{2ab^2(a-2)}$ take out common factors $3ab$ and $2ab^2$

$= \dfrac{3ab\,(a-2)}{2ab^2\,(a-2)}$ divide numerator and denominator by $ab(a-2)$

$= \dfrac{3}{2b}$

(ii) $\dfrac{y^2 - y - 6}{y^2 - 10y + 21}$

$= \dfrac{(y-3)(y+2)}{(y-3)(y-7)}$ factorize numerator and denominator

$= \dfrac{y+2}{y-7}$ divide numerator and denominator by $(y-3)$

(iii) $\dfrac{(x+y)^2 - z^2}{4x + 4y + 4z}$

$= \dfrac{(x+y-z)(x+y+z)}{4(x+y+z)}$ Num: factorize using Method 6
Den: take out common factor, 4

$= \dfrac{(x+y-z)}{4}$ divide num. and den. by $(x+y+z)$

(iv) $\dfrac{(a-b)^2}{3} \times \dfrac{1}{(a^2 - b^2)}$

$= \dfrac{(a-b)(a-b)}{3} \times \dfrac{1}{(a-b)(a+b)}$ since $(a-b)^2$
$= (a-b)(a+b)$
by Method 6

$= \dfrac{a-b}{3(a+b)}$ divide num. and den. by $(a-b)$

(v) $\dfrac{t^2 - t - 20}{t^2 - 25} \div \dfrac{t+1}{t^2 + 5t}$

$= \dfrac{(t-5)(t+4)}{(t-5)(t+5)} \div \dfrac{t+1}{t(t+5)}$

$= \dfrac{(t-5)(t+4)}{(t-5)(t+5)} \times \dfrac{t(t+5)}{t+1}$ divide num. and den. by $(t-5)(t+5)$

$= \dfrac{t(t+4)}{(t+1)}$

(vi) $\dfrac{c+1}{1-c} + \dfrac{c-1}{1+c}$

$= \dfrac{(c+1)(1+c) + (c-1)(1-c)}{(1-c)(1+c)}$ take $(1-c)(1+c)$ as the common denominator

$= \dfrac{(c+1)(c+1) + (c-1) \times -(c-1)}{1^2 - c^2}$

$= \dfrac{(c+1)^2 - (c-1)^2}{1 - c^2}$

$= \dfrac{c^2 + 2c + 1 - (c^2 - 2c + 1)}{1 - c^2}$ special binomial products

$= \dfrac{c^2 + 2c + 1 - c^2 + 2c - 1}{1 - c^2}$ simplify

$= \dfrac{4c}{1 - c^2}$

(vii) $5 - \dfrac{3m - n}{3m + n}$

$= \dfrac{5(3m + n) - (3m - n)}{3m + n}$ take $(3m + n)$ as the common denominator

$= \dfrac{15m + 5n - 3m + n}{3m + n}$ distributive law

$= \dfrac{15m - 3m + 5n + n}{3m + n}$

$= \dfrac{12m + 6n}{3m + n}$ simplify

(viii) $\dfrac{p^2}{p^2+3p+2} - \dfrac{2p}{p+2}$

$$= \dfrac{p^2}{(p+1)(p+2)} - \dfrac{2p}{p+2} \qquad \text{factorize}$$

$$= \dfrac{p^2 - 2p(p+1)}{(p+1)(p+2)} \qquad \begin{array}{l}\text{take } (p+1)(p+2) \text{ as the}\\ \text{common denominator}\end{array}$$

$$= \dfrac{p^2 - 2p^2 - 2p}{(p+1)(p+2)} \qquad \text{distributive law}$$

$$= \dfrac{-p^2 - 2p}{(p+1)(p+2)} \qquad \text{simplify}$$

$$= \dfrac{-p(p+2)}{(p+1)(p+2)} \qquad \text{factorize}$$

$$= \dfrac{-p}{(p+1)} \qquad \begin{array}{l}\text{divide num. and den. by}\\ (p+2)\end{array}$$

Extension C: Further products involving surds

Readers who wish to refresh their memory about surds and their basic operating rules should refer back to (*Section 3*) of Chapter 3.

EXAMPLE E4.3
Expand and then simplify the following algebraic expressions.

(i) $\sqrt{3}\,(5\sqrt{7} + 3)$

(ii) $(\sqrt{7} + 2\sqrt{3})\,(3\sqrt{2} - 5)$

(iii) $(\sqrt{8} - 2\sqrt{3})^2$

(iv) $(6 - \sqrt{3})(6 + \sqrt{3})$

(v) $(\sqrt{7} + \sqrt{5})\,(\sqrt{7} - \sqrt{5})$

(vi) $(2\sqrt{3} - \sqrt{5})\,(2\sqrt{3} + \sqrt{5})$

Solution

(i) $\sqrt{3}\,(5\sqrt{7} + 3) = (5\sqrt{7} \times \sqrt{3}) + (3 \times \sqrt{3}) \quad \text{distributive law}$

$\qquad\qquad\qquad = (5\sqrt{21}) + 3\sqrt{3} \quad \text{since } \sqrt{a} \times \sqrt{b} = \sqrt{ab}$

(ii) $(\sqrt{7} + 2\sqrt{3})\,(3\sqrt{2} - 5)$

$\quad = \sqrt{7}\,(3\sqrt{2} - 5) + 2\sqrt{3}\,(3\sqrt{2} - 5)$

$\quad = (3\sqrt{2} \times \sqrt{7} - 5\sqrt{7}) + (2\sqrt{3} \times 3\sqrt{2} - 5 \times 2\sqrt{3}) \quad \begin{array}{l}\text{distributive}\\ \text{law}\end{array}$

$\quad = 3\sqrt{14} - 5\sqrt{7} + 6\sqrt{6} - 10\sqrt{3}$

(iii) $(\sqrt{8} - 2\sqrt{3})^2$

$\quad = (\sqrt{8})^2 - 2(\sqrt{8} \times 2\sqrt{3}) + (2\sqrt{3})^2 \quad \begin{array}{l}\text{since } (a-b)^2\\ = a^2 - 2ab + b^2\end{array}$

$\quad = 8 - 4\sqrt{24} + 4 \times 3 \quad \text{since } (\sqrt{a})^2 = a$

$\quad = 8 + 12 - 4\sqrt{4} \times \sqrt{6} \quad \text{since } \sqrt{ab} = \sqrt{a} \times \sqrt{b}$

$\quad = 20 - 4 \times 2 \times \sqrt{6}$

$\quad = 20 - 8\sqrt{6}$

(iv) $(6 - \sqrt{3})(6 + \sqrt{3})$

$\quad = 6^2 - (\sqrt{3})^2 \quad \text{since } (a-b)(a+b) = a^2 - b^2$

$\quad = 36 - 3 \quad \text{since } (\sqrt{a})^2 = a$

$\quad = 33$

Note: When multiplying two binomial factors of the form $(a - b)(a + b)$, where '*a*' and/or '*b*' are surds, the product is **always** a rational number and the factors $(a - b)$ and $(a + b)$ are called **conjugate surds**.

Hence, $(6 - \sqrt{3})$ and $(6 + \sqrt{3})$ above are **conjugate surds**.

(v) $(\sqrt{7} + \sqrt{5})\,(\sqrt{7} - \sqrt{5})$

$\quad = (\sqrt{7})^2 + (\sqrt{5})^2 \quad \text{since } (a-b)(a+b) = a^2 - b^2$

$\quad = 7 - 5$

$\quad = 2$

Here, the factors $(\sqrt{7} + \sqrt{5})$ and $(\sqrt{7} - \sqrt{5})$ are **conjugate surds** as they are of the form $(a - b)$ and $(a + b)$.

(vi) $(2\sqrt{3} - \sqrt{5})\,(2\sqrt{3} + \sqrt{5})$

$\quad = (2\sqrt{3})^2 - (\sqrt{5})^2$

$\quad = 4 \times 3 - 5$

$\quad = 12 - 5$

$\quad = 7$

Here, the factors $(2\sqrt{3} - \sqrt{5})$ and $(2\sqrt{3} + \sqrt{5})$ are **conjugate surds**.

Rationalizing the denominator

Whenever a fraction has an irrational denominator such as $\dfrac{1}{\sqrt{2}}$, $\dfrac{\sqrt{3}}{\sqrt{5}}$, $\dfrac{2}{3\sqrt{7}}$, $\dfrac{3\sqrt{5}}{5\sqrt{3}}$, it is usually more convenient to form an equivalent expression that has a *rational* denominator. This process of 'rationalizing the denominator' is accomplished by multiplying both the numerator and the denominator by the surd in the denominator.

EXAMPLE E4.4
Using the examples from above,

(i) $\dfrac{1}{\sqrt{2}} = \dfrac{1}{\sqrt{2}} \times \dfrac{\sqrt{2}}{\sqrt{2}} = \dfrac{\sqrt{2}}{2}$

(ii) $\dfrac{\sqrt{3}}{\sqrt{5}} = \dfrac{\sqrt{3}}{\sqrt{5}} \times \dfrac{\sqrt{5}}{\sqrt{5}} = \dfrac{\sqrt{15}}{5}$

(iii) $\dfrac{2}{3\sqrt{7}} = \dfrac{2}{3\sqrt{7}} \times \dfrac{\sqrt{7}}{\sqrt{7}} = \dfrac{2\sqrt{7}}{3 \times 7} = \dfrac{2\sqrt{7}}{21}$

(iv) $\dfrac{3\sqrt{5}}{5\sqrt{3}} = \dfrac{3\sqrt{5}}{5\sqrt{3}} \times \dfrac{\sqrt{3}}{\sqrt{3}} = \dfrac{3\sqrt{15}}{5 \times 3} = \dfrac{\sqrt{15}}{5}$

Note: In parts (iii) and (iv) above, the numerator and denominator are multiplied by the **surd** in the denominator only, rather than by the complete term. That is, in (iii), multiply by $\sqrt{7}$ rather than $3\sqrt{7}$ and in (iv), multiply by $\sqrt{3}$ rather than $5\sqrt{3}$.

A fraction where the denominator is a binomial expression involving surds, such as $\dfrac{3}{\sqrt{2}-1}$, may also be written as a fraction with a *rational* denominator by multiplying the numerator and the denominator by the **conjugate** of the denominator.

EXAMPLE E4.5
Rationalize the denominator in each of the following expressions.

(i) $\dfrac{3}{\sqrt{2}-1}$

(ii) $\dfrac{\sqrt{8}+\sqrt{3}}{\sqrt{5}-\sqrt{8}}$

(iii) $\dfrac{2\sqrt{5}-\sqrt{3}}{6+\sqrt{2}}$

Solution

(i) $\dfrac{3}{\sqrt{2}-1} = \dfrac{3}{\sqrt{2}-1} \times \dfrac{\sqrt{2}+1}{\sqrt{2}+1}$ since $(\sqrt{2}+1)$ is the conjugate of $(\sqrt{2}-1)$

$= \dfrac{3(\sqrt{2}+1)}{(\sqrt{2}-1)(\sqrt{2}+1)}$

$= \dfrac{3\sqrt{2}+3}{(\sqrt{2})^2-(1)^2}$ since $(a-b)(a+b)=a^2-b^2$

$= \dfrac{3\sqrt{2}+3}{2-1}$ since $(\sqrt{a})^2=a$

$= \dfrac{3\sqrt{2}+3}{1}$

$= 3\sqrt{2}+3$

(ii) $\dfrac{\sqrt{8}+\sqrt{3}}{\sqrt{5}-\sqrt{8}}$

$= \dfrac{\sqrt{8}+\sqrt{3}}{\sqrt{5}-\sqrt{8}} \times \dfrac{\sqrt{5}+\sqrt{8}}{\sqrt{5}+\sqrt{8}}$

since $(\sqrt{5}+\sqrt{8})$ is the conjugate of $(\sqrt{5}-\sqrt{8})$

$= \dfrac{(\sqrt{8}+\sqrt{3})(\sqrt{5}+\sqrt{8})}{(\sqrt{5}-\sqrt{8})(\sqrt{5}+\sqrt{8})}$

$= \dfrac{\sqrt{8}(\sqrt{5}+\sqrt{8})+\sqrt{3}(\sqrt{5}+\sqrt{8})}{(\sqrt{5})^2-(\sqrt{8})^2}$

by distributive law and since $(a-b)(a+b)=a^2-b^2$

$= \dfrac{(\sqrt{8}\times\sqrt{5})+(\sqrt{8})^2+(\sqrt{3}\times\sqrt{5})+(\sqrt{3}\times\sqrt{8})}{5-8}$

by distributive law and since $(\sqrt{a})^2=a$

$= \dfrac{\sqrt{40}+8+\sqrt{15}+\sqrt{24}}{-3}$

since $\sqrt{a}\times\sqrt{b}=\sqrt{ab}$

$= \dfrac{\sqrt{4\times10}+8+\sqrt{15}+\sqrt{4\times6}}{-3}$

$= \dfrac{2\sqrt{10}+8+\sqrt{15}+2\sqrt{6}}{-3}$

since $\sqrt{ab}=\sqrt{a}\times\sqrt{b}$

(iii) $\dfrac{2\sqrt{5}-\sqrt{3}}{6+\sqrt{2}}$

$= \dfrac{2\sqrt{5}-\sqrt{3}}{6+\sqrt{2}} \times \dfrac{6-\sqrt{2}}{6-\sqrt{2}}$

since $(6-\sqrt{2})$ is the conjugate of $(6+\sqrt{2})$

$= \dfrac{(2\sqrt{5}-\sqrt{3})(6-\sqrt{2})}{(6+\sqrt{2})(6-\sqrt{2})}$

$= \dfrac{2\sqrt{5}(6-\sqrt{2})-\sqrt{3}(6-\sqrt{2})}{(6)^2-(\sqrt{2})^2}$

by distributive law and since $(a-b)(a+b)=a^2-b^2$

$= \dfrac{12\sqrt{5}-2\sqrt{10}-6\sqrt{3}+\sqrt{6}}{36-2}$

expand and simplify

$= \dfrac{12\sqrt{5}-2\sqrt{10}-6\sqrt{3}+\sqrt{6}}{34}$

Summary

This chapter has built upon your knowledge of basic algebra gained from Chapter 2 by examining more complex algebraic expressions and discovering ways of both expanding and factorizing them, with the assistance of some rules and properties.

Important terminology
(see p36 for definitions)
You should be able to recognise and write an example of each of the following:

    monomial

    multinomial

    binomial

    trinomial

    coefficient

    constant

    polynomial

    degree

Definition
The product of two binomial expressions is called a **binomial product** and is most easily determined using the distributive law.

$$(x+a)(y+b) = x(y+b)+a(y+b)$$
$$= xy+xb+ay+ab$$

Special binomial products

1. $(a+b)^2 = a^2+2ab+b^2$ ⎫ perfect squares
2. $(a-b)^2 = a^2-2ab+b^2$ ⎭

3. $(a-b)(a+b) = a^2-b^2$ difference between two squares

Definition

> **Factorization** is the reverse of the process of *expansion* and involves determining the factors that, when multiplied together, give the algebraic expression in question.

Factorization methods

1. $ab + ac = a(b + c)$ common factor
2. $ab + ac + pb + pc = (a + p)(b + c)$
3. $x^2 + (a + b)x + ab = (x + a)(x + b)$
                        common factors
4. $a^2 + 2ab + b^2 = (a + b)^2$
5. $a^2 - 2ab + b^2 = (a - b)^2$
6. $a^2 - b^2 = (a - b)(a + b)$

Numerical substitution was discussed again in this chapter as it is very important to be able to substitute given numbers into complex algebraic expressions and then to evaluate them with or without a calculator.

The **extensions** included in this chapter are particularly useful for those students undertaking disciplines with a strong mathematical orientation. They elaborate upon your knowledge of binomial products, algebraic fractions and surds (discussed in this and previous chapters) covering topics such as multinomial products, the simplification of more complex algebraic fractions using factorization methods and the determination of products involving surds using the distributive law, the notion of conjugate surds and the process of rationalization of the denominator.

Note:

(i) Two binomial factors of the form $(a - b)$ and $(a + b)$ where 'a' and/or 'b' are surds are called **conjugate surds**.

(ii) The product of conjugate surds is *always* a **rational** number.
   e.g. $(\sqrt{2} + 3)(\sqrt{2} - 3) = (\sqrt{2})^2 - 3^2 = 2 - 9 = -7$

4.1 Simplify each of the following:
  (a) $(\sqrt{3} + \sqrt{2})(\sqrt{3} + \sqrt{5})$
  (b) $(2\sqrt{5} - 3)(3 + 2\sqrt{3})$
  (c) $(\sqrt{5} + \sqrt{2})^2$
  (d) $(\sqrt{3} + \sqrt{2})(\sqrt{3} - \sqrt{2})$
  (e) $(2\sqrt{7} + 3\sqrt{2})(2\sqrt{7} - 3\sqrt{2})$

4.2 Simplify the following expressions:
  (a) $(a - b)(a + b)$
  (b) $(2a + 3)(3a - 2)$
  (c) $(x - 5)(5 - 2x)$
  (d) $(x + 2)(4x^3 + 3x + 2)$
  (e) $(3x^2 - x - 2)(-2x^2 - 3x + 4)$

4.3 Factorise each of the following expressions:
  (a) $x^2y - xy^2$
  (b) $x + 3xyz$
  (c) $12x^3 + 13x^2 + 24x$
  (d) $3x^4 + 15x^2$
  (e) $25 - 16x^2$
  (f) $0.09a^2 - 0.25$
  (g) $a^{2n} - 1$
  (h) $(x + y)^2 - 1$

4.4 Factorise
  (a) $x^2 + 3x - 4$
  (b) $m^2 + 5m - 24$
  (c) $(ax)^2 - (ax) - 6$
  (d) $(t + n)^2 + 5(t + n) - 14$
  (e) $3x^2 + 5x - 2$

4.5 Simplify each of the following expressions:
  (a) $\dfrac{x^2 + 7x + 12}{x + 3}$
  (b) $\dfrac{10a^2 + 7a - 12}{2a + 3}$
  (c) $\dfrac{x^2 - 16}{4 + x}$
  (d) $\dfrac{12a^2 + 6a - 18}{2a + 3}$

4.6 Rationalise the denominator of each of the following:
  (a) $\dfrac{1}{\sqrt{3}}$
  (b) $\dfrac{2}{3\sqrt{7}}$
  (c) $\dfrac{3}{\sqrt{7} + 2}$

4.6 *Cont.*

(d) $\dfrac{\sqrt{6}}{\sqrt{5}+\sqrt{2}}$

(e) $\dfrac{\sqrt{3}+\sqrt{2}}{\sqrt{3}-\sqrt{2}}$

*(f) $\dfrac{\sqrt{7}-3\sqrt{5}}{3\sqrt{5}-2\sqrt{2}}$

4.7 Simplify $\dfrac{a^2\sqrt{b}}{\sqrt[3]{c}}$, given that $a = 10$, $b = 100$ and $c = 1000$.

4.8 If $a = 4$, $b = -3$, $c = 5$ and $d = -6$, find the value of:

(a) $2a + 3b - c$

(b) $6a - 3b - 2d$

(c) $5a - \dfrac{cd}{2} + \dfrac{abc}{3}$

4.9 Show that $x^2 - 7x + 12 = 0$ if $x = 3$ and $x = 4$. What is the value of the expression $x^2 - 7x + 12$ when $x = -2$?

*4.10 Show that, when $x = 9$ and $y = 3$, the two expressions $4(x - y) + 5(x + y)$ and $7(x + y) + 2x - 6y$, are equal.

*4.11 Show that, when $a = 1$, 3 or 4, the expressions $a^3 + 19a$ and $4(2a^2 + 3)$ are equal. Which expression is greater when $a = 2$?

4.12 When x has the values -1, 4, -7 and 10, find the respective values of the expression $16 - x + x^2$.

5 Equations

Introduction

In the preceding chapters we have concentrated on the manipulation of both simple and complex arithmetic and algebraic expressions, presenting many rules, laws and techniques designed to enhance and simplify these manipulations.

Now we must move on and begin our examination of several mathematical topics all of which will be grouped together under the heading **equations**.

Before we proceed, let us pause here and define what we mean by an *equation*.

> An **equation** is a mathematical statement that tells us that two expressions are *equal*. These two expressions are separated by the *equality sign*, '='.

For example:

(i) $x + 3 = 5$

(ii) $2a + 8 = a - 3$

(iii) $y^2 + 2y + 1 = (y + 1)^2$

(iv) $\dfrac{b}{b - 4} = 2c + 3$

(v) $2z^2 + 1 = 0$

An *equation* usually contains at least one *variable*. In the above example, equations (i), (ii), (iii) and (v) are equations with one variable only, whereas equation (iv) contains two variables.

Note: You will generally find that equation (i) above would be referred to as an equation in x, while equation (iv) would be called an equation in b and c.

Almost every student, regardless of their chosen discipline, will at some stage be faced with having to 'solve an equation', and the underlying aim of this chapter is to introduce techniques to accomplish this task.

Of course, one obvious question arising from this is "what do we mean by *'solve'*?"

In *solving* an equation we will attempt to determine value(s) of the variable(s) that will result in the given mathematical statement being true (that is, so that the left-hand side **equals** the right-hand side). You will see as we progress through this and later chapters, that there are three possible results:

(i) A **unique** solution

e.g. $x + 3 = 5$

So, $x = 2$

(ii) **No** solution

e.g. $2z^2 + 1 = 0$

(since we cannot find $\sqrt{-\dfrac{1}{2}}$)

(iii) An **infinite number** of solutions

e.g. $y^2 + 2y + 1 = (y + 1)^2$

This statement is true for *all* values of y.

This chapter consists of five sections in the core, together with some extensions. The *first* section discusses methods of solving one linear equation only, while the *second* presents techniques for solving two linear equations simultaneously.

Section 3 examines procedures for solving what are known as quadratic equations, with *Section 4* interpreting the meaning of the phrase 'changing the subject of an equation' and seeing how this can be best accomplished.

Finally, *Section 5* presents some applications of the material discussed in *Sections 1* to *4* within a problem-solving framework.

The **extensions** to this chapter discuss three rather more complex topics on equations and hence are considered necessary only for those readers wishing to undertake studies in any of the mathematical and/or scientific disciplines. However, other readers are certainly encouraged to study the material for interest.

The additional topics include

(i) equations reducible to quadratic or linear form

(ii) absolute-value equations, and

(iii) equations involving algebraic fractions.

5.1 Solving linear equations

Definition 1

A **linear** equation is an equation that contains one or more variables, each being of the *first* degree only.

Remember: The *degree* of a variable is the power to which that variable is raised (see Chapter 4: Algebra revisited).

For example:

(i) $x + 3 = 5$

(ii) $x + y = 4$

(iii) $2(x + 3) = x - 2$

In general, the solution of a linear equation requires the formation of one or more **equivalent** equations.

Definition 2

An **equivalent** equation is formed by applying to both sides of the original equation any combination of arithmetic operations (excluding division by zero) using the same number or algebraic expression.

EXAMPLE 5.1

Consider the original equation

$x + 3 = 5$

Then some possible equivalent equations are

(i) $x + 3 - 3 = 5 - 3$ (subtract 3 from both sides)

(ii) $2(x + 3) = 2 \times 5$ (multiply both sides by 2)

(iii) $\dfrac{x + 3 + y}{2} = \dfrac{5 + y}{2}$ (add y to both sides and then divide by 2)

Now, keeping in mind that *equivalent equations* must be formed in order to solve a given linear equation, the most efficient method of reaching a solution (if possible) is to simplify the original equation by performing a series of inverse operations.

Remember:

Operation	+	−	x	÷	a^n	$\sqrt[n]{a}$	log
Inverse operation	−	+	÷	x	$\sqrt[n]{a}$	a^n	antilog

EXAMPLE 5.2

Solve for x, $\qquad x + 3 = 5$

Solution

$$x + 3 = 5$$
$$x + 3 - 3 = 5 - 3 \quad \text{(subtract 3 from both sides)}$$
$$\therefore \ x = 2$$

EXAMPLE 5.3

Solve for y, $\qquad 4y - 5 = y + 10$

Solution

$$4y - 5 = y + 10$$
$$4y - 5 + 5 = y + 10 + 5 \quad \text{(add 5 to both sides)}$$
$$4y = y + 15 \quad \text{(simplify both sides)}$$
$$4y - y = y + 15 - y \quad \text{(subtract } y \text{ from both sides)}$$
$$3y = 15 \quad \text{(simplify)}$$
$$\frac{3y}{3} = \frac{15}{3} \quad \text{(divide both sides by 3)}$$
$$\therefore \ y = 5$$

EXAMPLE 5.4

Solve for z, $\qquad \dfrac{2z}{z + 2} = 4$

Solution

$$\frac{2z}{z + 2} = 4$$
$$\frac{2z}{(z + 2)} \times (z + 2) = 4 \times (z + 2) \quad \text{(multiply both sides by } (z + 2))$$
$$2z = 4(z + 2) \quad \text{(simplify)}$$
$$\frac{2z}{2} = \frac{4(z + 2)}{2} \quad \text{(divide both sides by 2)}$$
$$z = 2(z + 2) \quad \text{(simplify)}$$
$$z = 2z + 4 \quad \text{(remove brackets)}$$
$$z - 2z = 2z + 4 - 2z \quad \text{(subtract } 2z \text{ from both sides)}$$
$$-z = 4 \quad \text{(simplify)}$$
$$-z \times -1 = 4 \times -1 \quad \text{(multiply both sides by } -1)$$
$$\therefore \ z = -4$$

EXAMPLE 5.5

Solve for x, $\qquad \sqrt{x^2 + 33} - x = 3$

Solution

$$\sqrt{x^2 + 33} - x = 3$$
$$\sqrt{x^2 + 33} - x + x = 3 + x \quad \text{(add } x \text{ to both sides)}$$
$$\sqrt{x^2 + 33} = x + 3 \quad \text{(simplify)}$$
$$\left(\sqrt{x^2 + 33}\right)^2 = (x + 3)^2 \quad \text{(square both sides)}$$
$$x^2 + 33 = x^2 + 6x + 9 \quad \text{(expand)}$$

$$x^2 + 33 - x^2 = x^2 + 6x + 9 - x^2$$

(subtract x^2 from both sides)

$$33 = 6x + 9$$

(simplify)

$$33 - 9 = 6x + 9 - 9$$

(subtract 9 from both sides)

$$24 = 6x$$

(simplify)

$$\frac{24}{6} = \frac{6x}{6}$$

(divide both sides by 6)

$$\therefore \ 4 = x \ \text{ or } \ x = 4$$

Very important:
Whenever mathematical procedures and/or operations are involved the possibility of error must be considered. Therefore, it is **always** very important to utilise any available techniques or mechanisms to check that the results obtained are accurate or at least reasonable.

Hence, in examples 5.2 to 5.5 above we should go one step further and substitute the resulting value back into the original equation, thus checking that the equality is true.

EXAMPLE 5.6
Check that the solution obtained in each of examples 5.2 to 5.5 above is accurate by substituting the solution back into the original equation.

Solution
(a) In example 5.2 we found that the given equation $x + 3 = 5$ had a unique solution, $x = 2$.

   Check:

   LHS: $2 + 3 = 5$

   RHS: 5

   $\therefore$ LHS = RHS

   Therefore, $x = 2$ is the correct solution to the given equation, $x + 3 = 5$.

(b) In example 5.3 we found that the given equation $4y - 5 = y + 10$ had a unique solution, $y = 5$.

   Check:

   LHS: $(4 \times 5) - 5 = 15$

   RHS: $5 + 10 = 15$

   $\therefore$ LHS = RHS

   Therefore, $y = 5$ is the correct solution to the given equation, $4y - 5 = y + 10$.

(c) In example 5.4 we found that the given equation $\dfrac{2z}{z + 2} = 4$ had a unique solution, $z = -4$.

   Check:

   LHS: $\dfrac{2 \times -4}{-4 + 2} = \dfrac{-8}{-2} = 4$

   RHS: 4

   $\therefore$ LHS = RHS

   Therefore, $z = -4$ is the correct solution to the given equation, $\dfrac{2z}{z + 2} = 4$.

(d) In example 5.5 we found that the given equation $\sqrt{x^2 + 33} - 3 = 3$ had a unique solution, $x = 4$.

   Check:

   LHS: $\sqrt{x^2 + 33} - x = \sqrt{4^2 + 33} - 4$

   $= \sqrt{16 + 33} - 4$

   $= \sqrt{49} - 4$

   $= 7 - 4$

   $= 3$

   RHS: 3

   $\therefore$ LHS = RHS

   Therefore, $x = 4$ is the correct solution to the given equation, $\sqrt{x^2 + 33} - x = 3$.

YOUR TURN

5.1   *Check* that $x = 0$ is a solution to (i.e. satisfies) the equation
$$x(7 + x) - 2(x + 1) - 3x = -2.$$

5.2   *Check* that $y = 2$ and $y = -4$ are both solutions to the equation
$$2y + y^2 - 8 = 0.$$

In Exercises 5.3 to 5.5 write down the operations that must be applied to the first equation in order to obtain the second equivalent equation.

5.3 $8x - 4 = 16$; $x - \dfrac{1}{2} = 2$

5.4 $\dfrac{2}{y - 2} + y = y^2$; $2 + y(y - 2) = y^2(y - 2)$

5.5 $2t^2 - 9 = t$; $t^2 - \dfrac{1}{2}t = \dfrac{9}{2}$

In Exercises 5.6 to 5.10, solve the given equation for the unknown variable. *Remember* to check that the solution you obtain is correct by substituting it back into the original equation.

5.6 $\quad 3 - 2x = 4$

5.7 $\quad 7x + 7 = 2(x + 1)$

5.8 $\quad \dfrac{2y - 3}{4} = \dfrac{6y + 7}{3}$

5.9 $\quad t = 2 - 2[2t - 3(1 - t)]$

5.10 $\quad \dfrac{3}{2}(4a - 3) = 2\{a - (4a - 3)\}$

In Exercises 5.11 to 5.15, solve each of the given equations and then check that each solution is correct.

5.11 $\quad \dfrac{x + 3}{x} = \dfrac{2}{5}$

5.12 $\quad \dfrac{1}{p - 1} = \dfrac{2}{p - 2}$

5.13 $\quad 6 - \sqrt{2y + 5} = 0$

5.14 $\quad (t - 3)^{\frac{3}{2}} = 8$

5.15 $\quad \sqrt{a^2 - 9} = 9 - a$

5.2 Solving simultaneous linear equation

Definition 3

Simultaneous linear equations is the name given to a set of two or more equations in the first degree where the same solution for each of the unknown variables satisfies all the equations in the set.

For example:

(i) $\quad 2x + y = 7$

$\quad 3x - 2y = 8$

This is a set of two linear equations in two unknowns (i.e. variables x and y)

(ii) $\quad 3x + 2y + 4z = 7$

$\quad x - y + 2z = 5$

$\quad 2x + 4y - z = 8$

This is a set of three linear equations in three unknowns (i.e. variables x, y and z).

Note:

Solving a set of two or more linear equations **simultaneously** involves determining a solution for each of the unknown variables that satisfies **all** the equations in the set at the same time.

5.2.1 Methods of solution

Although several methods for solving simultaneous linear equations are available, we will concentrate on two only in this chapter – namely, the **elimination** method and the **substitution** method.

Furthermore, using each of these two methods, we will restrict our discussion to the simplest case of determining the solution to a set of *two* equations in two unknowns.

However, it should be noted that the procedures involved in solving the simple case above can be readily generalised to more complex problems requiring the simultaneous solution of n equations in n unknowns.

Note: In Chapter 7 the **graphical** method of solving simultaneous equations will be discussed, while in Chapter 11 the solution of simultaneous equations involving **matrix algebra** will be presented.

A. Solving a set of two linear equations in two unknowns using the *elimination method*

As the heading suggests, *the first step* of this solution method involves the 'elimination' of *one* of the two variables, thus allowing the value of the remaining variable to be determined. *The second step* involves substituting this value back into either of the original equations in order to calculate the value of the 'eliminated' variable. As always, the *final step* is to check that the solutions obtained do satisfy both equations.

EXAMPLE 5.7

Solve, using the *elimination method*,

$$2x + y = 8$$
$$3x - y = 7$$

Solution

$$2x + y = 8 \qquad\qquad \text{—(1)}$$
$$3x - y = 7 \qquad\qquad \text{—(2)}$$

Step 1: Add together term by term equations **(1)** and **(2)** to eliminate y. Then solve for x.

$$2x + y = 8 \qquad\qquad \text{—(1)}$$
$$3x - y = 7 \qquad\qquad \text{—(2)}$$
$$5x = 15 \ (\text{since } y + (^-y) = 0)$$
$$\therefore x = 3$$

Step 2: Substitute $x = 3$ back into equation **(1)** and solve for y.

$$2 \times 3 + y = 8$$
$$\therefore y = 8 - 6$$
$$\text{So,} \quad y = 2$$

Step 3: Check that the solution $x = 3$, $y = 2$ satisfies both equations.

$$LHS: 2 \times 3 + 2 = 8 = RHS \quad —(1)$$
$$LHS: 3 \times 3 - 2 = 7 = RHS \quad —(2)$$

Therefore, the required solution is $(x = 3, y = 2)$.

EXAMPLE 5.8

Solve, using the *elimination method*,

$$2x + 3y = 19$$
$$4x - y = 3$$

Solution

$$2x + 3y = 19 \quad —(1)$$
$$4x - y = 3 \quad —(2)$$

Note: Before any variable can be eliminated it must have the **same** numerical coefficient in *both* equations. This may involve some adjustments to one or both equations to form the appropriate equivalent equations.

In this example there are two possibilities:

(i) *Multiply* equation **(1)** by 2 and then *subtract* equation **(2)** from equation **(1)** to eliminate x.

(ii) *Multiply* equation **(2)** by 3 and then *add* the equations together to eliminate y.

In fact, whenever we wish to solve two linear equations simultaneously, there will always be two possible approaches depending upon which variable is to be eliminated. In general, the variable to be eliminated will be the one that requires the smaller number of adjustments.

In example 5.7 above, we eliminated y as *no* adjustments were necessary. Had we chosen to eliminate x, both equations would have required adjustment until the coefficient of x was the same in both.

In this example, 5.8, since elimination of either x or y requires the same number of adjustments, the choice is left to the individual. We will eliminate x, but you should re-do the problem yourself, eliminating y, to check that you obtain the same solution set.

Step 1: Multiply each term of **(1)** by 2, subtract **(2)** to eliminate x, and then solve for y.

$$2x + 3y = 19 \quad —(1)$$
$$4x - y = 3 \quad —(2)$$
$$\textbf{(1)} \times 2 \quad 4x + 6y = 38 \quad —(3)$$
$$\textbf{(3)} - \textbf{(2)} \quad 4x - y = 3$$
$$7y = 35 \text{ (since } 4x - 4x = 0$$
$$\text{and } 6y - (^-y) = 7y)$$
$$\therefore \quad y = 5$$

Step 2: Substitute $y = 5$ back into equation **(2)** and solve for x.

$$4x - 5 = 3$$
$$4x = 8$$
$$\therefore \quad x = 2$$

Step 3: Check that the solution $x = 2$, $y = 5$ satisfies both equations.

$$LHS: 2 \times 2 + 3 \times 5 = 4 + 15 = 19 = RHS \ —(1)$$
$$LHS: 4 \times 2 - 5 = 8 - 5 = 3 = RHS \quad —(2)$$

Therefore, the required solution is $(x = 2, y = 5)$.

EXAMPLE 5.9

Solve, using the *elimination method*,

$$2x + 3y = 8$$
$$3x - 2y = -7$$

Solution

$$2x + 3y = 8 \quad —(1)$$
$$3x - 2y = -7 \quad —(2)$$

Step 1: To eliminate y, multiply equation **(1)** by 2 and equation **(2)** by 3 to give the 'common' coefficient 6, and then add the two new equations together.

$$2x + 3y = 8 \quad —(1)$$
$$3x - 2y = -7 \quad —(2)$$
$$\textbf{(1)} \times 2 \quad 4x + 6y = 16 \quad —(3)$$
$$\textbf{(2)} \times 3 \quad 9x - 6y = -21 \quad —(4)$$
$$\textbf{(3)} + \textbf{(4)} \quad 13x = -5$$
$$\therefore \quad x = \frac{-5}{13}$$

Step 2: Substitute $x = \frac{-5}{13}$ back into **(1)** and solve for y.

$$2 \times \frac{-5}{13} + 3y = 8$$
$$3y = 8 + \frac{10}{13}$$
$$= 8\frac{10}{13}$$
$$\therefore \quad y = \frac{114}{13} \times \frac{1}{3}$$
$$= \frac{38}{13}$$

Step 3: Check that the solution $x = \frac{-5}{13}$, $y = \frac{38}{13}$ satisfies both original equations.

LHS: $2 \times \dfrac{-5}{13} + 3 \times \dfrac{38}{13} = \dfrac{-10 + 114}{13}$ —(1)

$= \dfrac{104}{13} = 8 = RHS$

LHS: $3 \times \dfrac{-5}{13} - 2 \times \dfrac{38}{13} = \dfrac{-15 - 76}{13}$ —(2)

$= \dfrac{-91}{13} = -7 = RHS$

Therefore, the required solution is $(x = \dfrac{-5}{13}, y = \dfrac{38}{13})$.

YOUR TURN

In Exercises 5.16 to 5.20, solve the given simultaneous linear equations using the *elimination* method.

5.16 $x + 2y = 10$
$x - y = 1$

5.17 $4x - y = 0$
$12x - y = -2$

5.18 $2x + y = 4$
$5x + 2y = 9$

5.19 $x + 3y = 3$
$y - 2x = 8$

5.20 $2x - 3y = 5$
$5x + 2y = -16$

B. Solving a set of two linear equations in two unknowns using the *substitution method*

This method involves taking one of the two equations and making one of the two variables 'the subject' (i.e. writing the chosen variable in terms of the remaining one). The resulting expression is then substituted into the second equation in place of the chosen variable, thus allowing a solution for the remaining variable to be obtained.

With the value of one variable determined, the value of the second is easily obtained by substituting back into one of the original equations.

Once again, *always* be sure to check that your final solution set satisfies both equations.

Note: Making a variable 'the subject' of an equation usually involves forming an *equivalent* equation where that variable is isolated on one side of the '=' sign.

For example:

(i) $2x + y = 3$
$\therefore y = 3 - 2x$

Here y is made the subject of the given equation and it is usual to say that y is written in terms of x.

(ii) $2x + y = 3$
$\therefore x = \dfrac{3 - y}{2}$

Here x is the subject of the given equation, implying that x is written in terms of y.

EXAMPLE 5.10

Solve, using the *substitution method*, the simultaneous equations given in example 5.7, namely

$2x + y = 8$
$3x - y = 7$

Solution

The *substitution method* requires some decisions to be made initially – based on the question, 'which variable will I make the subject of which equation?' Clearly there are 4 possible choices and, as with anything mathematical, the best advice always is to follow the simplest path even though all routes provide the same answer.

In this first example we will take two routes just to prove that the solution set remains the same.

Given: $2x + y = 8$ —(1)
$3x - y = 7$ —(2)

Route 1: Make y the subject of equation (1)

$y = 8 - 2x$

Now substitute $(8 - 2x)$ for y in equation (2)

$3x - (8 - 2x) = 7$

$\therefore 3x - 8 + 2x = 7$ (remove brackets)

$\therefore 5x = 15$ (simplify)

$\therefore x = 3$

Now substitute $x = 3$ into equation (1)

$2 \times 3 + y = 8$

$\therefore y = 8 - 6$

$= 2$

Check: solution set $(x = 3, y = 2)$

LHS: $2 \times 3 + 2 = 8 = RHS$ —(1)

LHS: $3 \times 3 - 2 = 7 = RHS$ —(2)

Therefore, the required solution is $(x = 3, y = 2)$.

(Compare with the solution to example 5.7.)

Route 2: Make x the subject of equation (2)

$x = \dfrac{7 + y}{3}$

Now substitute $\dfrac{7 + y}{3}$ for x in equation (1)

$$2\frac{(7 + y)}{3} + y = 8$$

$$2(7 + y) + 3y = 24 \quad \text{(multiply all terms by 3)}$$

$$14 + 2y + 3y = 24 \quad \text{(remove brackets)}$$

$$\therefore \quad 5y = 10 \quad \text{(simplify)}$$

$$\therefore \quad y = 2$$

Now substitute $y = 2$ into equation **(2)**

$$3x - 2 = 7$$

$$\therefore \quad 3x = 9$$

$$\therefore \quad x = 3$$

Check: solution set $(x = 3, y = 2)$

$LHS: 2 \times 3 + 2 = 8 = RHS$ —**(1)**

$LHS: 3 \times 3 - 2 = 7 = RHS$ —**(2)**

Therefore, the required solution is $(x = 3, y = 2)$.

(Again, compare this with the solution to example 5.7 and to *Route 1* of this example.)

EXAMPLE 5.11

Solve, using the *substitution method*, the simultaneous equations given in example 5.8, namely

$$2x + 3y = 19$$

$$4x - y = 3$$

Solution

Given: $2x + 3y = 19$ —**(1)**

$\qquad\quad 4x - y = 3$ —**(2)**

On inspection of **(1)** and **(2)**, the simplest route would be to make y the subject of equation **(2)**.

$$-y = 3 - 4x$$

$$\therefore \quad y = 4x - 3 \quad \text{(multiply both sides by -1)}$$

Now substitute $(4x - 3)$ for y in equation **(1)**

$$2x + 3(4x - 3) = 19$$

$$\therefore \quad 2x + 12x - 9 = 19 \quad \text{(remove brackets)}$$

$$\therefore \quad 14x = 28 \quad \text{(simplify)}$$

$$\therefore \quad x = 2$$

Now substitute $x = 2$ back with equation **(1)**

$$(2 \times 2) + 3y = 19$$

$$\therefore \quad 3y = 19 - 4$$

$$= 15$$

$$\therefore \quad y = 5$$

Check: solution set $(x = 2, y = 5)$

$LHS: 2 \times 2 + 3 \times 5 = 19 = RHS$ —**(1)**

$LHS: 4 \times 2 - 5 = 3 = RHS$ —**(2)**

Therefore, the required solution is $(x = 2, y = 5)$.

(Compare this result with that obtained in example 5.8.)

EXAMPLE 5.12

Using the *substitution method*, solve the simultaneous equations given in Example 5.9, namely

$$2x + 3y = 8$$

$$3x - 2y = -7$$

Solution

Given: $2x + 3y = 8$ —**(1)**

$\qquad\quad 3x - 2y = -7$ —**(2)**

In this case, making x the subject of equation **(1)**, we have $x = \dfrac{8 - 3y}{2}$.

Now substitute $\left(\dfrac{8 - 3y}{2}\right)$ for x in equation **(2)**,

$$3\left(\frac{8 - 3y}{2}\right) - 2y = -7$$

$$\therefore \quad 3(8 - 3y) - 4y = -14 \quad \text{(multiply throughout by 2)}$$

$$\therefore \quad 24 - 9y - 4y = -14 \quad \text{(remove brackets)}$$

$$\therefore \quad 38 = 13y \quad \text{(simplify)}$$

$$\therefore \quad y = \frac{38}{13}$$

Now substitute $y = \dfrac{38}{13}$ back into equation **(1)**, giving

$$2x + 3\left(\frac{38}{13}\right) = 8$$

$$\therefore \quad 2x = 8 - \frac{114}{13}$$

$$= \frac{104 - 114}{13}$$

$$= \frac{-10}{13}$$

$$\therefore \quad x = \frac{-5}{13}$$

Check: solution set $(x = \dfrac{-5}{13}, y = \dfrac{38}{13})$

$LHS: \quad 2 \times \dfrac{-5}{13} + 3 \times \dfrac{38}{13} = \dfrac{-10 + 114}{13}$

$$= \frac{104}{13}$$

$$= 8 = RHS \quad —\textbf{(1)}$$

LHS: $3 \times \dfrac{-5}{13} + 2 \times \dfrac{38}{13} = \dfrac{-15 - 76}{13}$

$= \dfrac{-91}{13}$

$= -7 = RHS$ ——(2)

Therefore, the required solution is

$(x = \dfrac{-5}{13}, y = \dfrac{38}{13})$. (Again, compare this result with that obtained in example 5.9.)

YOUR TURN

5.21 Solve the simultaneous equations given in Example 5.11 using the *substitution method*, but this time making *x* the subject of equation (1). Check that your solution agrees with the one already obtained.

In Exercises 5.22 to 5.26, use the *substitution method* to solve the simultaneous equations given in Exercises 5.16 to 5.20. Check that your solution here agrees with that obtained using the *elimination method*.

5.22 $x + 2y = 10$

$x - y = 1$

5.23 $4x - y = 0$

$12x - y = -2$

5.24 $2x + y = 4$

$5x + 2y = 9$

5.25 $x + 3y = 3$

$y - 2x = 8$

5.26 $2x - 3y = 5$

$5x + 2y = -16$

5.3 Solving quadratic equations

In Chapter 2 we learnt that a **polynomial** is an expression in *one* variable of degree n. When $n = 2$, the expression is called a **quadratic**.

For example:

(i) $x^2 + 2x + 1$

(ii) $\dfrac{4}{3}x - 6x^2 + 4$

Hence, the following definition is a logical extension.

Definition 4

A **quadratic equation** is an equation in *one* variable of degree *two*, with general form $ax^2 + bx + c = 0$, where a, b and c are constants and $a \neq 0$.

For example:

(i) $x^2 + 2x + 1 = 0$

(ii) $x^2 + 3x = 0$

(iii) $x^2 - x - 12 = 0$

(iv) $5 - 9x - 4x^2 = 0$

(v) $2x^2 = 8$

Note: Every quadratic equation has **at most** two solutions.

5.3.1 Methods of solution

There are *two* methods of solving quadratic equations.

A. The **quadratic formula** that can be used to solve **all** quadratic equations.

B. **Factorisation**, provided the quadratic expression has factors.

A. Solution using the quadratic formula

If the general form of a quadratic equation is $ax^2 + bx + c = 0$, then the **quadratic formula**, appropriate for solving *all* quadratics, is given by

$$x = \dfrac{-b \pm \sqrt{b^2 - 4ac}}{2a}.$$

Note: Although the proof of the above statement is well beyond the scope of this text, interested readers are urged to consult a more advanced mathematical text.

EXAMPLE 5.13

Use the *quadratic formula* to solve

$$x^2 + 7x + 12 = 0$$

Solution

Given: $\qquad x^2 + 7x + 12 = 0$

Here $\qquad a = 1, b = 7$ and $c = 12$.

Using the *quadratic formula*, we know that

$x = \dfrac{-b \pm \sqrt{b^2 - 4ac}}{2a}$

$= \dfrac{-7 \pm \sqrt{49 - 4 \times 1 \times 12}}{2 \times 1}$

$= \dfrac{-7 \pm \sqrt{1}}{2}$

$= \dfrac{-7 \pm 1}{2}$

$= \dfrac{-6}{2}, \dfrac{-8}{2}$ or $-3, -4$

Check:

(i) When $x = -3$,

LHS: $x^2 + 7x + 12 = 9 - 21 + 12 = 0 = RHS$

(ii) When $x = -4$,

LHS: $x^2 + 7x + 12 = 16 - 28 + 12 = 0 = RHS$

Therefore, the solution to $x^2 + 7x + 12 = 0$ is

$x = -3$ and $x = -4$.

EXAMPLE 5.14

Use the *quadratic formula* to solve, correct to 2 decimal places,

$$2x^2 + 4x - 7 = 0$$

Solution

Given: $\quad 2x^2 + 4x - 7 = 0$

Here $\quad a = 2, b = 4$ and $c = -7$.

Using the *quadratic formula*, we have

$$x = \frac{-b \pm \sqrt{b^2 - 4ac}}{2a}$$

$$= \frac{-4 \pm \sqrt{16 - (4 \times 2 \times -7)}}{4}$$

$$= \frac{-4 \pm \sqrt{16 + 56}}{4}$$

$$= \frac{-4 \pm \sqrt{72}}{4}$$

$$= \frac{-4 \pm 8.485}{4}$$

$$= \frac{-12.485}{4}, \frac{4.485}{4} \quad \text{or} \quad -3.12, 1.12$$

Therefore, the approximate solution to $2x^2 + 4x - 7 = 0$ is $x = -3.12$ and $x = 1.12$.

EXAMPLE 5.15

Solve, using the *quadratic formula*,

$$x^2 - 2x - 5 = 0$$

Solution

Given: $\quad x^2 - 2x - 5 = 0$

Here $\quad a = 1, b = -2$ and $c = -5$.

Using the *quadratic formula*, we have

$$x = \frac{-b \pm \sqrt{b^2 - 4ac}}{2a}$$

$$= \frac{-(-2) \pm \sqrt{(-2)^2 - (4 \times 1 \times -5)}}{2 \times 1}$$

$$= \frac{2 \pm \sqrt{4 + 20}}{2}$$

$$= \frac{2 \pm \sqrt{24}}{2}$$

$$= \frac{2 \pm 4.899}{2}$$

$$= \frac{-2.899}{2}, \frac{6.899}{2} \quad \text{or} \quad -1.45, 3.45$$

Therefore, the approximate solution to $x^2 - 2x - 5 = 0$ is

$x = -1.45$ and $x = 3.45$.

YOUR TURN

In each of Exercises 5.27 to 5.31, solve the given quadratic equation using the *quadratic formula*. Calculate your solution correct to 2 decimal places.

5.27 $x^2 - 7x - 3 = 0$

5.28 $3x^2 + 9x + 5 = 0$

5.29 $4x^2 - 2x - 3 = 0$

5.30 $2x^2 + 5x = 2$

5.31 $6x^2 + 5x = 6$

B. Solution using factorisation

Any quadratic equation where the quadratic expression has factors may be solved using the following procedure:

(i) factorise the quadratic expression

(ii) set each factor equal to zero and solve for the unknown variable.

Note: The above solution method depends upon the following fact:

If $a \times b = 0$, then either $a = 0$ or $b = 0$.

For example: If $(x - 1)(x - 2) = 0$,

then $(x - 1) = 0$ or $(x - 2) = 0$,

giving $x = 1$ or $x = 2$.

EXAMPLE 5.16

Use the *factorisation method* to solve the quadratic equation given in example 5.13, namely

$$x^2 + 7x + 12 = 0$$

Solution

Factorising $x^2 + 7x + 12$ we find that

$$(x + 4)(x + 3) = 0$$

$\therefore$ we have $(x + 4) = 0$ or $(x + 3) = 0$

$\therefore$ the required solution to $x^2 + 7x + 12 = 0$ is $(x = -4$ or $x = -3)$, as was found in example 5.13 using the *quadratic formula*.

EXAMPLE 5.17

Use the *factorisation method* to solve

$$3x^2 - 3x - 18 = 0$$

Solution

Factorising $3x^2 - 3x - 18$ we find that

$$(3x + 6)(x - 3) = 0$$

$\therefore$ $(3x + 6) = 0$ or $(x - 3) = 0$, implying that

$$x = -\frac{6}{3} = -2 \text{ or } x = 3.$$

Check:

(i) When $x = -2$,

LHS: $3x^2 - 3x - 18 = 3 \times (-2)^2 - 3(-2) - 18$
$= 12 + 6 - 18 = 0 = RHS$

(ii) When $x = 3$,

LHS: $3x^2 - 3x - 18 = 3 \times 9 - 3(3) - 18$
$= 27 - 27 = 0 = RHS$

Therefore, the required solution to $3x^2 - 3x - 18 = 0$ is ($x = -2$, $x = 3$).

EXAMPLE 5.18

Use the *factorisation method* to solve

$$x^2 = 6x$$

Solution

Re-write the given equation as

$$x^2 - 6x = 0$$

Factorising $x^2 - 6x$ we find

$$x(x - 6) = 0$$

$\therefore$ $x = 0$ or $x = 6$

Check:

(i) When $x = 0$,

$LHS = 0 = RHS$

(ii) When $x = 6$,

$LHS = 6^2 = 36 = 6 \times 6 = RHS$

Therefore, the required solution to $x^2 = 6x$ is

($x = 0$, $x = 6$).

YOUR TURN

In each of Exercises 5.32 to 5.35, solve the given quadratic equation using the *factorisation method*. Check that your solution satisfies the equation.

5.32 $x^2 - 9x + 14 = 0$

5.33 $2x^2 + 12x + 16 = 0$

5.34 $x^2 + 25 = 10x$

5.35 $2x^2 - 7x = -6$

(*Hint*: In 5.34 and 5.35, begin by forming an equivalent equation equal to zero.)

Which method?

Although the *factorisation method* is usually quicker, provided the factors can be easily determined, not all quadratic equations have factors. In the latter case, or when the factors are difficult to extract, the *quadratic formula* is the preferred method of solution.

YOUR TURN AGAIN

Solve each of the following quadratic equations using the method you consider the simpler or more appropriate. Whenever the solutions are not exact, give an approximate answer correct to two decimal places.

5.36 $x^2 - 5x = 2$

5.37 $x^2 + 6x = 72$

5.38 $x^2 - 7x + 3 = 0$

5.39 $4x^2 - 36x = 0$

5.40 $x^2 - 10x = 75$

5.41 $(2x - 1)^2 = 36$

5.42 $3x^2 + x - 1 = 0$

5.43 $5x^2 - 9x + 2 = 0$

5.44 $(2x + 3)^2 = (x + 4)(x + 3)$

5.45 $18x - 24 - 3x^2 = 0$

5.4 Changing the subject of an equation

In *Section 2* we saw how to make a simple change of subject in order to solve simultaneous linear equations. In this section we will endeavour to extend this procedure, by means of examples, to cover more complicated equations involving several variables.

However, it must always be remembered that changing the subject of any equation, no matter how complex, involves isolating the variable of interest and re-writing an equivalent equation in terms of all other variables so that the original equality is maintained.

EXAMPLE 5.19

Consider the equation $A = l \times b$.

Express:

(a) l in terms of A and b

(b) b in terms of A and l.

Solution

(a) If $A = l \times b$, then

$l = \dfrac{A}{b}$ (divide both sides by b and simplify)

(b) If $A = l \times b$, then
$$b = \frac{A}{l}.$$

EXAMPLE 5.20
Consider the equation $V = \pi r^2 h$.
Express:

(a) h in terms of V, π and r

(b) r in terms of V, π and h.

Solution

(a) If $V = \pi r^2 h$, then

$$h = \frac{V}{\pi r^2} \qquad \text{(divide both sides by } \pi r^2\text{)}$$

(b) If $V = \pi r^2 h$, then

$$r^2 = \frac{V}{\pi h} \qquad \text{(divide both sides by } \pi h\text{)}$$

$$\therefore r = \pm\sqrt{\frac{V}{\pi h}} \text{ (take the square root of both sides)}$$

EXAMPLE 5.21
Consider the equation $S = P + Prt$, a very useful equation in financial mathematics.
Solve the equation for P. That is, make P the subject of the equation.

Solution

If $S = P + Prt$, then

$$S = P(1 + rt) \text{ (factoring the } RHS\text{)}$$

$$\therefore \frac{S}{1 + rt} = P \qquad \text{(divide both sides by } (1 + rt)\text{)}$$

$$\text{or} \quad P = \frac{S}{1 + rt}$$

EXAMPLE 5.22
Solve $(a + c)x + x^2 = (x + a)^2$ for x.
That is, make x the subject of
$$(a + c)x + x^2 = (x + a)^2$$

Solution

Given: $(a + c)x + x^2 = (x + a)^2$

$$\therefore ax + cx + x^2 = x^2 + 2ax + a^2 \qquad \text{(expand)}$$

$$ax + cx = 2ax + a^2 \qquad \text{(subtract } x^2\text{)}$$

$$\therefore cx - ax = a^2 \qquad \text{(subtract } 2ax\text{)}$$

$$\therefore x(c - a) = a^2 \qquad \text{(factorise)}$$

$$\therefore x = \frac{a^2}{c - a}$$

5.46 Make q the subject of $p = -3q + 6$.

5.47 Make r the subject of $S = P(1 + rt)$.

5.48 In the following equation, express m in terms of all other variables.
$$r = \frac{2ml}{B(n + 1)}$$

5.49 Make a_1 the subject of $S = \frac{n}{2}(a_1 + a_n)$.

5.5 Practical applications

One of the questions often asked by budding mathematics students is 'why do we learn the various techniques? Is there any practical use or is it all just a mental exercise?'

The answer is that almost all mathematical techniques have some practical application in the 'real' world and the main purpose of this section is to present a few realistic examples that utilise some of the theoretical procedures discussed in this and previous chapters.

The best way to approach 'practical applications' is by example and here we will present problems that involve two different methods of solution, at least initially.

(i) The easier of the two methods requires us to solve a practical problem where the appropriate equation(s) is (are) actually given.

(ii) The more difficult method requires us to solve a given practical problem 'from scratch'. This involves actually translating the practical 'word' problem into mathematical symbolic terms using algebra, the language of mathematics. The resulting equation(s) can then be solved using algebraic techniques, and the mathematical solution obtained allows for the solution of the 'real' problem.

EXAMPLE 5.23
Use the equation $P = 2l + 2b$ to find the length l of a rectangle whose perimeter P is 960 metres and whose breadth, b, is 120 metres.

Solution
Since we are asked to find l, given $P = 2l + 2b$, the first step is to make l the subject.

So, if $P = 2l + 2b$

then $P - 2b = 2l$

$$\therefore l = \frac{P - 2b}{2}$$

Given $P = 960$ and $b = 120$ we find, on substitution, that

$$l = \frac{960 - 2(120)}{2}$$

$$= \frac{960 - 240}{2}$$

$$= \frac{720}{2}$$

$$= 360$$

Therefore, the length of the given rectangle is 360 metres.

EXAMPLE 5.24

A group of zoologists was studying the effect on the body weight of rats of varying the amount of yeast in the diet.

By changing the percentage P of yeast in the diet, the average weight gain, g (in grams), over time was estimated to be

$$g = -200P^2 + 200P + 20$$

What percentage of yeast would you expect to give an average weight gain of 70 grams?

Solution

Here we wish to solve

$$g = -200P^2 + 200P + 20 \quad \text{for } P, \text{ given } g = 70.$$

With a small amount of re-arranging we can write the given equation in the general form of a quadratic and then solve for P using the *quadratic formula*. With $g = 70$, we have

$$70 = -200P^2 + 200P + 20$$

or $\qquad 200P^2 - 200P + 50 = 0$

or $\qquad 4P^2 - 4P + 1 = 0$ (divide by 50)

Now, using the *quadratic formula* with $a = 4$, $b = -4$ and $c = 1$, we have

$$P = \frac{-b \pm \sqrt{b^2 - 4ac}}{2a} = \frac{4 \pm \sqrt{16 - 16}}{8}$$

$$= \frac{4}{8} = \frac{1}{2}$$

Therefore, for an average weight gain of 70 grams, we would expect $\frac{1}{2}$ % of yeast in the diet.

EXAMPLE 5.25

Suppose that Mary travels a certain distance on the first day and twice as far on the second day. If her total distance travelled is 60 km, how far did she travel on the first day.

Solution

Let $\qquad x =$ distance travelled on day 1.

Then $\qquad 2x =$ distance travelled on day 2.

Given $x + 2x = 60$, solve for x.

Here $\qquad 3x = 60$

$\qquad \therefore x = 20$

Therefore, on the first day Mary travelled 20 km.

EXAMPLE 5.26

Joy, Pam, Sandra and Lesley each make a donation ($) to the Guide Dogs Association.

Sandra gives *twice* as much as Lesley, Pam gives *three* times as much as Sandra and Joy gives *four* times as much as Pam.

If their *total* gift is $132, find the amount of Lesley's donation and hence the value of the remaining three.

Solution

Let $x =$ Lesley's donation.

Then $\qquad$ Sandra donates $2x$,

$\qquad$ Pam donates $3 \times 2x = 6x$

and $\qquad$ Joy donates $4 \times 6x = 24x$

We know that the total gift is $132.

Hence, $x + 2x + 6x + 24x = 132$

$\qquad \therefore 33x = 132$

$\qquad \therefore x = 4$

Donations:	Lesley:	$ 4.00
	Sandra:	$ 8.00
	Pam:	$24.00
	Joy:	$96.00

EXAMPLE 5.27

The tickets for an ice-skating display were being sold at $5 for adults and $2 for children.

If 101 tickets were sold altogether for a total cost of $394, find the number of children's tickets sold.

Solution

Let $x =$ number of adult tickets sold

and $y =$ number of children's tickets sold.

From the information given we can construct *two* equations.

Total tickets sold $\qquad x + y = 101 \qquad$ —(1)

Total cost of tickets $\qquad 5x + 2y = 394 \qquad$ —(2)

Using the *elimination method*, double equation (1) and then subtract equation (2) to eliminate x.

(1) x 2 $\qquad 2x + 2y = 202 \qquad$ —(3)

$\qquad\qquad 5x + 2y = 394 \qquad$ —(2)

(3) – (2) $\qquad -3x = -192$

$\qquad\qquad \therefore x = 64$

Now substitute $x = 64$ into **(1)**

$$64 + y = 101$$
$$\therefore \ y = 37$$

Check that the solution set (64, 37) satisfies both equations.

(i) *LHS*: $64 + 37 = 101 = RHS$

(ii) *LHS*: $320 + 74 = 394 = RHS$

Therefore, of the 101 tickets sold, 64 are adults and 37 are children.

EXAMPLE 5.28

A farmer wishes to build a temporary pen for new lambs. The pen will be a rectangle of area 40 square metres and enclosed by 26 metres of fencing. Find the length of the longer side of the pen.

Solution

Let y = length of longer side.

Since 26 metres of fencing enclose the rectangle, then 13 metres must go half way round – representing a long and a short side.

Since y = length of long side,

then $13 - y$ = length of short side.

Because the area of a rectangle equals length times breadth, we know that

$$y(13 - y) = 40 \text{ square metres.}$$

As you will see on expansion, this is a quadratic equation that, in this case, can be solved by *factorisation*.

$$13y - y^2 = 40$$
$$\therefore \quad y^2 - 13y + 40 = 0 \quad \text{(subtract 40 from both sides and multiply by } -1)$$
$$\therefore \quad (y - 8)(y - 5) = 0$$

Thus, $y = 5$ or $y = 8$.

Check:

(i) When $y = 5$,

   LHS: $5(13 - 5) = 40 = RHS$

(ii) When $y = 8$,

   LHS: $8(13 - 8) = 40 = RHS$.

Therefore, we have that the longer side of the rectangle is 8 metres while the shorter side is 5 metres.

YOUR TURN

5.50 Using the equation $A = \frac{1}{2} bh$, find the height h (in cms) of a triangle whose base b is 15 cm and whose area A is 75 cm^2.

5.51 The number, n, of prey eaten by a particular predator over a given period of time has been found to be

$$n = \frac{10d}{1 + 0.1d}$$

where d is the prey density (i.e. number of prey per unit of area).
Find the prey density necessary for the survival of a predator who must eat 50 prey over a given time period.

5.52 Suppose that the height h (metres) of an object thrown straight up from the ground is given by

$$h = 44.1t - 4.9t^2$$

where t is the elapsed time (in seconds).
(a) After how many seconds does the object hit the ground ($h = 0$)?
(b) Find the elapsed time, t, when the object reaches a height of 88.2 metres.

In Exercises 5.53 to 5.55, write the given statements as equations and then solve the equations.

5.53 Tom is 4 years older than Susan. If their combined ages total 28, how old are Tom and Susan?

5.54 One number subtracted from another number twice as big leaves 36. Find both numbers.

5.55 Five times a certain number is divided by three. If the result is ten, find the number.

5.56 At a sale, Lorraine bought 4 blouses and one pair of slacks for $94.46. At the same sale, Joy bought 3 blouses of the same type and two pairs of slacks of the same type for $113.97.
(a) What was the sale price of a blouse?
(b) What was the sale price of a pair of slacks?

Suggested solution steps:
(i) Let x be the cost of ....
(ii) Let y be the cost of ....
(iii) Write one equation for Lorraine's shopping and a second equation for Joy's shopping.
(iv) Solve the equations and check your solutions.

5.57 A man is 3 times as old as his son. Four years ago he was 4 times as old as his son. Determine the present ages of father and son.
(Construct two equations and solve simultaneously).

60 | EQUATIONS

5.58 In a special pen, a farmer keeps rabbits and ducks. Together they have 98 feet and 40 heads.
How many rabbits and how many ducks are there?
(Again, construct two equations and solve simultaneously).

5.59 A used-car dealer bought 2 cars, a Toyota and a Mazda, for a total of $14 500. He sold the Toyota at a 20% *profit* and the Mazda at a 5% *loss*, while still making a total profit of $1 700 on the overall transaction. What did he pay for each car?

5.60 A small oil field contains 20 wells and produces 4 000 barrels of oil daily. A recent report estimates that for each additional well drilled, daily production at each well would decrease by five barrels. If the oil company wishes to increase oil production to 4 420 barrels a day, how many new wells should they drill?

[*Hint*: Let W = number of new wells drilled. Then form a quadratic equation stating that the total desired daily production (4 420) may be calculated as the product of the total number of wells and the daily output per well.]

Extensions to chapter 5

Extension A: Equations reducible to the quadratic or linear form

While linear and quadratic equations are relatively easy to solve, the solutions to equations of degree 3 or higher are far more complicated. However, some equations, initially of a more complex nature, may be re-written in a simpler form after a suitable substitution has been made. This generally results in a reduction to the quadratic form.

EXAMPLE E5.1
Solve the following equation using whatever method is available
$$x^4 - 10x^2 + 9 = 0$$

Solution
If we let $a = x^2$, then $x^4 = a^2$ and $10x^2 = 10a$.

∴ the equation $x^4 - 10x + 9 = 0$ may be written as
$$a^2 - 10a + 9 = 0 \text{ in quadratic form or}$$
$$(a - 9)(a - 1) = 0 \text{ by factoring.}$$
Therefore, $a = 9$ or $a = 1$.

However, since $a = x^2$, we have that
$$x^2 = 9 \text{ or } x^2 = 1$$
$$\therefore \qquad x = \sqrt{9} \text{ or } x = \sqrt{1},$$
implying that $x = \pm 3$ or $x = \pm 1$.
Therefore, $x^4 - 10x^2 + 9 = 0$ has solutions when $x = \pm 3$ or $x = \pm 1$.

Check:

(i) When $x = 3$, $x^4 - 10x^2 + 9 = 81 - 90 + 9 = 0$ *OK*

(ii) When $x = -3$, $x^4 - 10x^2 + 9 = 81 - 90 + 9 = 0$ *OK*

(iii) When $x = 1$, $x^4 - 10x^2 + 9 = 1 - 10 + 9 = 0$ *OK*

(iv) When $x = -1$, $x^4 - 10x^2 + 9 = 1 - 10 + 9 = 0$ *OK*

EXAMPLE E5.2
Solve the following equation for x.
$$(2x + 1)^4 - 5(2x + 1)^2 + 4 = 0$$

Solution
Here, if we let $a = (2x + 1)^2$, then $(2x + 1)^4 = a^2$ and $5(2x + 1)^2 = 5a$.

∴ the equation $(2x + 1)^4 - 5(2x + 1)^2 + 4 = 0$ may be written as $a^2 - 5a + 4 = 0$ where $a = (2x + 1)^2$.
The resulting equation in a is now in quadratic form and can be factorised.

That is $a^2 - 5a + 4 = 0$

or $(a - 4)(a - 1) = 0$ by factoring.

Therefore, $a = 4$ or $a = 1$.

However, since $a = (2x + 1)^2$, we have that

(i) $(2x + 1)^2 = 4$ or

(ii) $(2x + 1)^2 = 1$

In (i), $(2x + 1)^2 = 4$ implies that $(2x + 1) = \sqrt{4} = \pm 2$

Hence, $x = \dfrac{\pm 2 - 1}{2}$ or $x = \dfrac{1}{2}$ or $x = -1\dfrac{1}{2}$.

In (ii), $(2x + 1)^2 = 1$ implies that $(2x + 1) = \sqrt{1} = \pm 1$

Hence, $x = \dfrac{\pm 1 - 1}{2}$ or $x = 0$ or $x = -1$.

Therefore, the equation $(2x + 1)^4 - 5(2x + 1)^2 + 4 = 0$ has solutions when $x = \dfrac{1}{2}$, $x = -1\dfrac{1}{2}$ $x = 0$ or $x = -1$.

Check:

(i)   When $x = \dfrac{1}{2}$, $(2x + 1)^4 - 5(2x + 1)^2 + 4$
$$= 2^4 - 5(2)^2 + 4$$
$$= 16 - 20 + 4 = 0 \ OK$$

(ii)   When $x = -1\dfrac{1}{2}$, $(2x + 1)^4 - 5(2x + 1)^2 + 4$
$$= (-2)^4 - 5(-2)^2 + 4$$
$$= 16 - 20 + 4 = 0 \ OK$$

(iii) When $x = 0$, $(2x + 1)^4 - 5(2x + 1)^2 + 4$
$$= (1)^4 - 5(1)^2 + 4$$
$$= 1 - 5 + 4 = 0 \ OK$$

(iv) When $x = -1$, $(2x + 1)^4 - 5(2x + 1)^2 + 4$
$$= (-1)^4 - 5(-1)^2 + 4$$
$$= 1 - 5 + 4 = 0 \ OK$$

EXAMPLE E5.3
Find the value of x if
$$(36)^{2x - 4} = 6^{x + 4}.$$

Solution
Since $36 = 6^2$, we can re-write both sides of the equation as a power of 6
$$(36)^{2x - 4} = 6^{x + 4}$$
$$\therefore (6^2)^{2x - 4} = 6^{x + 4}$$
$$\therefore 6^{4x - 8} = 6^{x + 4}$$

Using the rule,

$$\boxed{\text{if } a^m = a^n, \text{ then } m = n}$$

we have that
$$4x - 8 = x + 4$$
$$\therefore 3x - 8 = 4 \qquad \text{subtract } x \text{ from both sides}$$
$$\therefore 3x = 12 \qquad \text{add 8 to both sides}$$
$$\therefore x = 4 \qquad \text{divide both sides by 3}$$

Therefore, $(36)^{2x - 4} = 6^{x + 4}$ when $x = 4$.

Check:
LHS: $36^{8 - 4} = 36^4 = 1679616$
RHS: $6^{4 + 4} = 6^8 = 1679616$ $\qquad \therefore LHS = RHS$

Extension B: Absolute-value equations
In chapter 1 we defined the **absolute value** of a number x, denoted $|x|$, as its distance from zero.

However, now that we have completed 2 chapters (namely Chapters 2 and 4) that together cover all the basic algebraic concepts, we are able to present a more formal algebraic definition of absolute value.

Definition 5

> **(i)** If $a > 0$, then $|a| = a$ e.g. $|+9| = 9$
> **(ii)** If $a = 0$, then $|a| = 0$ e.g. $|0| = 0$
> **(iii)** If $a < 0$, then $|a| = -a$ e.g. $|-2| = -(-2) = 2$

Using the above definition we are now able to solve equations involving absolute values.

EXAMPLE E5.4
Solve for x, $|2x - 5| = 11$, and check that the solutions satisfy the original equation.

Solution
If $|2x - 5| = 11$ then, since $|-11| = 11$ and $|11| = 11$,
$$2x - 5 = 11 \quad \text{or} \quad 2x - 5 = -11$$
That is, $\qquad 2x = 16 \quad \text{or} \qquad 2x = -6$
Hence, $\qquad x = 8 \quad \text{or} \qquad x = -3$
Therefore, $|2x - 5| = 11$ when $x = 8$ or $x = -3$.
Check:

(i) When $x = 8$, $|2x - 5| = |16 - 5|$
$$= |11| = 11 \ OK$$

(ii) When $x = -3$, $|2x - 5| = |-6 - 5| = |-11|$
$$= -(-11) = 11 \ OK$$

EXAMPLE E5.5
Solve for x, $|2x - 4| = |3x + 16|$, and check the validity of your solutions.

Solution
From the definition of absolute value there appear to be four possibilities, namely
$$\pm(2x - 4) = \pm(3x + 16)$$

That is, $+(2x - 4) = +(3x + 16)$ $\qquad$ —(1)
$\qquad +(2x - 4) = -(3x + 16)$ $\qquad$ —(2)
$\qquad -(2x - 4) = +(3x + 16)$ $\qquad$ —(3)
$\qquad -(2x - 4) = -(3x + 16)$ $\qquad$ —(4)

However, equations **(1)** and **(4)** are equivalent as are equations **(2)** and **(3)**.

Therefore, all solutions to the given absolute value equation may be found by solving equations **(1)** and **(2)** only.

Thus, $+(2x + 4) = +(3x + 16)$ **or** $+(2x - 4) = -(3x - 16)$
$\qquad 2x - 4 = 3x + 16 \qquad\qquad 2x - 4 = -3x + 16$
$\qquad\quad -x = 20 \qquad\qquad\qquad\quad 5x = 20$
$\qquad\quad\ x = -20 \qquad\qquad\qquad\ x = 4$

Therefore, we must check that
$$|2x - 4| = |3x + 16| \text{ when } x = -20 \text{ or } x = 4.$$
Check:

(i) When $x = -20$,
$$LHS = |2x - 4| = |-40 - 4| = |-44|$$
$$= -(-44) = 44$$
$$RHS = |3x + 16| = |-60 + 16| = |-44|$$
$$= -(-44) = 44$$
$$\therefore LHS = RHS \quad \therefore x = -20 \text{ is a solution.}$$

(ii) When $x = 4$,
$$LHS = |2x - 4| = |8 - 4| = |4| = 4$$
$$RHS = |3x + 16| = |12 + 16| = |28| = 28$$
$$\therefore LHS \neq RHS \quad \therefore x = 4 \text{ is } not \text{ a solution.}$$

Therefore, $x = -20$ satisfies the equation

$$|2x - 4| = |3x + 16|.$$

EXAMPLE E5.6

Solve for x, $|6x - 5| = 5x + 27$, and check the validity of each solution.

Solution

Since $|6x - 5| = 5x + 27$, we know that

$$(6x - 5) = (5x + 27) \text{ or } (6x - 5) = -(5x + 27)$$

That is, $\quad 6x - 5 = 5x + 27 \quad$ or $\quad 6x - 5 = -5x - 27$

$$\therefore \quad x = 32 \qquad \text{or} \ \ 11x = -22 \text{ or } x = -2$$

Now we must check whether each solution, $x = 32$ and $x = -2$, satisfies the original equation.

Check:

(i) When $x = 32$,

$$\begin{aligned} LHS &= |6x - 5| = |6 \times 32 - 5| = |192 - 5| \\ &= |187| = 187 \end{aligned}$$

$$\begin{aligned} RHS &= 5x + 27 = 5 \times 32 + 27 \\ &= 160 + 27 = 187 \end{aligned}$$

So, $RHS = LHS \quad \therefore \ x = 32$ is a valid solution.

(ii) When $x = -2$,

$$LHS = |6x - 5| = |-12 - 5| = |-17|$$
$$= -(-17) = 17$$

$$RHS = 5x + 27 = -10 + 27 = 17$$

So, $RHS = LHS \therefore \quad x = -2$ is a valid solution.

Therefore, $x = -2$ and $x = 32$ both satisfy the original equation $|6x - 5| = 5x + 27$.

Extension C: Solving equations involving algebraic fractions

The first step in solving equations involving fractions is to multiply both sides by the **lowest common denominator** (*LCD*) in order to remove the fractions. The resulting equivalent equations can then be solved using an appropriate procedure.

EXAMPLE E5.7

Solve $\dfrac{x}{2} - \dfrac{x}{5} = 1$ and check the validity of your solution.

Solution

In the equation $\dfrac{x}{2} - \dfrac{x}{5} = 1$, the *LCD* is $(2 \times 5) = 10$.

Thus, to remove the fractions from the *LHS* we must multiply throughout by 10.

Hence, $\dfrac{x}{2} \times 10 - \dfrac{x}{5} \times 10 = 1 \times 10$

That is, $\quad 5x - 2x = 10$

$$\text{or} \quad 3x = 10$$

$$\therefore \ \ x = \frac{10}{3} \text{ or } 3\frac{1}{3}$$

Therefore, $\dfrac{x}{2} - \dfrac{x}{5} = 1$ when $x = \dfrac{10}{3}$.

Check:

When $x = \dfrac{10}{3}$,

$$LHS = \frac{10}{3} \times \frac{1}{2} - \frac{10}{3} \times \frac{1}{5} = \frac{5}{3} - \frac{2}{3} = \frac{3}{3} = 1 = RHS \ OK$$

EXAMPLE E5.8

Solve $\dfrac{3x - 2}{2x + 3} = \dfrac{3x - 1}{2x + 1}$ and check that your solution satisfies the equation.

Solution

Since the first step is to remove the fractions from the equation, we must multiply both sides by the *LCD* – namely, $(2x + 3)(2x + 1)$.

Thus, $\dfrac{(3x - 2)(2x + 3)(2x + 1)}{2x + 3}$

$$= \frac{(3x - 1)(2x + 3)(2x + 1)}{2x + 1}$$

or $\quad (3x - 2)(2x + 1) = (3x - 1)(2x + 3)$

or $\ 6x^2 - 4x + 3x - 2 = 6x^2 - 2x + 9x - 3$

or $\qquad 6x^2 - x - 2 = 6x^2 + 7x - 3$

$$\therefore -x - 2 = 7x - 3 \ \text{ subtract } 6x^2 \text{ from both sides}$$

$$\therefore \ \ 1 = 8x$$

$$\text{or} \ \ x = \frac{1}{8}$$

Therefore, $\dfrac{3x - 2}{2x + 3} = \dfrac{3x - 1}{2x + 1}$ when $x = \dfrac{1}{8}$.

Check: When $x = \dfrac{1}{8}$,

$$LHS: \quad \frac{3x - 2}{2x + 3} \qquad\qquad RHS: \quad \frac{3x - 1}{2x + 1}$$

$$= \frac{3 \times \frac{1}{8} - 2}{2 \times \frac{1}{8} + 3} \qquad\qquad = \frac{3 \times \frac{1}{8} - 1}{2 \times \frac{1}{8} + 1}$$

$$= \frac{-1\frac{5}{8}}{3\frac{1}{4}} \qquad\qquad\qquad = \frac{-\frac{5}{8}}{1\frac{1}{4}}$$

$$= \frac{-\frac{13}{8}}{\frac{13}{4}} \qquad\qquad\qquad = \frac{-\frac{5}{8}}{\frac{5}{4}}$$

$$= -\frac{13}{8} \times \frac{4}{13} \qquad\qquad = -\frac{5}{8} \times \frac{4}{5}$$

$$= -\frac{1}{2} \qquad \therefore LHS = RHS \qquad = -\frac{1}{2}$$

Therefore, $x = \dfrac{1}{8}$ satisfies the equation

$$\frac{3x-2}{2x+3} = \frac{3x-1}{2x+1}.$$

EXAMPLE E5.9

Solve $\dfrac{3x+4}{x+2} - \dfrac{3x-5}{x-4} = \dfrac{12}{x^2-2x-8}$ and check that your solution is valid.

Solution

We know that to remove the fractions from the equation we must multiply each term by the *LCD*

On the *LHS* of the equation the *LCD* is
$$(x+2)(x-4).$$

On the *RHS*, by factorising, we find that
$$x^2 - 2x - 8 = (x+2)(x-4).$$

Thus the *LCD* across both sides of the equation is
$$(x+2)(x-4).$$

Multiplying each term on both sides of the equation, by the *LCD*, we have

$$\frac{(3x+4)(x+2)(x-4)}{x+2} - \frac{(3x-5)(x+2)(x-4)}{x-4}$$

$$= \frac{12(x+2)(x-4)}{(x+2)(x-4)}$$

Simplifying, $(3x+4)(x-4) - (3x-5)(x+2) = 12$

$$3x^2 + 4x - 12x - 16 - (3x^2 - 5x + 6x - 10) = 12$$

$$3x^2 - 8x - 16 - (3x^2 + x - 10) = 12$$

$$3x^2 - 8x - 16 - 3x^2 - x + 10 = 12$$

$$-9x - 6 = 12$$

$$-9x = 18$$

$$\therefore x = -2$$

Check:

When $x = -2$, the first term $\dfrac{3x+4}{x+2}$ gives $\dfrac{-2}{0}$ (an invalid result). Therefore, although $x = -2$ is a solution to the equivalent equation $(3x+4)(x-4) - (3x-5)(x+2) = 12$, it is **not** a solution of the *original* equation $\dfrac{3x+4}{x+2} - \dfrac{3x-5}{x-4} = \dfrac{12}{x^2-2x-8}$, since the original equation is *not defined* for $x = -2$ (as we cannot divide by zero).

Summary

In this chapter we have discussed in some detail a very broad and very important area of mathematics entitled **equations**.

Remember: An equation is just a mathematical statement telling us that two expressions are **equal** and is denoted:

$$\text{expression 1} = \text{expression 2}$$

In *Section 1* we learned how to solve *linear* equations by forming one or more *equivalent* equations by carrying out a series of *inverse operations*. By definition,

(i) a *linear* equation is one that contains one or more variables of the first degree only.

(ii) an *equivalent* equation is formed whenever the same arithmetic operation is applied to both sides of the original equation.

(iii) *inverse operations*

+	×	a^n	logarithm
−	÷	$\sqrt[n]{a}$	exponential

Section 2 presented two methods for solving *simultaneous linear equations*.

Remember: Simultaneous linear equations refers to two or more equations in the first degree where the same solution for each unknown satisfies all the equations at the same time.

Method A: **Solving two linear equations in two unknowns** *–the elimination method*

This involved:

(i) forming equivalent equations so that on either addition or subtraction of these (as appropriate) one of the variables will be eliminated, allowing the remaining one to be determined.

(ii) substituting back into one of the original equations to determine the value of the eliminated variable.

(iii) checking that the solutions obtained satisfy the original equations.

Method B: **Solving two linear equations in two unknowns** *– the substitution method*

This involved:

(i) taking one of the equations and making one variable the subject (by writing it in terms of the other).

(ii) substituting this resulting expression into the second equation, allowing for a solution for the remaining variable to be obtained.

(iii) substituting back into one of the original equations to determine the value of the second variable.

(iv) checking that the final solution set satisfies the original equations.

Section 3 entitled *solving quadratic equations*, began by defining a quadratic equation to be an equation in *one* variable of degree *two*, usually written as
$$ax^2 + bx + c = 0 \text{ (for } a \neq 0).$$

To solve a quadratic equation, we discussed *two* different methods:

Method A: Using the *quadratic formula*
The solution to *all* equations of the form

$ax^2 + bx + c = 0$ is given by $x = \dfrac{-b \pm \sqrt{b^2 - 4ac}}{2a}$

Method B: Using *factorisation*
Provided the quadratic expression $ax^2 + bx + c$ has factors, then

(i) factorise the expression

(ii) set each factor equal to zero and solve – using the rule

> if $a \times b = 0$, then either $a = 0$ or $b = 0$.

Section 4 had another look at **changing the subject** of a given equation.

This involves isolating the variable of interest and rewriting the equivalent equation in terms of all other variables so that the original equality is maintained.

Section 5 presented some practical 'real-life' problems involving equations and their solutions.

This is very important as it is essential that prospective students are able to 'put into practice' the theoretical techniques and procedures learnt.

The **extensions** to this chapter examined, by means of examples, three further topics on equations, considered to be rather more complicated than those dealt with in the core sections. For this reason they are in general only necessary for those readers who are intending to head along a mathematical and/or scientific path, although others are certainly encouraged to read this material for interest.

The three topics included a discussion about solving

(i) equations reducible to the quadratic or linear form

(ii) absolute-value equations

(iii) equations involving algebraic fractions.

PRACTICE PROBLEMS

5.1 Solve each of the following equations:
  (a) $2x - 3 = 15$
  (b) $8y - 1 = 39$
  (c) $6a = 4a + 16$
  (d) $19 = 6a - 5$
  (e) $6x - 3 = 2x + 21$
  (f) $9y - 4 = 36 - y$
  (g) $3(x - 8) = 8$
  (h) $8(x - 2) = 20$
  (i) $7(x - 3) = 2(x + 7)$
  (j) $\dfrac{2x}{5} = 4$
  (k) $\dfrac{2x - 6}{4} = 6$
  (l) $\dfrac{3x}{2} = \dfrac{2x}{3} + 5$
  (m) $\dfrac{3x}{10} = 4 - \dfrac{x}{2}$
  (n) $\dfrac{x - 3}{3} = \dfrac{x + 2}{4}$

5.2 Solve the following systems of equations in 2 unknowns:
  (a) $x + y = 5$
    $2x + y = 7$
  (b) $x - y = 1$
    $x + 2y = 4$
  (c) $3a + b = 6$
    $2a + 3b = 11$
  (d) $y = 3x$
    $x + y = 8$
  (e) $m = 1 - 2n$
    $2m + 3n = 3$

5.3 Solve for x each of the following:
  (a) $(x - 6)(x - 2) = 0$
  (b) $(x - 4)(x + 7) = 0$
  (c) $(x - 5)(x - 2) = 0$
  (d) $x(x + 5) = 0$
  (e) $4x(2x - 1) = 0$
  (f) $x^2 - 7x + 12 = 0$
  (g) $x^2 + 2x - 15 = 0$
  (h) $x^2 + 5x = 0$
  (i) $2x^2 = 32 - 12x$
  (j) $4x^2 - 9 = 0$
  (k) $(x - 5)^2 = 36$
  (l) $(x + 2)(x + 3) = 2$

5.4 Using the *quadratic formula*, solve for x each of the following equations. Where applicable, leave your answer in surd form.
  (a) $x^2 + 6x = 7$
  (b) $x^2 + 6x + 4 = 0$
  (c) $5x^2 - 2x - 4 = 0$
  (d) $x^2 - 7x + 3 = 0$

5.4 *Cont.*

(e) $3x^2 = 2x + 2$

(f) $3x^2 - x - 1 = 0$

(g) $(x-3)^2 = 9$

(h) $x^2 - 6x - 2 = 0$

(i) $2x^2 + 3x = 5$

5.5 Solve for x each of the following equations.

(a) $5x = 20$

(b) $6x + 8 = 44$

(c) $15x = 6x + 36$

(d) $\dfrac{3x}{2} = 4 - \dfrac{x}{2}$

(e) $4(x + 1) = 2x$

(f) $\dfrac{x + 6}{5} = 3$

(g) $\dfrac{5x + 1}{3} = x + 4$

(h) $\dfrac{x}{4} = \dfrac{2}{3}$

(i) $\dfrac{2x}{5} + 5 = 3$

(j) $\dfrac{x}{4} = \dfrac{x}{6} + \dfrac{1}{12}$

5.6 Solve for x the following equations:

(a) $|8x| = 24$

(b) $|2x - 1| = 5$

(c) $\left|\dfrac{x}{2}\right| = 4$

(d) $\left|\dfrac{x + 1}{4}\right| = 1$

(e) $|9x - 4| = 5$

*5.7 Solve for x each of the following equations by first making a substitution so that the equation reduces to a quadratic.

(a) $x^4 - 5x^2 + 4 = 0$

(b) $x^6 - 9x^3 + 8 = 0$

(c) $(x + 1)^4 - 6(x + 1)^2 + 8 = 0$

(d) $\left(x + \dfrac{1}{x}\right)^2 - 5\left(x + \dfrac{1}{x}\right) + 6 = 0$

(e) $(5^x)^2 - 26(5^x) + 25 = 0$

(f) $(2x + 1)^4 - (2x + 1)^2 - 12 = 0$

6 Inequalities

Introduction

In the preceding chapter we discussed at some length the concept of a mathematical equation and presented in detail several methods of solving different types of equations.

Remember: An equation is just a mathematical statement telling us that two expressions are **equal.**

However, there will be many instances in your mathematical career when the two expressions of interest are *not necessarily* equal.

This then brings us to the topic for discussion in this chapter – namely, **inequalities** (often called **inequations** - the 'opposite of' equations).

Consider the following:

If a and b are any two real numbers on the number line, then one of the following statements must be true.

Either　**(i)**　a lies to the left of b

or　　**(ii)**　a lies to the right of b

or　　**(iii)**　a coincides exactly with b.

From this it follows, quite logically, that

(i)　if 'a lies to the left of b', then 'a is *less than* b' or equivalently 'b is *greater than* a'.

(ii)　if 'a lies to the right of b', then 'a is *greater than* b' or equivalently 'b is *less than* a'.

(iii)　if 'a coincides exactly with b', then 'a equals b'.

The above statements lead us now to the simple but very comprehensive definition of an inequality.

> An **inequality** describes the positional relationship between two numbers.

This chapter is divided into three sections, as follows:

(i)　*Section 1* covers the notation relevant to inequalities,

(ii)　*Section 2* lists the three basic rules involving inequalities, while

(iii)　*Section 3*, entitled 'solving equalities',

presents this topic by means of several examples, including some practical applications.

The **extensions** to this chapter involve *two* concepts only - namely, *solving inequalities whose unknown is in the denominator* and *solving absolute value inequalities*. As with several previous chapters, these more complicated problems are dealt with in the **extensions** because, in general, this knowledge is required only by those readers who are intending to study and/or work in the more mathematical or scientific disciplines. However, those of you who have little difficulty with the core sections of the chapter are encouraged to read the **extensions**, if only from an interest perspective.

6.1　Notation

The majority of the notation relevant to inequalities is most efficiently presented in tabular form.

In words	Diagrammatic representation	Symbolic notation	Examples
a is equal to b		$a = b$	$3 = 3$ $x + 4 = 6$ when $x = 2$
a is less than b or b is greater than a		$a < b$ $b > a$	$7 < 9$ $8 + 4 > 2 + 7$
a is greater than b or b is less than a		$a > b$ $b < a$	$x + 1 > 2$ for $x > 1$ $-6 < -4$

There are three additional inequality symbols not included in the above table as they require some brief explanation.

(i) The symbol '≤' is read 'is less than or equal to' and is defined $a \le b$ if and only if $a < b$ or $a = b$.

(ii) The symbol '≥' is read 'is greater than or equal to' and is defined $a \ge b$ if and only if $a > b$ or $a = b$.

(iii) Suppose that $a < b$ and that x is somewhere in between a and b. That is,

Then, not only is $a < x$ but also $x < b$, and, *by convention*, we write this as $a < x < b$.

For example:

$6 < 9 < 15$: 9 lies between 6 and 15

$4 < x < 10$: x lies between 4 and 10

$0 \le y \le 1$: y lies between 0 and 1 *inclusive*

Note:

An inequality of the form above, i.e. $a < x < b$, is sometimes referred to as a **dual inequality**, since it is made up of two distinct inequalities ($a < x$ and $x < b$).

6.2 Basic rules involving inequalities

These need to be mentioned here because it will make the solution strategies in *Section 3* much easier to follow and understand.

> **Rule 1**: If the same number is added to or subtracted from both sides of an inequality, then the **sign** remains **unchanged**.

For example:

(i) $\boxed{10 > 6}$

$10 + 3 > 6 + 3$ add 3 to both sides

$10 - 5 > 6 - 5$ subtract 5 from both sides

(ii) $\boxed{-5 < -3}$

$-5 + 3 < -3 + 3$ add 3 to both sides

$-5 - 2 < -3 - 2$ subtract 2 from both sides

> **Rule 2**: If both sides of an inequality are multiplied or divided by the same **positive** number, then the **sign** remains **unchanged**.

For example:

(i) $\boxed{2 < 6}$

$2 \times 5 < 6 \times 5$ multiply both sides by 5

$2 + 2 < 6 + 2$ divide both sides by 2

(ii) $\boxed{-10 > -15}$

$-10 \times 2 > -15 \times 2$ multiply both sides by 2

$-10 + 4 > -15 + 4$ divide both sides by 4

> **Rule 3**: If both sides of an inequality are multiplied or divided by the same **negative** number, then the **sign** is **reversed**.

For example:

(i) $\boxed{4 < 6}$

$4 \times -2 > 6 \times -2$ multiply both sides by -2

$-8 > -12$ and reverse the sign

$4 + -4 > 6 + -4$ divide both sides by -4 and

$-1 > -5$ reverse the sign

(ii) $\boxed{-3 < -2}$

$-3 \times -4 > -2 \times -4$ multiply both sides by -4

$12 > 8$ and reverse the sign

$-3 + -2 > -2 + -2$ divide both sides by -2

$1.5 > 1$ and reverse the sign

(iii) $\boxed{1 > -5}$

$1 \times -3 < -5 \times -3$ multiply both sides by -3

$-3 < 15$ and reverse the sign

$1 + -2 < -5 + -2$ divide both sides by -2

$-0.5 < 2.5$ and reverse the sign

6.3 Solving inequalities

In the preceding chapter we presented in some detail several methods for solving various types of **equations**. In fact, these methods carry over to inequalities and, as you will see in the following examples, the techniques used to solve **inequalities** are very similar to those used to solve **equations**, *except that* all the arithmetic operations performed *must* follow the three basic rules for inequalities stated in *Section 2*.

A. Solving inequalities using rules 1 and 2 – no fractions

EXAMPLE 6.1

Solve for x, $6x > 24$, and draw the solution on the number line.

Solution

$$6x > 24$$

$$\therefore \frac{6x}{6} > \frac{24}{6} \quad \text{divide both sides by 6}$$

$$\therefore \quad x > 4$$

Therefore, $6x > 24$ when $x > 4$.

Note: The solution $x > 4$ does *not* include $x = 4$. This is denoted ⊕ on the diagram.

EXAMPLE 6.2
Solve for y, $3y - 1 \leq 14$, and draw the solution on the real number line.

Solution

$$3y - 1 \leq 14$$
$$\therefore 3y - 1 + 1 \leq 14 + 1 \quad \text{add 1 to both sides}$$
$$\therefore 3y \leq 15$$
$$\therefore \frac{3y}{3} \leq \frac{15}{3} \quad \text{divide both sides by 3}$$
$$\therefore y \leq 5$$

Therefore, $3y - 1 \leq 14$ when $y \leq 5$.

Note: The solution $y \leq 5$ *does* include $y = 5$. This is denoted • on the diagram.

EXAMPLE 6.3
Solve for r, $-9 \leq 3r < 18$, and draw your solution on the real number line.

Solution

$$-9 \leq 3r < 18$$
$$\therefore \frac{-9}{3} \leq \frac{3r}{3} < \frac{18}{3} \quad \text{divide each part by 3}$$
$$\therefore -3 \leq r < 6$$

Therefore, $-9 \leq 3r < 18$ when $-3 \leq r < 6$.

Note: Here the solution includes $r = -3$ but *not* $r = 6$.

EXAMPLE 6.4
Solve for x, $7 < 4x + 3 \leq 15$, and draw your solution on the real number line.

Solution

$$7 < 4x + 3 \leq 15$$
$$\therefore 7 - 3 < 4x + 3 - 3 \leq 15 - 3 \quad \text{subtract 3 from all parts}$$
$$\therefore 4 < 4x \leq 12$$
$$\therefore \frac{4}{4} < \frac{4x}{4} \leq \frac{12}{4} \quad \text{divide all parts by 4}$$
$$\therefore 1 < x \leq 3$$

Therefore, $7 < 4x + 3 \leq 15$ when $1 < x \leq 3$.

Note: Here the solution $1 < x \leq 3$ does not include $x = 1$ but does include $x = 3$.

EXAMPLE 6.5
Solve for t, $8t - 5 > 5t + 7$, and draw your solution on the real number line.

Solution

$$8t - 5 > 5t + 7$$
$$8t - 5 - 5t > 5t + 7 - 5t \quad \text{subtract } 5t \text{ from both sides}$$
$$\therefore 3t - 5 > 7$$
$$\therefore 3t - 5 + 5 > 7 + 5 \quad \text{add 5 to both sides}$$
$$\therefore 3t > 12$$
$$\therefore \frac{3t}{3} > \frac{12}{3} \quad \text{divide both sides by 3}$$
$$\therefore t > 4$$

Therefore, $8t - 5 > 5t + 7$ when $t > 4$.

Note again that the solution $t > 4$ does not include $t = 4$.

Solve the following inequalities in terms of the unknown variable and draw each solution on the real number line.

6.1 $8x < 32$
6.2 $4y + 3 > y + 15$
6.3 $24 \leq 4t \leq 30$
6.4 $2x - 3 \leq 4 + 7x$
6.5 $-8 < 3x - 2 < 4$
6.6 $5(y - 3) \geq 2(y + 7)$

B. Solving inequalities using rules 1 and 2 – with fractions

EXAMPLE 6.6
Solve for y, $3(y - 1) \leq \dfrac{2y + 9}{2}$, and draw your solution on the real number line.

Solution

$$3(y-1) \le \frac{2y+9}{2}$$

$\therefore \ 6(y-1) \le 2y+9$ multiply both sides by 2

$\therefore \ 6y-6 \le 2y+9$

$\therefore \ 4y-6 \le 9$ subtract $2y$ from both sides

$\therefore \ 4y \le 15$ add 6 to both sides

$\therefore \ y \le 3\frac{3}{4}$ divide both sides by 4

Therefore, $3(y-1) \le \frac{2y+9}{2}$ when $y \le 3\frac{3}{4}$.

Note: Here the solution $y \le 3\frac{3}{4}$ does include $y = 3\frac{3}{4}$.

EXAMPLE 6.7

Solve for t, $\frac{3t-5}{3} \ge \frac{2-t}{2}$, and draw your solution on the real number line.

Solution

$$\frac{3t-5}{3} \ge \frac{2-t}{2}$$

$\therefore \ \frac{6(3t-5)}{3} \ge \frac{6(2-t)}{2}$ multiply both sides by 6 to remove the fractions

$\therefore \ 2(3t-5) \ge 3(2-t)$

$\therefore \ 6t-10 \ge 6-3t$ expand

$\therefore \ 9t-10 \ge 6$ add $3t$ to both sides

$\therefore \ 9t \ge 16$ add 10 to both sides

$\therefore \ t \ge 1\frac{7}{9}$ divide both sides by 9

Therefore, $\frac{3t-5}{3} \ge \frac{2-t}{2}$ when $t \ge 1\frac{7}{9}$.

YOUR TURN

Solve the following inequalities and draw each solution on the real number line.

6.7 $\frac{x+6}{2} \le x$

6.8 $2y-1 \ge \frac{9y+1}{4}$

6.9 $\frac{2t}{5} + 5 > 3$

6.10 $-1 < \frac{2x}{3} < 3$

6.11 $2(3p-6) < \frac{5p+3}{2}$

6.12 $\frac{3(2r-2)}{2} > \frac{6r-3}{5} + \frac{r}{10}$

C. Solving inequalities using rule 3

EXAMPLE 6.8

Solve for x, $6 - 5x > 16$, and draw your solution on the real number line.

Solution

$$6 - 5x > 16$$

$\therefore \ -5x > 10$ subtract 6 from both sides

$\therefore \ \frac{-5x}{-5} < \frac{10}{-5}$ divide both sides by –5 and reverse the inequality sign (see *rule 3* in *Section 2*)

$\therefore \ x < -2$

Therefore, $6 - 5x > 16$ when $x < -2$.

Note: Here the solution $x < -2$ does not include $x = -2$.

EXAMPLE 6.9

Solve for y, $9 \le 3 - 2y < 6$, and draw your solution on the real number line.

Solution

$$9 \le 3 - 2y < 6$$

$\therefore \ 9-3 \le 3-2y-3 < 6-3$ subtract 3 from both sides

$\therefore \ 6 \le -2y < 3$

Since the solution involves dividing by –2 and thus reversing the signs of the inequality (by *rule 3*) it will be easier at this stage to treat each inequality separately and then combine them again on completion, if appropriate.

Hence, we have

$$6 \le -2y \text{ or } -2y < 3$$

$\therefore \ \frac{6}{-2} \ge \frac{-2y}{-2}$ or $\frac{-2y}{-2} > \frac{3}{-2}$ divide both parts by –2 and reverse the sign

$\therefore \ -3 \ge y$ or $y > -1\frac{1}{2}$

Therefore, $9 \le 3 - 2y < 6$ when either

$y \le -3$ or $y > -1\frac{1}{2}$. (See over for diagram.)

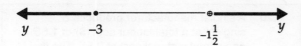

y −3 $-1\frac{1}{2}$ y

EXAMPLE 6.10

Solve for q, $-8 < 2(4 - 3q) \le 14$, and draw your solution on the real number line.

Solution

$$-8 < 2(4 - 3q) \le 14$$

$$\therefore \ -4 < 4 - 3q \le 7 \quad \text{divide each part by 2}$$

$$\therefore \ -8 < -3q \le 3 \quad \text{subtract 4 from each part}$$

As with the previous example we will again treat each part of the dual inequality separately since the next step of the solution requires a division by −3 resulting in a reversal of the signs (see *rule 3* in *Section* 2).

Hence, we have

$$-8 < -3q \text{ and } -3q \le 3$$

$$\therefore \ \frac{-8}{-3} > \frac{-3q}{-3} \text{ and } \frac{-3q}{-3} \ge \frac{3}{-3} \quad \begin{array}{l}\text{divide by −3 and}\\\text{reverse the sign}\end{array}$$

$$\therefore \quad 2\frac{2}{3} > q \text{ and } q \ge -1$$

Therefore, the solution to $-8 < 2(4 - 3q) \le 14$ is that $q \ge -1$ and $q < 2\frac{2}{3}$.

Combining these two parts to form a dual inequality, we can write that

$$-8 < 2(4 - 3q) \le 14 \text{ when } -1 \le q < 2\frac{2}{3}.$$

$$\underline{\bullet\oplus}$$
     −1 q $2\frac{2}{3}$

Note: Here the solution $-1 \le q < 2\frac{2}{3}$ includes −1 but does not include $2\frac{2}{3}$.

YOUR TURN

Solve the following inequalities and draw each solution on the real number line.

6.13 $-5x \le 20$

6.14 $13 - 3y \ge 19$

6.15 $3 - 2(s - 1) < 8$

6.16 $9 \le \dfrac{2 - 0.01t}{0.2}$

6.17 $-21 < 3 - 3x \le 6$

D. Applications of inequalities

As was mentioned in the previous chapter, all mathematical concepts and procedures are far easier to understand if they can be seen to have some practical application.

The following examples are just two of the many practical situations that may give rise to inequalities.

EXAMPLE 6.11   Purchase or rent?

Mr. Smith must decide whether to rent or buy a certain piece of farm machinery. Comparing costs he finds:

(i)   To rent: $500 per month rental fee (on an annual basis), $50 for each day of use to cover petrol, oil and driver.

(ii)   To buy: $3500 fixed annual cost, $75 for each day of use to cover operating and maintenance costs.

Determine the least number of days Mr. Smith must use the machine per year to justify renting rather than buying.

Solution

Strategy:

(i)   Find expressions for rental cost and purchase cost.

(ii)   Determine when rental cost is less than purchase cost.

Let d = number of days per year the machine is used.

Then, rental cost = annual rental fees + daily charges

$$= 12 \times 500 + 50 \times d$$

$$= 6000 + 50d$$

purchase cost = annual fixed cost + daily costs

$$= 3500 + 75d$$

We want

rental cost < purchase cost

$$\therefore \ 6000 + 50d < 3500 + 75d$$

or $\quad\quad\quad 2500 < 25d$

or $\quad\quad\quad 100 < d$

Therefore, Mr. Smith must use the machinery more than 100 days per year to justify renting it.

EXAMPLE 6.12   Profit

For a manufacturer of hearing aid batteries, the combined labour and materials cost is $4 per battery. The fixed costs (i.e. those incurred over a given time period regardless of output) is $45 000. If the selling price of each battery is $7.00, how many must be sold for the company to earn a profit?

Solution

Strategy: Remember that,

profit = total revenue (sales) – total cost

Therefore, if we find expressions to represent total revenue and total cost, then all that remains is to determine when their difference is *positive*.

Let q = the number of batteries that must be sold. Since the cost of producing q batteries is $4q$, then the total cost for the company is

$$4q + 45\,000$$

The total revenue (sales) for the company will be $7q$ (if q batteries are sold).

Now, since **profit** implies that total revenue exceeds total production costs, we want

total revenue – total cost > 0

i.e. $7q - (4q + 45\,000) > 0$

or $7q - 4q - 45\,000 > 0$

or $3q > 45\,000$

$\therefore q > 15\,000$

Therefore, more than 15 000 batteries must be sold for the company to make a profit.

Extensions to chapter 6

There are only two extensions being presented here, both of which may be considered to be rather difficult and hence probably more relevant to those readers who intend to pursue a mathematics and/or science degree.

The first deals with the situation where the unknown variable in the inequality occurs in the denominator, while the second continues with our treatment of absolute value — a topic dealt with in the extensions of several previous chapters.

Extension A: Solving inequalities with the unknown in the denominator

In order to solve an inequality where the unknown, say x, is in the denominator, one must multiply both sides of the inequality by x (or by the expression involving x).

However, this poses a special problem because we do not know whether x (and hence the denominator) is positive or negative.

Remember that whenever the solution to an inequality involves multiplication by a **negative** number, then the **direction** of the inequality is reversed.

Therefore, when solving an inequality whose unknown is in the denominator, two cases must be considered before a complete solution can be obtained:

(i) when the denominator is positive

(ii) when the denominator is negative.

EXAMPLE E6.1

Solve for y, $\dfrac{1}{y} < \dfrac{1}{4}$

Solution

Case (i):

If $y > 0$, then multiplying both sides by $4y$ gives

$$4 < y,$$

that is, $y > 4$.

Now, the conditions $y > 0$ and $y > 4$ must both be true **simultaneously**. Clearly, this will occur whenever $y > 4$.

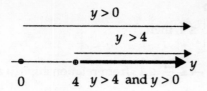

Case (ii):

If $y < 0$, then multiplying both sides by $4y$ gives

$$4 > y,$$

that is, $y < 4$.

Again the conditions $y < 0$ and $y < 4$ must both be true **simultaneously**. From the diagram below this will occur whenever $y < 0$.

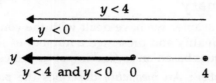

Therefore, combining the individual solutions to each of *cases* (i) and (ii), we find that the complete solution to $\frac{1}{y} < \frac{1}{4}$ is $y < 0$ or $y > 4$.

EXAMPLE E6.2

Solve for t, $\dfrac{4}{t-3} > 5$

Solution

Case (i):

If $t - 3 > 0$, i.e. if $t > 3$, then multiplying both sides by $t - 3$ gives

$$4 > 5t - 15$$

That is, $\quad 19 > 5t$

or $\qquad 3\frac{4}{5} > t$

$\therefore \ t < 3\frac{4}{5}$

Now, for the conditions $t > 3$ and $t < 3\frac{4}{5}$ to *both* be

true, we must have $3 < t < 3\frac{4}{5}$.

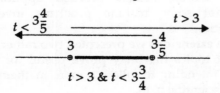

Case (ii):

If $t - 3 < 0$, i.e. if $t < 3$, then multiplying both sides by $t - 3$ gives

$$4 < 5t - 15$$

That is, $\quad 19 < 5t$

or $3\frac{4}{5} < t$

$\therefore \ t > 3\frac{4}{5}$

Clearly, the conditions $t < 3$ *and* $t > 3\frac{4}{5}$ cannot *both* be true at the same time. Hence, we say that there is *no* solution when $t < 3$.

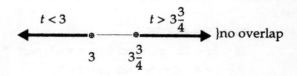

Therefore, combining our results of each of *cases* (i) and (ii), we find that the complete solution to $\frac{4}{t-3} > 5$ is $3 < t < 3\frac{3}{4}$.

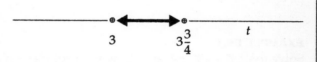

Extension B: Solving absolute – value inequalities

The solution of inequalities surrounding absolute values is most easily determined if the absolute value of any expression is defined in terms of its distance from zero (as it was in Chapter 1). Thus,

(i) if $|x| = 5$, then x is *exactly* 5 units from zero. That is, $x = +5$ or $x = -5$ (usually written as $x = \pm 5$).

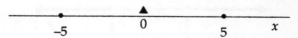

(ii) if $|x| < 5$, then x is *less than* 5 units from zero. That is, x lies between -5 and $+5$, written as $-5 < x < 5$.

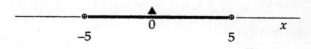

(iii) if $|x| > 5$, then x is *more than* 5 units from zero. That is, x is less than –5 *or* x is more than 5; written as $x < -5$ or $x > 5$.

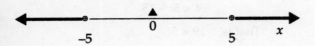

Using the above relationships (i), (ii) and (iii) we will now consider several examples requiring the solution of inequalities involving absolute values.

EXAMPLE E6.3

Solve for x, $|3x + 6| < 18$, and draw your solution on the real number line.

Solution

Since $|3x + 6| < 18$, from relationship (ii) above we know that $(3x + 6)$ lies between –18 and +18.
That is,

$$-18 < (3x + 6) < 18$$

$\therefore$ $\;-24 < 3x < 12$ subtract 6 from each part

$\therefore$ $\;-8 < x < 4$ divide each part by 3

Therefore, the solution to $|3x + 6| < 18$ is $-8 < x < 4$.

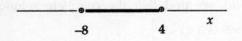

EXAMPLE E6.4

Solve for y, $|7y - 3| \geq 4$, and draw your solution on the real number line.

Solution

Since $|7y - 3| \geq 4$, then

$7y - 3 \geq 4$ or $7y - 3 \leq -4$ (from relationship (iii) above)

$\therefore$ $\;7y \geq 7$ or $7y \leq -1$ add 3 to both sides

$\therefore$ $\;y \geq 1$ or $y \leq -\dfrac{1}{7}$ divide both sides by 7

Therefore, the solution to $|7y - 3|$ is $y \geq 1$ or $y \leq -\dfrac{1}{7}$.

EXAMPLE E6.5

Solve for t, $\left|\dfrac{2t}{5}\right| \leq 4$, and draw your solution on the real number line.

Solution

Since $\left|\dfrac{2t}{5}\right| \leq 4$, then

$$-4 \leq \frac{2t}{5} \leq 4 \quad \text{from relationship (ii) above}$$

$\therefore$ $\;-20 \leq 2t \leq 20$ multiply all parts by 5

$\therefore$ $\;-10 \leq t \leq 10$ divide all parts by 2

Therefore, the solution to $\left|\dfrac{2t}{5}\right| \leq 4$ is $-10 \leq t \leq 10$.

Summary

In this chapter we have dealt with the concept of an **inequality** and presented a number of solution techniques based on several important rules.

Remember: An *inequality* describes the *positional relationship* between two numbers.

Section 1 listed the notation relevant to this topic, **inequalities,** namely

$a < b$	(*a* less than *b*)
$b > a$	(*b* greater than *a*)
$a \leq b$	(*a* less than or equal to *b*)
$b \geq a$	(*b* greater than or equal to *a*)
$a < x < b$	(*x* lies between *a* and *b*)
$a \leq x \leq b$	(*x* lies between *a* and *b* inclusive).

Section 2 presented the three basic rules involving inequalities.

Rule 1: Adding or subtracting the same number from both sides of an inequality leaves the sign unchanged.

Rule 2: Multiplying or dividing both sides of an inequality by the same **positive** number leaves the sign unchanged.

Rule 3: Multiplying or dividing both sides of an inequality by the same **negative** number **reverses** the sign.

Section 3 discussed a number of examples covering techniques for solving different types of inequality problems using the three basic rules. The final part of this section, headed practical applications, looked at some real-life problems involving inequalities.

In the **extensions** we presented two rather more complex topics that may prove useful for those readers intending to follow a more mathematical and/or scientific path.

A: Solving inequalities when the unknown is in the denominator

These types of examples require special treatment due to the fact that the unknown, and hence the denominator, may be either *positive* (sign remains unchanged) or *negative* (sign is reversed).

B: Solving absolute-value inequalities

As we found, these types of problems are most easily solved if we define the absolute value of an expression in terms of its distance from zero.

PRACTICE PROBLEMS

In problems 6.1 to 6.16 below, solve for x the given inequalities.

6.1 $-5x \geq 40$

6.2 $7x - 2 < 19$

6.3 $9x < 3x + 27$

6.4 $4(x + 1) \geq 2x$

6.5 $6x - 5 \leq 9x + 11$

6.6 $\dfrac{x + 6}{2} \leq 2x$

6.7 $\dfrac{5x + 1}{3} > x + 4$

6.8 $\dfrac{2x}{5} + 5 < 3$

6.9 $\dfrac{x + 6}{3} < 5$

6.10 $\dfrac{x}{4} \leq \dfrac{x}{6} + \dfrac{1}{3}$

6.11 $\dfrac{1}{x} > \dfrac{1}{4}$

6.12 $\dfrac{3}{x - 1} < \dfrac{5}{2}$

6.13 $\dfrac{12}{3x + 2} > 4$

6.14 $|2x - 1| < 5$

6.15 $\left|\dfrac{x}{2}\right| < 5$

6.16 $\left|\dfrac{3x - 1}{2}\right| > 1$

6.17 A manufacturer has 3000 units of product in stock. The product is selling at $4.50 per unit. Next month a rise of $0.50 in the price of the product will occur. The manufacturer wants to sell 2500 units for no less than $12000. What is the maximum number of units that can be sold this month?

*6.18 A company invests a total of $50000 of excess funds at two annual rates of interest: 5.5% and 4%. The company wishes to make an annual yield of at least 5%. What is the least amount of money that it must invest at the 5.5% p.a. rate?

*6.19 A business person is deciding whether to hire or buy a car. The costs associated with hiring are $135 per month with operating costs of $0.05 per kilometre. To buy the car the fixed annual cost is $1000 and all other costs would amount to $0.10 per kilometre. What is the least number of kilometres driven so that renting would be no more expensive than buying?

*6.20 A sales position is being offered under two choices of payment of the annual salary. The first method pays $25000 plus a bonus of 2% of your yearly sales. The other method pays a straight 10% commission on your sales. For what yearly sales level is it better to choose the first method?

7 Functions and graphs

Introduction

During our day-to-day life we often come across quantities or variables that we believe are totally unrelated, implying that, at any given time, knowledge of the value of one variable has absolutely no bearing on the possible value of the second variable.

For example: It is difficult to imagine that a woman's height could help determine or in any way influence her annual interest earned in a savings account.

However, this does not alter the fact that there are many very familiar variables where the value of one does in fact depend upon or is determined, at least in part, by the value of the other.

For example: If you drive your car at a constant 70 km per hour then the distance covered depends upon the length of time you drive. This implies that if you are told the travelling time, t, then the distance travelled, d, can be calculated.

In mathematics, however, rather than saying "distance covered depends on travelling time", you are more likely to say that "distance, d, is a function of time, t."

Although we will present a formal definition in the first section, the term 'function' can be thought of as simply a name given to a special 'input/output' relationship that indicates how the value of one variable (the output) is determined by the value of another variable (the input).

Hence, in our above example, distance, d, would represent the 'output' whose value is a function of (depends upon) the travelling time, t, the 'input'.

Functional relationships such as this are usually specified by a simple rule, equation or formula that indicates what operations must be performed on the input in order to calculate the associated output.

Additional or alternative methods of expressing a functional relation involve the use of tables and/or graphs, with a graph being particularly useful as it provides a visual representation of a function and its behaviour.

In the first half of this chapter we will discuss the concept of mathematical functions together with some related issues including dependent versus independent variables and the domain and the range of a function.

The second half of the chapter covers graphs of functions, first discussing some plotting rules and techniques before presenting several of the more common graphs.

As far as this topic – namely, **functions and graphs** – is concerned, there is no additional material specific to particular discipline areas that needs to be included as an 'extension' to this chapter.

7.1 Defining functions

> A **function** is a mathematical rule that assigns to each and every input a **unique** output.

Notation: The conventional notation used to represent a function is $y = f(x)$.

This is read 'y is a function of x', where $f(x)$ produces an output, y, for any given input, x.

Note: Although any letter can be used to represent variables, e.g. s, t, p, q, d, etc., the most commonly used are x and y because functions are often graphed and, as you will see later in this chapter, graphs involve the x and y axes.

Also, f is not the only representation of a function rule. Other letters commonly used to denote a function of x, say, $H(x)$, $F(x)$, $g(x)$, $G(x)$, etc., clearly are necessary whenever several functions are involved.

For example: $y = g(x) = 2x + 3$ defines y as a function of x, where the function rule g is 'double the input and add 3'.

To symbolically indicate that we wish to determine the output, y, when the input x equals -1, say, we write $y = g(-1) = 2x(-1) + 3 = 1$, or just $g(-1) = 1$ (read 'g of -1 equals 1'.) Similarly, $g(3) = 2 \times 3 + 3 = 9$ and $g(t + 1) = 2(t + 1) + 3 = 2t + 5$.

EXAMPLE 7.1

Consider the following:

Let d = distance travelled (in km)

s = travelling speed (in km/hour).

Then, if we wish to determine the distance travelled after 2 hours, we can say that distance equals *twice* the speed of travel, **or** symbolically,

$d = f(s) = 2s.$

Hence, we can see here that distance is a function of speed.

Furthermore, if $d = 2s$ then the following table shows the output, d, for some given input values, s.

Input, s	Function $f(s) = 2s$	Output, d
60 km/hr	$f(60) = 2 \times 60$	120 km
75 km/hr	$f(75) = 2 \times 75$	150 km
100 km/hr	$f(100) = 2 \times 100$	200 km
120 km/hr	$f(120) = 2 \times 120$	240 km

EXAMPLE 7.2

Consider the following:

$y = f(x) = x^2 + 3$

Here y is a function of x and the above statement $y = f(x) = x^2 + 3$ is usually abbreviated to either:

(i) $y = x^2 + 3$, or

(ii) $f(x) = x^2 + 3$

The following table shows the output, y, that can be determined by evaluating the function at various values of the input, x.

Input, x	Function $f(x) = x^2 + 3$	Output, y
2	$f(2) = 2^2 + 3$	7
–1	$f(-1) = (-1^2) + 3$	4
0	$f(0) = 0^2 + 3$	3
a	$f(a) = a^2 + 3$	$a^2 + 3$
$b + 2$	$f(b + 2) = (b + 2)^2 + 3$	$b^2 + 4b + 7$

Beware!! $f(x)$ does *not* mean f times x (that is, $f \times x$ or $f \cdot x$), it is *always* interpreted as 'function of x'.

7.2 Dependent and independent variables

In the introduction to this chapter we saw that the distance a car travels is dependent upon the travelling time, provided the speed remains constant. Hence, it would be reasonable to refer to distance, d, as the 'dependent' variable.

On the other hand travelling time, t, whose possible values can be freely selected and then used to calculate the corresponding distances, is termed the 'independent' variable.

The concept of dependent versus independent variables can now be formalised with the following definition.

A variable that represents the input for some given function is called the **independent** variable, whilst a variable that represents the output from the function is called the **dependent** variable because its value is determined by or depends upon the particular input. Hence, we can say that the dependent variable is a function of the independent variable.

From the above definition, it follows that:

(a) In example 7.1, where $d = f(s) = 2s$,

    (i) distance, d, is the dependent variable, whilst

    (ii) speed, s, is the independent variable.

(b) In example 7.2, where $y = f(x) = x^2 + 3$,

    (i) y is the dependent variable, whilst

    (ii) x is the independent variable.

EXAMPLE 7.3

If $y = f(x) = 3x - 1$, find

(i) $f(5)$

(ii) $f(0)$

(iii) $f(-3)$

Solution

(i)   Given $f(x) = 3x - 1$,

    then $f(5) = (3 \times 5) - 1$

               $= 15 - 1$

               $= 14$

(ii) If $f(x) = 3x - 1$,

    then $f(0) = (3 \times 0) - 1$

               $= 0 - 1$

               $= -1$

(iii) If $f(x) = 3x - 1$,

    then $f(-3) = (3 \times -3) - 1$

               $= -9 - 1$

               $= -10$

EXAMPLE 7.4

If $H(x) = 2^x + 2^{-x}$,

(a) find
   (i) $H(1)$
   (ii) $H(-3)$

(b) show that $H(2) = H(-2)$

Solution

(a) (i) Given $H(x) = 2^x + 2^{-x}$,

   then $H(1) = 2^1 + 2^{-1}$

   $= 2 + \dfrac{1}{2}$

   $= 2\dfrac{1}{2}$

   (ii) Given $H(x) = 2^x + 2^{-x}$,

   then $H(-3) = 2^{-3} + 2^{-(-3)}$

   $= \dfrac{1}{2^3} + 2^3$

   $= \dfrac{1}{8} + 8$

   $= 8\dfrac{1}{8}$

(b) Given $H(x) = 2^x + 2^{-x}$,

   then $H(2) = 2^2 + 2^{-2}$

   $= 4 + \dfrac{1}{4}$

   $= 4\dfrac{1}{4}$

   and $H(-2) = 2^{-2} + 2^{-(-2)}$

   $= \dfrac{1}{2^2} + 2^2$

   $= \dfrac{1}{4} + 4$

   $= 4\dfrac{1}{4}$

   $\therefore H(2) = H(-2)$

EXAMPLE 7.5

If $\quad g(x) = x - \dfrac{1}{x}$

(a) show that $g(\dfrac{1}{2}) = g(-2)$

(b) find a, if $g(a) = 0$

Solution

(a) Given $\quad g(x) = x - \dfrac{1}{x}$,

   then $g(\dfrac{1}{2}) = \dfrac{1}{2} - \dfrac{1}{1/2}$

   $= \dfrac{1}{2} - \dfrac{1}{1} \times \dfrac{2}{1}$

   $= \dfrac{1}{2} - 2$

   $= -1\dfrac{1}{2}$

   and $\quad g(-2) = -2 - \dfrac{1}{-2}$

   $= -2 + \dfrac{1}{2}$

   $= -1\dfrac{1}{2}$

   Therefore $g(\dfrac{1}{2}) = g(-2)$

(b) Given $\quad g(x) = x - \dfrac{1}{x}$,

   then $\quad g(a) = a - \dfrac{1}{a}$,

   But, $\quad g(a) = 0$,

   $\therefore \quad a - \dfrac{1}{a} = 0$

   or $a = \dfrac{1}{a}$

   Therefore, multiplying both sides by a, we have $a^2 = 1$ or $a = \pm 1$.

   Hence, $g(a) = 0$ when $a = 1$ or $a = -1$.

YOUR TURN

7.1 If $f(x) = 3x^2 + 1$, find
   (i) $f(1)$
   (ii) $f(-3)$
   (iii) $f(0)$.

7.2 If $F(x) = \dfrac{3x + 2}{4x - 1}$, find
   (i) $F(1)$
   (ii) $F(\dfrac{1}{2})$
   (iii) $F(-2)$.

7.3 If $f(x) = 3x + |x|$, find $f(-2)$.

7.4 If $g(x) = x^2 + 3 + \dfrac{1}{x^2}$, show that
   (i) $g(a) = g(\dfrac{1}{a})$
   (ii) $g(a) = g(-a)$.

7.3 The domain and range of a function

An additional concept necessary for an understanding of functions involves the notion that any given function may be defined only for particular values of the **independent** variable.

> The **domain** of a function is the set of all possible values of the independent (input) variable for which the function is defined.

For example: We could have

(i) $y = f(x) = x^3$, for all real x.

(ii) $y = f(x) = x^2 + 5$, for $-3 < x < 3$.

(iii) $y = f(x) = \sqrt{1 - x^2}$.

Sometimes, as in (ii) above, the domain is deliberately restricted either for convenience or for practical reasons in a given physical situation.

Also, as in (iii) above, it is quite common to describe a function without specifying the domain. Whenever this occurs, the domain is assumed to be the set of real numbers for which the function is

meaningful. Actually, $y = f(x) = \sqrt{1 - x^2}$ is only defined for $-1 \le x \le 1$.

Why? For all $x > 1$ or $x < -1$, $(1 - x^2) < 0$ and we cannot find the square root of a negative number.

> The **range** of a function is the set of all corresponding values of the dependent (output) variable, $y\ (= f(x))$, obtained from all values of the independent (input) variable, x, in the domain.

For example: The range of $y = f(x) = x^2$ includes all *positive* real numbers and zero because for *all* values of x (positive or negative), the value of the function cannot be a negative number.

EXAMPLE 7.6

For each of the functions given in column 1 of the following table, determine the *range*, R, of the dependent (output) variable, $y = f(x)$, associated with the *domain*, D, of the independent (input) variable given in column 2.

Note: In each case, the appropriate range is given in column 3 of the table.

Function	D: *Domain* of the independent variable, x	R: *Range* of the dependent variable, $y = f(x)$
(a) $f(x) = \dfrac{1}{x+1}$	D: All real numbers except $x = -1$. **Why?** If $x = -1$, then the denominator, $x + 1 = 0$, and division by zero is undefined.	R: All real numbers except $f(x) = 0$. **Why?** There is no value of x for which $y = f(x) = 0$.
Note: In examples similar to (a) above, the domain and range are often written in terms of the *restrictions* only. So, in (a), $D: x \ne -1$ and $R : y = f(x) \ne 0$.		
(b) $y = x^2 + 4$	D: all real numbers, x	R: $y = f(x) \ge 4$
(c) $g(t) = \sqrt{3t - 1}$	D: all $t \ge \dfrac{1}{3}$	R: $g(t) \ge 0$
(d) $p = s + 4$, for $-4 \le s \le 4$	D: $-4 \le s \le 4$	R: $0 \le p \le 8$
(e) $s(t) = 2t - 5$	D: all real t	R: all real $s(t)$
(f) $h(x) = 6$ (called a *constant* function)	D: all real x	R: $h(x) = 6$
(g) $F(x) = \begin{cases} 8x^5, & \text{for } x \ge 7 \\ 2x + 3, & \text{for } x < 7 \end{cases}$	D: all real x	R: all real $F(x)$
In example (g), $F(x)$ is called a 'piece-wise' or 'compound' function as it defines two rules for $y = F(x)$ – namely, (i) $F(x) = 8x^5$, for $x \ge 7$, and (ii) $F(x) = 2x + 3$, for $x < 7$.		

7.5 In defining the following functions, write down any values of x that must be excluded from the domain.

(i) $y = \dfrac{8}{x}$

(ii) $f(x) = \dfrac{1}{x-3}$

(iii) $y = \dfrac{x}{x^2 - 9}$

(iv) $g(x) = \sqrt{1 - x^2}$

7.6 What happens to the expression $\sqrt{4 - x^2}$ if $x > 2$? Hence state the domain of
$f(x) = \sqrt{4 - x^2}$.

7.7 Suppose you are given the function $y = f(x) = x + 5$ for the domain $-5 \le x \le 5$. What is the range of this function?

7.8 State the domain to be assumed for the following functions.

(i) $F(x) = \sqrt{x - 5}$

(ii) $g(x) = \sqrt{x + 4}$

7.9 Determine the range of each of the following functions over the given domain.

(i) $f(x) = -2x^2$, for all real x

(ii) $g(x) = x^2 - 5$, for all real x.

7.4 Graphs of functions

A mathematical function, $y = f(x)$, is very often graphed in order to provide a useful visual representation of its behaviour – that is, whether it is increasing, decreasing etc.

The graph of any function is constructed by plotting the ordered pairs (x, y) that belong to the function $y = f(x)$.

7.4.1 Plotting graphs

The graph of a given function is plotted on what is called **the real number plane**.

This number plane consists of two intersecting straight lines – one **horizontal**, called the **x-axis** and the other **vertical**, called the **y-axis**. The two axes (the x-axis and the y-axis) intersect (or cross each other) at the point (0,0), called the **origin**, as shown in Figure 7.1 following.

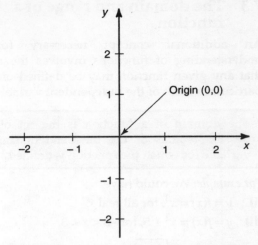

Fig 7.1: The x- and y-axes intersect at the origin.

Steps involved in plotting points

For example: Plot the point (3,4) on the real number plane.

Step 1: Locate $x = 3$ on the horizontal x-axis.

Step 2: Locate $y = 4$ on the vertical y-axis.

Step 3: From the point $x = 3$, draw an imaginary line vertically up 4 units whilst at the same time drawing an imaginary line horizontally 3 units to the right from the point $y = 4$. These two imaginary lines will meet at the point (3,4) and this junction should be marked on the graph with a dot or a small cross.

The plotting procedure is summarised in Figure 7.2 below.

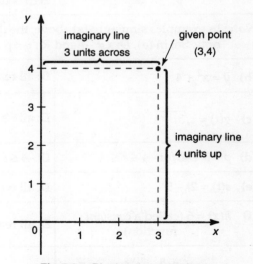

Fig 7.2: Plotting the point (3,4)

Note:

The point, $P(3,4)$, is *not* the same as the point, $Q(4,3)$, as the 'order' in which the numbers making up the ordered pair occur differs from P to Q.

Remember that, by convention, an ordered pair is always written (x,y) – that is, the x-value always precedes the y-value.

Hence, at the point $P(3,4)$, $x = 3$ and $y = 4$, whilst at the point $Q(4,3)$, $x = 4$ and $y = 3$.

The difference between P and Q is shown in Figure 7.3 below.

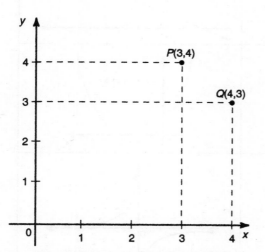

Fig 7.3: Showing the difference between $P(3,4)$ and $Q(4,3)$.

Note:

The *scale* of each axis, that is, the values to be marked and the interval between them, does *not* have to be the same for each axis. What is important is that, for either axis, the interval from one marked value to the next must be *the same* throughout the length of the axis.

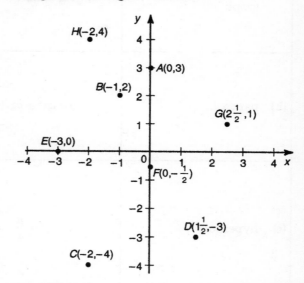

EXAMPLE 7.7

Draw a single set of x- and y-axes, label the axes, mark an appropriate scale of values along each and then plot the following points on the number plane.

(i) $A(0,3)$

(ii) $B(-1,2)$

(iii) $C(-2,-4)$

(iv) $D(1.5,-3)$

(v) $E(-3,0)$

(vi) $F(0,-\frac{1}{2})$

(vii) $G(2\frac{1}{2},1)$

(viii) $H(-2,4)$

Solution

Step 1: Draw and label the horizontal x-axis and the vertical y-axis so that they intersect at the origin, $(0,0)$.

Step 2: Mark in appropriate values of x and y along their respective axes after referring back to the question in order to determine the range of values to be included. Here we need x-values from -3 to $+3$ (to include E and G) and y-values from -4 to $+4$ (to include C and H).

7.5   Some common graphs

The table below is designed solely to acquaint the reader with the various types of graphs and their associated symbolic representations. Each will be explained in more detail in the following sub-sections.

Common name of graph	Symbolic representation	Pictorial representation
(a) linear (straight line) graph	$y = mx + b$	
(b) parabola	$y = ax^2 + bx + c$	
(c) hyperbola	$y = \dfrac{a}{x}$	
(d) logarithmic	$y = \log_a n$	
(e) exponential	$y = a^x$	
(f) Circle (with radius r and centred at (0,0))	$x^2 + y^2 = r^2$	
(g) Circle (with radius r and centre (h,k))	$(x - h)^2 + (y - k)^2 = r^2$	

7.5.1 The linear (or straight line) graph.

From the previous table we know that, symbolically, the general form of a linear graph is

$$y = mx + b$$

where m is called the *gradient* (or slope) of the line relative to the horizontal axis

b is called the *y-intercept* (that is, the point where the line cuts the y-axis, or the value of y when $x = 0$).

EXAMPLE 7.8

Graph the linear function, $y = x + 2$.

Solution

Step 1: Construct a table of ordered (x,y) pairs by selecting at least three x-values and calculating the corresponding y-values for the given function.

x	–1	0	1
$y = x + 2$	1	2	3

Step 2: Draw the x- and y-axes and mark on some reasonable values using the table in Step 1 as a guide.

Take the three (x,y) pairs from the table in Step 1, plot these points on the graph and then join the points with a straight line, labelling the axes and the graph itself, where appropriate.

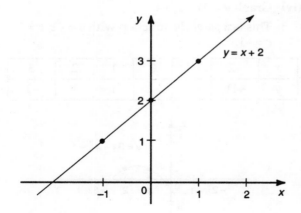

Note: In this example, with $y = x + 2$, the y-intercept is 2 and the slope is 1.

EXAMPLE 7.9

Graph the linear function, $y = 3 - 2x$.

Solution

x	–1	1	2
$y = 3 - 2x$	5	1	–1

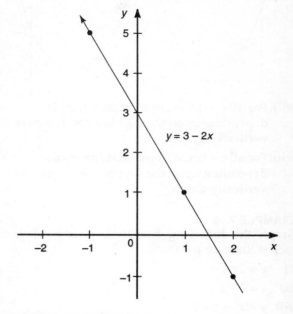

Note: In this example, with $y = 3 - 2x$, the y-intercept is 3 and the gradient or slope is –2.

7.5.2 The parabola

From the table given at the beginning of this section we know that, symbolically, the general form of a parabola is

$$y = ax^2 + bx + c$$

where a, b and c are constants.

The effect on the graph of a parabola, as the sign of the constants changes from positive to negative is shown in the following diagrams.

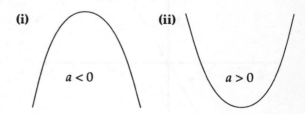

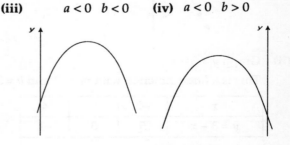

(v) $a > 0$ $b < 0$ **(vi)** $a > 0$ $b > 0$

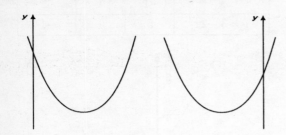

(vii) For all $c > 0$, the appropriate parabola, depending upon the sign of a and b, moves **vertically up**.

(viii) For all $c < 0$, the appropriate parabola, depending upon the sign of a and b, moves **vertically down**.

EXAMPLE 7.10
Graph the following functions, stating whether each is linear or parabolic.

(i) $y = 2x$

(ii) $y = 3 - x$

(iii) $y = x^2 - x + 2$

(iv) $y = 3x - x^2 - 1$

Solution

(i)   Graph $y = 2x$

This is a *linear* function with $m = 2$ and $b = 0$.

x	-1	0	1
$y = 2x$	-2	0	2

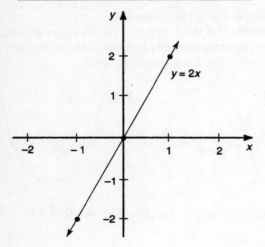

(ii)   Graph $y = 3 - x$

This is a *linear* function with $m = -1$ and $b = 3$.

x	-2	0	4
$y = 3 - x$	5	3	-1

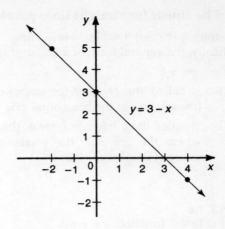

(iii) Graph $y = x^2 - x + 2$

This is a *parabolic* function with $a = 1$, $b = -1$, $c = 2$.

x	-2	-1	0	1	2
y	8	4	2	2	4

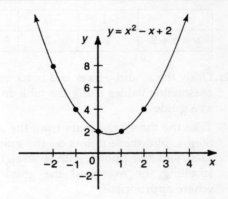

(iv) Graph $y = 3x - x^2 - 1$

This is a *parabolic* function with $a = -1$, $b = 3$, $c = -1$.

x	-2	-1	0	1	2	3
y	-11	-5	-1	1	1	-1

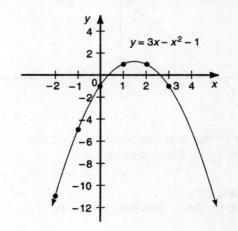

7.5.3 The hyperbola

Again, referring back to the table at the beginning of this section, we see that the general form of a hyperbolic function is

$$y = \frac{a}{x}$$

where a is a constant
$$x \neq 0.$$

EXAMPLE 7.11
Graph each of the following hyperbolic functions.

(i) $\quad y = \frac{2}{x}$ $\qquad$ (ii) $\quad y = \frac{-1}{x}$

Solution

(i) $\quad y = \frac{2}{x}$, for $x \neq 0$

x	-8	-6	-4	-2	-1	1	2	4	6	8
y	$-\frac{1}{4}$	$-\frac{1}{3}$	$-\frac{1}{2}$	-1	-2	2	1	$\frac{1}{2}$	$\frac{1}{3}$	$\frac{1}{4}$

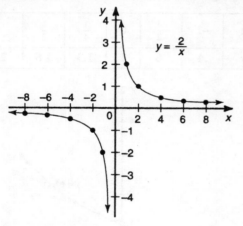

(ii) $\quad y = \frac{-1}{x}$, for $x \neq 0$

x	-8	-4	-2	-1	1	2	4	8
y	$\frac{1}{8}$	$\frac{1}{4}$	$\frac{1}{2}$	1	-1	$-\frac{1}{2}$	$-\frac{1}{4}$	$-\frac{1}{8}$

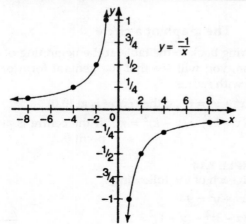

7.5.4 The logarithmic graph

As given in the table at the beginning of this section, the general form of the logarithmic graph is

$$y = \log_a x$$

where $a > 0$ and $x > 0$.

EXAMPLE 7.12
Graph each of the following logarithmic functions.

(i) $\quad y = \log_{10} x$
(ii) $\quad y = \log_e x$
(iii) $\quad y = \log_2 x$

Solution

(i) $\quad y = \log_{10} x$

x	0.1	1	10	100	1000
y	-1	0	1	2	3

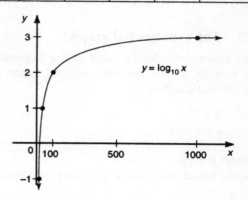

(ii) $\quad y = \log_e x = \ln x$

x	0.25	0.5	1	1.5	2
y	-1.4	-0.7	0	0.4	0.7

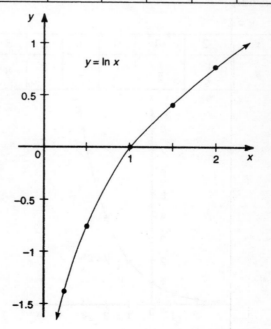

(iii) $y = \log_2 x$

x	0.5	1	2	4	8
y	−1	0	1	2	3

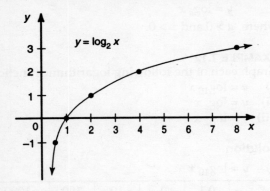

Note: All logarithmic graphs of the same form as (i), (ii) and (iii) above, pass through the point $(1, 0)$.

7.5.5 The exponential graph

Once again, refer to the table at the beginning of this section to see that the general form of an exponential function is

$$y = a^x$$

for $a \neq 0$ or 1.

EXAMPLE 7.13

Graph each of the following exponential functions.

(i) $y = 2^x$

(ii) $y = 2^{-x}$

(iii) $y = e^x$

Solution

(i) $y = 2^x$

x	−2	−1	0	1	2	3
y	$\frac{1}{4}$	$\frac{1}{2}$	1	2	4	8

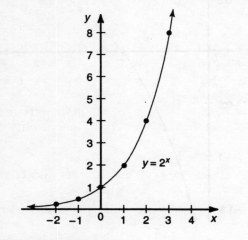

(ii) $y = 2^{-x}$

x	−2	−1	0	1	2	3
y	4	2	1	$\frac{1}{2}$	$\frac{1}{4}$	$\frac{1}{8}$

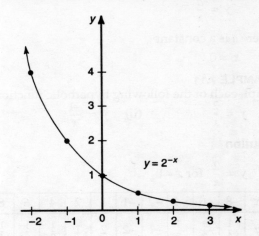

(iii) $y = e^x$

x	−1	$-\frac{1}{2}$	0	$\frac{1}{4}$	$\frac{1}{2}$	1
y	.4	.6	1	1.3	1.6	2.7

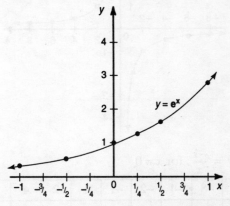

Note: All exponential graphs of the same form as (i), (ii) and (iii) above, pass through the point $(0,1)$

7.5.6 The graph of a circle

Referring back to the table at the beginning of this section, you will see that the general form of the circle with radius r is

(i) $x^2 + y^2 = r^2$, if the centre is at $(0,0)$

(ii) $(x - h)^2 + (y - k)^2 = r^2$, if the centre is at the point (h,k).

EXAMPLE 7.14

Graph each of the following

(i) $x^2 + y^2 = 9$

(ii) $(x - 3)^2 + (y + 1)^2 = 1$

Solution

(i) $x^2 + y^2 = 9$

This is the graph of a circle, centred at the origin and with radius, $r = 3$.

Therefore, the circle must go through the points $(0,3)$, $(-3,0)$, $(0,-3)$, $(3,0)$.

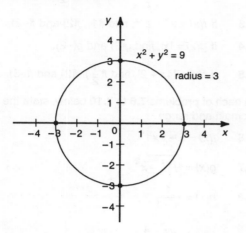

(ii) $(x - 3)^2 + (y + 1)^2 = 1$

This is the graph of a circle, centred at $(3,-1)$, and with radius, $r = 1$. Therefore, the circle must go through the points $(3,0)$, $(3,-2)$, $(2,-1$, $(4,-1)$.

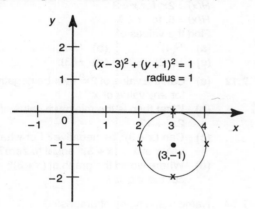

Important note:

The graphs presented in 7.5.1 to 7.5.5 above are all graphs of functions because they can be expressed in the form $y = f(x)$, where x is the independent variable and y is the **unique** dependent variable.

However, this is *not* the case in the final sub-section, 7.5.6, where the graph of a circle is presented. In fact, the symbolic representation of a circle is *not* a function because for each value of the variable, x, there are **two** values of the variable, y, thus contradicting the uniqueness requirement in the definition of a function.

YOUR TURN

7.11 Construct a table of (x,y) pairs for the function $y = 3x + 2$, given the x-values:

$$-2, -1, 0, 1, 2.$$

Plot your (x,y) pairs and draw the graph.

7.12 On the one set of axes, draw the graphs
 (i) $y = 2 - x$
 (ii) $y = 2x - 1$,

for values of x from -4 to 4.

Where do these two graphs meet?

7.13 Graph the following parabolas from $x = -4$ to $x = 4$.
 (i) $y = x^2 + x - 6$
 (ii) $y = x^2 + 2$

7.14 Graph the hyperbolic function $y = \dfrac{x}{8}$, for

x-values of ± 8, ± 4, ± 2, ± 1, $\pm \dfrac{1}{2}$.

7.15 In the function $y = \dfrac{12}{x - 1}$, which x-values

must be omitted from the domain? In drawing the graph, where do you think some care is needed? Try to choose suitable x-values to plot. *For example*, the function is easy to calculate and plot when $x = 7$, but not so when $x = 6$.

7.16 Graph the curve represented by each of the following equations and state which curves are functions and which are not.
 (i) $x^2 + y^2 = 25$
 (ii) $y = 4^x - 1$
 (iii) $y = x^3 + 1$
 (iv) $(x + 1)^2 + (y - 2)^2 = 9$

Summary

This chapter discusses the very important topic of mathematical 'functions' and shows that any given function has two representations – namely,

(i) a **symbolic** representation in the form of an equation, and

(ii) a **pictorial** representation in the form of a graph.

In *section 1* we defined a function, $y = f(x)$, as a rule that assigns to each input, x, a unique output, y.

Note:

(i) $F(x)$, $G(x)$, $g(x)$, $H(x)$ etc. are also commonly used to denote functions of x.

(ii) For a given function, $y = g(x)$, a particular

output, y, for a given input, say $x = -3$, is written as $g(-3)$ and is calculated by substituting $x = -3$ into the function, $g(x)$.

In *section 2* we introduced the concept of dependent and independent variables, stating that for a given function, $y = f(x)$, the *input* variable, x, is called the *independent variable* while the *output* variable, y, is known as the *dependent variable* since its value depends upon the particular input.

Section 3, entitled '**The domain and range of a function**', looks at the notion that for a particular function, $y = f(x)$, both the input and the output values may be restricted in some way.

The **domain** gives the set of values of the independent variable, x, for which the function is defined, while the **range** is the set of all corresponding values of the dependent variable, $y = f(x)$.

Note: The domain of a function may be restricted

(i) for convenience or practical reasons, or

(ii) because there are certain values of x for which the function has no meaning (that is, where $f(x)$ is undefined).

Sections 4 and *5* covered the graphical segment of this chapter – that is, the pictorial representation of a function.

In *section 4* we had a brief discussion on how to construct a graph by plotting a set of ordered (x,y) pairs belonging to the given function, $y = f(x)$, onto the real number plane.

Remember that the *real number plane* consists of two straight lines, the horizontal x-axis and, the vertical y-axis, that intersect at the *origin* $(0,0)$.

In *section 5* we looked at some of the more common functions that you are likely to meet during your studies, giving both the symbolic and the graphical representations of each.

These functions included

(i) the straight line

(ii) the parabola

(iii) the hyperbola

(iv) the logarithmic function

(v) the exponential function.

Finally, we presented both the symbolic and pictorial representations of a circle, noting carefully that

(i) $x^2 + y^2 = r^2$ (a circle centred at $(0,0)$)

and (ii) $(x - h)^2 + (y - k)^2 = r^2$ (a circle centred at (h,k))

are **not** functions, because for each value of x there are **two** values of y, contradicting the uniqueness requirement that defines a function.

7.1 If $f(x) = 9 - x^2$, find $f(2)$ and $f(-2)$. For what value(s) of x does $f(x) = 0$?

7.2 If $H(x) = x - \dfrac{1}{x}$, show that $H(\tfrac{1}{2}) = H(-2)$.

7.3 If $f(x) = 2^x + 2^{-x}$, find $f(1)$, $f(2)$ and $f(-2)$.

7.4 If $g(x) = |x|$, find $g(3)$ and $g(-2)$.

7.5 If $f(x) = |4x + 2|$, find $f(\tfrac{1}{2})$, $f(2)$ and $f(-3)$.

In each of problems **7.6** to **7.10** below, state the domain and range.

7.6 $f(x) = 5 - 2x$

7.7 $g(x) = \sqrt{16 - x^2}$

7.8 $h(x) = \dfrac{1}{x - 1}$

7.9 $g(x) = \dfrac{\sqrt{x - 4}}{2x + 10}$

7.10 $f(x) = 2^x$

7.11 A function $H(x)$ is defined as:
$H(x) = 2x$, for $x < 3$
$H(x) = 6$, for $x \geq 3$
Find the values of
(a) $H(4)$ **(b)** $H(0)$
(c) $H(-2)$ **(d)** $H(3)$.

7.12 **(a)** Can the value of $2x^2$ ever be negative for *any* value of x?
(b) What then is the minimum value of $2x^2 + 1$?
(c) Can $(x + 3)^2$ be negative? For what value of x is $(x + 3)^2$ equal to zero?
(d) Where would the graph of $(x + 3)^2 - 1$ cut the y-axis?

7.13 Name these types of graphs:
(a) $y = \dfrac{4}{x}$ **(b)** $y = 2x - 1$
(c) $y = 3x^2 + 2x - 1$ **(d)** $y = \log_5 x$
(e) $y = 3^x$

7.14 Sketch the following:
(a) $y = 2x + 1$, $-4 \leq x \leq 4$
(b) $y = -\dfrac{2}{x}$, $-2 \leq x \leq 4$
(c) $y = |x|$, $-4 \leq x \leq 4$
(d) $y = x^2 - 1$, $-3 \leq x \leq 3$

7.15 Show that if $f(x) = x^3 + 3x$, then $f(-a) = -f(a)$ for all a.

7.16 Show that if $f(x) = x^4$, then $f(a) = f(-a)$ for all a.

*7.17 Which of the graphs below best describes the following statement:

The distance a walker travels after a certain number of hours, starting at base camp.

(I)

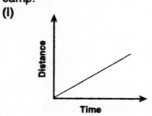

(ii)

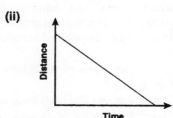

(iii)

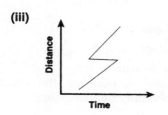

*7.18 The following graph shows the journey taken by a motorist.

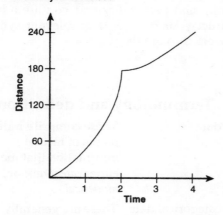

(a) How many kilometres did the motorist travel?

(b) What was his average speed for the journey?

(c) The motorist stopped for lunch. How long was this lunch break?

(d) What was the fastest stage of the journey?

*7.19 The following graph shows the amount of petrol in a car's petrol tank during a trip. Describe what you think the graph shows.

7.19 *Cont.*

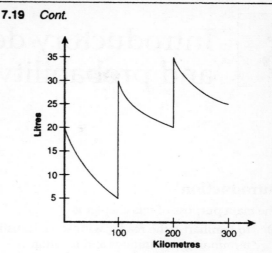

*7.20 The length of a spring with a mass attached to its lower end can be represented by the equation $l = 3m + 10$ where l centimetres is the length of the spring and m kilograms is the mass attached to it. The following figure shows the graph of the equation.

(a) How long is the spring before the mass is attached?

(b) Find the length of the spring when a mass of 8 kg is attached to it. How much did the spring extend when this mass was attached?

(c) If the spring has length 28 cm, what mass is attached?

(d) If the spring is extended by 9 cm, what mass is attached?

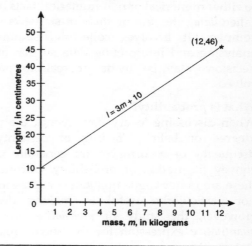

8 Introductory descriptive statistics and probability

Introduction

The main purpose of this chapter is

(i) to familiarise the reader with some statistical terminology, definitions and notation,

(ii) to introduce several simple procedures for summarising and describing data, and

(iii) to discuss some of the basic concepts involved in the study of probability.

As with other chapters, it is *not* the purpose of this text to delve into the mathematical theory or the derivation of formulae underlying various concepts, but rather to present some of the simpler techniques for determining the more widely-used statistical and probabilistic measures, from a logical and practical standpoint.

Before we proceed it is important to define the meaning of *statistics* and *probability* in this context.

What is statistics?

While the *term* **statistics**, in everyday usage, refers to either numerical or non-numerical facts (usually called *data*), the *field* or *study* of **statistics** is more complex. It involves collecting, summarising, analysing and interpreting data so that effective decisions may be made or real-life problems solved.

What is probability?

When discussing 'every day' events, the phrases 'degree of belief', 'feeling of certainty' and 'frequency of occurrence' are generally used to convey the notion of probability. Unfortunately, these are rather vague qualitative expressions and not particularly useful in making decisions. However, the study of *probability* provides a *quantitative* expression for the above qualitative statements.

That is,

> **probability** is the tool that allows us to quantify variability and uncertainty in many practical situations.

When dealing with basic statistics and probability

it is important to realise that, just as in any other area of mathematics or in any other subject areas such as chemistry, history, economics or biology, some 'jargon' or subject-specific terminology is necessary.

These important and relevant statistical terms are presented in *section 1*.

Sections 2, 3 and *4* cover the statistical aspects of this chapter – namely, summarising and describing data using

(i) tabular displays,

(ii) pictorial displays, and

(iii) numerical measures.

Sections 5 and *6* examine the concepts of elementary probability by discussing

(i) sets and events,

(ii) ideas and basic rules (or laws) of probability.

Finally, **notation** is very important in all areas of statistics and probability and you are advised to construct your own table of relevant symbols as you work through the chapter.

8.1 Terminology and definitions

(i)	data	These comprise individual pieces of factual information that may be either *categorical* or *numerical*.
(ii)	categorical data	These are generally non-numerical or qualitative, since each item is a description rather than a number, eg. colour, occupation, sex etc.
(iii)	numerical data	These are quantitative rather than qualitative, referring to measurements of some kind, and are thus expressed as numbers, eg. age, income, IQ, weight etc.

(iv) population — This is the complete set of all items (measurements) of interest in a particular study.

(v) sample — This is some part or subset of the measurements in the population of interest.

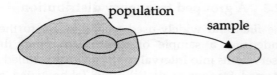

(vi) random sample — This is a sample selected in such a manner that every member of the population has an *equal* chance of being included.

(vii) variable — This refers to the characteristic actually being measured, assuming that this characteristic varies in either quality or quantity.

(viii) parameter — This is a numerical summary measure that describes some characteristic of a **population.**

(ix) statistic — This is a numerical summary measure that describes some characteristic of a **sample.**

(x) die (pl: dice) — This is a small 6-sided object used in many games of chance. Each of the sides displays dots, and the die is said to be fair if each side has an equal chance of appearing uppermost.

 a die

(xi) standard pack of cards — This contains 4 suits – clubs, hearts, diamonds and spades. The spades and clubs are black, whilst the diamonds and hearts are red. Each of the 4 suits consists of 13 cards – ace, 2, 3, 4, 5, 6, 7, 8, 9, 10, jack, queen and king. Hence, a full standard pack contains 52 cards.

8.2 Summarising data – tabular displays

In this section we will discuss, briefly, a few of the more common methods of presenting data in tabular form as this may affect

(i) the choice of pictorial display (the topic for *Section 3*), and

(ii) the method used to calculate numerical summaries (the topic for *Section 4*).

Generally, when data are collected in a practical situation, the observations themselves appear in a rather meaningless random order.

There are three main techniques often used for condensing data into a more comprehensible form. The simplest way to present these techniques is to take an example and then see how each may be applied to the given data.

EXAMPLE 8.1
Each day the sales manager of a store notes the total number of customers entering the store. These numbers, for a period of 40 days, are listed below:

47	64	69	44	65	74	49	66
83	53	47	72	54	64	37	72
66	56	44	74	77	66	69	70
84	83	70	84	84	56	74	56
65	63	77	56	66	35	35	47

Observations presented in the above form are often referred to as **raw data.**

It is difficult to obtain much information from the data in this form, except to notice that the numbers of customers all appear to be under 100.

8.2.1 An array

One simple technique for presenting data in a more meaningful way is to re-write the values as an **array** – that is, to arrange the data **in some order**.

The most common method is to arrange the data in order of magnitude from the smallest to the largest observation.

EXAMPLE 8.2
Table 8.1 following shows the raw data of Example 8.1 presented as an **array**.

Table 8.1: Array of customer numbers (from example 8.1)

35	47	54	63	66	69	74	83
35	47	56	64	66	70	74	83
37	47	56	65	66	70	74	84
44	49	56	65	66	72	77	84
44	53	56	65	69	72	77	84

8.2.2 A frequency distribution

A second technique for condensing data, going a step further than an array, is called a **frequency distribution.**

Now,

(i) the **frequency** of an observation is just the number of times that observation occurs in a given batch of data, and

(ii) a series of statistical observations is sometimes said to have a certain **distribution.**

So, logically it follows that

> a **frequency distribution** gives a listing of the *different* observations (in order of magnitude), each with its corresponding frequency alongside.

EXAMPLE 8.3
By condensing the **array** displayed in Table 8.1, we can construct Table 8.2. This table shows the raw data of Example 8.1 presented as a **frequency distribution.**

Table 8.2: Frequency distribution of customer numbers

No. of customers	35	37	44	47	49	53	54
Frequency	2	1	2	3	1	1	1
No. of customers	56	63	64	65	66	69	70
Frequency	4	1	1	3	4	2	2
No. of customers	72	74	77	83	84		
Frequency	2	3	2	2	3		

Note:
When constructing a frequency table, always check that the total frequency equals the number of observations in the raw data (e.g. 40 in the example above).

8.2.3 A grouped frequency distribution

One final and widely-used technique for further condensing a sample of data is to group the observations into **intervals** to form what is called a **grouped frequency distribution** (abbreviated to *G.F.D.*).

However, grouping data in this manner does have one fundamental disadvantage – some of the initial information will be lost.

For example, if it is known that there are six observations in an interval labelled 15 – 20, we cannot say whether they are all at one end of the interval or spread throughout it.

When constructing a *G.F.D.*, there are several **rules** that should be followed very carefully:

(a) The intervals should be **non-overlapping** – so that any one observation cannot belong to more than *one* interval.

(b) In most cases, the intervals should be the *same* width.

(c) If there are no observations in a particular interval, it should still be included to avoid a misleading impression of the data.

(d) The frequency of observations in each interval should be recorded in an adjacent column.

The **number** of intervals comprising the *G.F.D.* is, to some extent, **arbitrary.** However, when selecting the appropriate interval width, keep in mind that **too many** intervals will reduce the effectiveness of the *G.F.D.* as a tool for *condensing* data, while **too few** intervals may have the opposite effect, resulting in the loss of too much information. Also keep in mind that, conventionally, intervals are of width 2, 5, 10, 20, 50 or 100.

EXAMPLE 8.4
The customer numbers for the 40 days, given in Example 8.1, are condensed into the following *G.F.D.*:

Table 8.3: Grouped frequency distribution of customer numbers

Interval	Frequency
30 – 39	3
40 – 49	6
50 – 59	6
60 – 69	11
70 – 79	9
80 – 89	5
Total	**40**

Note:

The following **definitions** associated with a *G.F.D.* may be of interest.

(i) The **interval limits** are the largest and smallest observations that could belong to a given interval. So, for the interval 30 – 39 in the *G.F.D.* in Example 8.4, the **lower limit** is 30 and the **upper limit** is 39.

(ii) The **interval boundaries** are the dividing lines between the intervals. So, for the intervals 40 – 49 and 50 – 59 in the *G.F.D.* in Example 8.4, the boundaries are 39.5, 49.5, 59.5.

(iii) The **width** of a given interval is the difference between the upper and lower **boundaries**. So, the *real* width of the interval 60 – 69 in the *G.F.D.* in Example 8.4 is 69.5 – 59.5 = 10. (In fact, *all* intervals in that example are of width 10).

(iv) The **midpoint** of a certain interval is found by adding the upper and lower *limits* and dividing by 2.

So, for the interval 60 – 69 in the *G.F.D.* in Example 8.4, the *midpoint* is $\frac{1}{2}(60 + 69) = 64.5$.

YOUR TURN

8.1 Consider the following sample of raw data:

12	17	24	26	31	12	16	23
29	31	16	23	17	17	21	21
21	25	26	26	27	27	25	31

(a) Present the above data as an array, ordering the observations from smallest to largest.

(b) Summarise the above data using a **frequency distribution**.

8.2 Summarise the sample of data in exercise 8.1 by completing the **grouped frequency distribution** below:

Interval	Frequency
10 – 14	2
15 – 19	
20 – 24	
25 – 29	
30 – 34	
Total	**24**

8.3 Using your grouped frequency distribution in exercise 8.2, write down
(a) the upper and lower limits of the intervals 20 – 24 and 25 – 29
(b) the upper and lower boundaries of the intervals 20 – 24 and 25 – 29
(c) the width of each interval
(d) the midpoint of the intervals 10 – 14 and 15 – 19.

8.4 Again using your *G.F.D.* constructed in exercise 8.2, write down the number of observations with a value
(a) less than 25
(b) more than 14.

8.3 Summarising data – pictorial displays

The old saying "a picture is worth a thousand words" is, in fact, very true – much can be learned very quickly from one simple picture. This is especially true in statistics where we are often dealing with large amounts of information in the form of numbers. Hence, in order to make this information easier to understand, it is always advisable to summarise the data using tables and pictures.

8.3.1 Pictorial displays for categorical data

As mentioned in *Section 1*, categorical data arise when observations are classified into several distinct, non-overlapping categories.
For example:

(i) People being surveyed may be classified as male or female; and they may be 'for', 'against' or 'undecided about' some proposal.

(ii) A new method of migraine treatment may be successful or unsuccessful (depending upon whether or not the patient recovers quickly).

(iii) People may be classified according to occupation, country of birth, ethnic background, religion etc.

The two most common methods of displaying qualitative, categorical data are to construct

(i) **a pie chart**, or

(ii) **a bar chart**.

A brief description

(i) A pie chart

A **pie chart** is a circular display that partitions categorical data into sectors or pie pieces. The area of each sector reflects the percentage of the total count or the relative frequency assigned to each category. In general, the categories are ordered by size (largest to smallest) in a clockwise direction, beginning at 12 o'clock.

(ii) A bar chart

A **bar chart** consists of a series of labelled rectangular bases (one for each category) with

- the height of each box equal to the data frequency in that category,
- the width of each box the same for all boxes,
- even spacing between boxes,
- a scale of measurement on the left-hand side of the display (i.e. the vertical axis) showing data frequencies or percentage frequencies.

As in the previous section, we once again examine these two displays with the aid of an example.

EXAMPLE 8.5

Information was collected from several insurance companies on the number of road accidents involving red cars in each of the 4 quarters of 1992.

A summary of the data obtained is shown in Table 8.4 below.

Table 8.4: Road accidents involving red cars in 1992.

Quarter	1st	2nd	3rd	4th
No. of accidents	32	18	25	45

Present these data using

(i) a pie chart, and

(ii) a bar chart.

Solution

(i) A pie chart

Before we can construct a pie chart, we must first determine the percentage or relative frequency assigned to each category.

Since there are 120 cars altogether, for each category we have

$$\text{relative frequency} = \frac{\text{actual frequency}}{120} \times 100$$

Hence, for each quarter, we have

Quarter	Frequency	Relative frequency
1st	32	$\frac{32}{120} \times 100 = 26.7\%$
2nd	18	$\frac{18}{120} \times 100 = 15.0\%$
3rd	25	$\frac{25}{120} \times 100 = 20.8\%$
4th	45	$\frac{45}{120} \times 100 = 37.5\%$
Total	**120**	**100%**

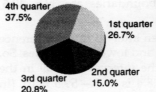

Pie chart showing the accident rate for red cars in each of the 4 quarters in 1992.

Note:

When describing a pie chart earlier, we said that, in general, the categories (and hence the sectors in the pie chart) are ordered from largest to smallest. However, when the categories themselves are intrinsically ordered, as in this example, then this ordering is retained so that fluctuations across the categories can be examined.

(ii) A bar chart

When constructing a bar chart, the vertical axis shows either the frequency *or* the relative frequency.

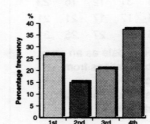

Bar chart showing the accident rate (%) for red cars in each of the 4 quarters in 1992.

Note:

As with the pie chart, the bars in the bar chart have not been ordered from largest to smallest because of the intrinsic time-ordering in the categories. One of the purposes of this bar chart is to indicate the fluctuation in accident rate across the four categories.

8.5 At a particular point on a certain main highway, the colour of the first 200 cars was noted. A summary of the results is shown in the following table.

Car colour	Number of cars
white	80
red	45
yellow	30
blue	25
other	20

Display the data in the above table using
(a) a bar chart,
(b) a pie chart (be careful to calculate the percentage frequency of each category first).

8.6 You are interested in the religious preferences of the other 80 members of a club to which you belong. Patiently you ask each member to indicate his/her preference – Roman Catholic, Protestant, Other, No preference. You find that your club consists of 16 Roman Catholics, 20 Protestants, 16 Others, 20 No preference and 8 who said "none of your business" – or something less restrained.

Summarise and display the above data using
(i) a pie chart,
(ii) a bar chart.

8.7 The consumer price index (*CPI*) is one method of measuring changes in the cost of living. The index is constructed by determining what percentage of their total expenditure an average family spends in a variety of areas. In one particular month in 1993, it was found that the average Australian family spent its money in the following way.

Area	Percentage expenditure
Food	20%
Clothing	7%
Housing	15%
Household maintenance	18%
Transportation	17%
Tobacco and alcohol	8%
Health	5%
Recreation and education	10%

Display the above data using
(i) a bar chart,
(ii) a pie chart.

8.3.2 Pictorial displays for numerical data

Numerical or quantitative data consist of counts or measurements and are said to be either **discrete** or **continuous**.

Discrete data usually consist of counts and take only whole-number values e.g. the number of pups produced by a pregnant 'mum', the number of members in a certain club, the number of cigarettes smoked per day, etc.

Continuous data are not restricted to whole numbers, but can take any value in a range of possible values, e.g. the birth weight of a baby is not restricted to a whole number of grams, but can take any decimal value. Other examples include the time taken to complete a certain task, the melting point of a particular chemical compound, etc.

The most widely-used technique for summarising and displaying quantitative data is to first construct a *G.F.D.* of the data and to then represent the information in the *G.F.D.* using what is called a **histogram**. Associated with any histogram is another graphical display called a **frequency polygon.**

We look first at how to construct a histogram and then follow this with a discussion of frequency polygons.

1. A **histogram** is a graphical display using **adjoining** rectangular boxes. The height of each box reflects the frequency of the data within that interval, and the width is the distance between the lower boundary of that interval and the lower boundary of the next interval above.

EXAMPLE 8.6

Refer back to Example 8.4 on page 92 where the customer numbers have been condensed into a *G.F.D.*

Display the information in this *G.F.D.* using a histogram.

Solution

Histogram of the data in the *G.F.D.* in EXAMPLE 8.4

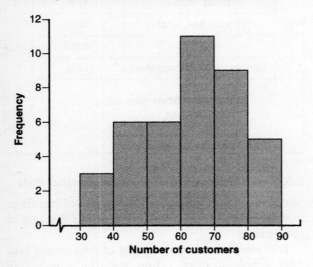

Note the following features:

(i) The frequency is shown on the 'vertical axis'. It always begins at zero and covers the range of all frequency values.

(ii) The 'horizontal axis' also always starts at zero. However, in this example the axis is broken ( ) since the lowest interval boundary is 30.

(iii) The histogram has a title.

(iv) The rectangles adjoin each other to show the continuity of the intervals.

(v) The lower interval boundaries are shown on the *x*-axis and indicate the end-points of each interval.

2. A **frequency polygon** is a straight line graph of a *G.F.D.* It is used to reinforce the continuous nature of the data and is superimposed on the histogram by joining the midpoints of the tops of the interval rectangles with a series of straight lines.

EXAMPLE 8.7

Draw the frequency polygon on to the histogram constructed in **example 8.6**.

Solution

Histogram and frequency polygon of customer numbers

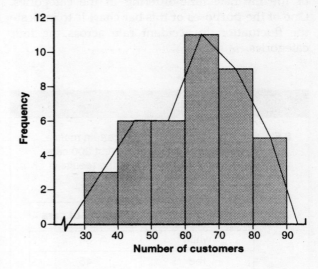

Note the following features:

(i) It is the midpoints (not the boundaries) of each interval that are joined.

(ii) Straight lines (not smooth curves) join the interval midpoints.

(iii) The endpoints are extended to meet the *x*-axis at the midpoints of the intervals below and above the first and last intervals respectively.

EXAMPLE 8.8

The diastolic blood pressure (measured in millimetres of mercury) obtained from a group of men during a routine medical check-up are shown below.

86	100	84	80	88	80	82	98	86	84
80	84	92	102	70	84	94	100	95	74
106	80	90	80	70	90	90	84	98	98
84	88	92	88	72	90	84	86	92	94
112	92	78	84	110	112	100	82	94	94

a) Summarise the above data by constructing a *G.F.D.* using intervals of width 10, (i.e. 70 – under 80, 80 – under 90, etc.).

b) Graph the information contained in the *G.F.D.* in **a)** above using

(i) a histogram

(ii) a frequency polygon.

Solution

(a) Grouped frequency distribution (*G.F.D.*) summarising the given blood pressure data obtained from 50 men.

Blood pressure	Frequency
70 – under 80	5
80 – under 90	21
90 – under 100	16
100 – under 110	5
110 – under 120	3
Total	**50**

(b) – (c) Histogram and frequency polygon

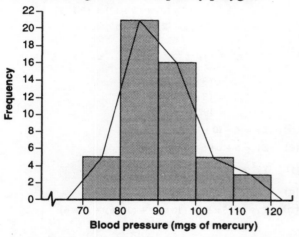

8.9   Refer back to exercise 8.2.

Graph the information contained in your *G.F.D.* using
(i)   a histogram,
(ii)   a frequency polygon.

8.10   The weights in kilograms of trout caught in a particular river are as follows:

1.3 0.2 0.1 2.7 3.0 0.7 1.0 0.5
0.6 1.2 2.1 0.9 1.7 2.2 0.4 1.3
0.4 0.1 1.3 0.8 2.3 1.1

(a)   Condense the data into a *G.F.D.* using intervals of width 0.5 (i.e. 0 – under 0.5, 0.5 – under 1, etc.).
(b)   To represent the data pictorially, construct
(i)   a histogram,
(ii)   a frequency polygon.

8.11   Using a histogram.

Consider the following histogram of patients' body temperature.

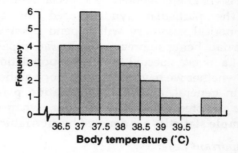

(a)   Fill in the missing interval boundaries. What interval width has been used?
(b)   How many patients were there altogether?
(c)   What were the lowest and the highest temperatures?
(d)   What was the most common temperature?
(e)   What does the gap between the two rectangles on the right tell you?
(f)   What proportion of patients have a temperature of 38°C or more?

YOUR TURN

8.8   The following data were collected by the research department of a particular insurance company. They give the 'life' in months of shop windows from the time of insurance to the time of the claim (i.e. when the window was broken).

Lifetime of shop windows

13	16	24	7	39	30	62	40	7
2	2	68	40	54	14	9	12	8
7	17	0	53	66	1	61	19	38
17	5	5	9	15	56	18	4	16

a)   Summarise the above data using a *G.F.D.* with intervals of width 10, (i.e. 0 – under 10, 10 – under 20, etc.).
b)   Graph the information contained in a) above, using
(i)   a histogram,
(ii)   a frequency polygon.

8.4   Summarising data – numerical measures

In *sections 3* and *4* we saw how tabular and pictorial displays may be constructed to summarise a sample of data. However, it is also necessary (especially if further analysis of the data is intended) to have concise *numerical* summaries.

In the statistical sense, most data sets (populations or samples) are effectively summarised by **two** numerical measures – namely,

(i) a measure of **central tendency** of the data (the subject of sub-section 8.4.1), and

(ii) a measure of **spread** or **variability** of the data (the subject of sub-section 8.4.2).

We have already discussed, in *section 1*, the difference between a **population** and a **sample**, and will now present **two** very important **definitions** that should be noted very carefully.

---

1. A **parameter** is a numerical summary used to summarise a **population** of data.

2. A **statistic** is a numerical summary used to summarise a **sample** of data.

You will see, as we progress, that the need for distinguishing between populations and samples becomes *very* important.

Each numerical summary calculated from a data set is generally denoted by a special symbol.

The particular symbol used for a given numerical summary will depend upon whether the data being summarised is a *whole population* or just a *sample* selected from some population – that is, whether we have a **parameter** or a **statistic.**

In general, almost all **population parameters** are denoted by a **Greek** letter, whereas most **sample statistics** are denoted by an **Arabic** letter.

Important note:

Many of the numerical summaries in statistics involve summing data or derived quantities. It will be assumed here that most readers are familiar with **summation notation** involving the Greek letter Σ, and can easily distinguish between and calculate quantities such as

$$\Sigma x, \ \Sigma x^2, \ (\Sigma x)^2, \ \Sigma(x-\bar{x})^2, \ \Sigma xy, \ \Sigma x \cdot \Sigma y, \ \text{etc.}$$

However, for those of you who feel the need to revise your knowledge of summation notation, the following brief review should be of assistance.

Summation notation

The Greek symbol, Σ (pronounced sigma), is mathematical shorthand and is used to indicate 'the sum of' or 'add up'.

If you wish to specify a particular range of x values to be summed, the lower and upper limits are shown at the bottom and top of the symbol respectively.

So,

$$\sum_{i=2}^{35} x_i$$ means the sum of x values starting with x_2 and finishing with x_{35}.

When all values are to be included in the sum, these limits are usually omitted. So, Σx means the sum of all x values.

The symbol may also be used to indicate the sum of a particular function of x values.

For example: $\sum_{j=2}^{6} (x_j^2 - 4)$ means the sum of (the square of each x value, minus 4) starting with x_2 and finishing with x_6.

EXAMPLE 8.9

Given: $x_1 = 5$, $x_2 = 10$, $x_3 = 1$, $x_4 = 7$, $x_5 = 0$, $x_6 = 5$

and $y_1 = 8$, $y_2 = 1$, $y_3 = 6$, $y_4 = 6$, $y_5 = 9$, $y_6 = 4$

(a) $\sum_{i=2}^{4} y_i = y_2 + y_3 + y_4 = 1 + 6 + 6 = 13$

(b) $\sum_{i=4}^{6} (x_i + 12) = (x_4 + 12) + (x_5 + 12) + (x_6 + 12)$

$$= (7 + 12) + (0 + 12) + (5 + 12)$$
$$= 19 + 12 + 17 = 48$$

(c) $\sum_{n=1}^{2} 3y_n{}^2 = 3y_1{}^2 + 3y_2{}^2$

$$= (3 \times 8^2) + (3 \times 1^2)$$
$$= (3 \times 64) + (3 \times 1)$$
$$= 192 + 3 = 195$$

(d) $\Sigma x = 5 + 10 + 1 + 7 + 0 + 5 = 28$

(e) $\Sigma y = 8 + 1 + 6 + 6 + 9 + 4 = 34$

(f) $\Sigma x^2 = 25 + 100 + 1 + 49 + 0 + 25 = 200$

(g) $(\Sigma x)^2 = 28^2 = 784$

(h) $\Sigma xy = (5 \times 8) + (10 \times 1) + (1 \times 6) + (7 \times 6)$

$$+ (0 \times 9) + (5 \times 4)$$
$$= 40 + 10 + 6 + 42 + 0 + 20 = 118$$

(i) $(\Sigma x)(\Sigma y) = 28 \times 34 = 952$

Note:

It is important to distinguish between **(f)** and **(g)** above, and also between **(h)** and **(i)**.

$$\sum x^2 \text{ is } not \text{ the same as } (\sum x)^2$$

and $\sum xy$ is *not* the same as $(\sum x)(\sum y)$.

Now that we have dealt with the preliminaries and definitions necessary for the successful calculation of numerical summaries, we can begin our discussion with the first of these – namely, *measures of central tendency.*

8.4.1 Measures of central tendency

Statisticians use 'measure of central tendency' to describe the general notion of a typical value. In principal, a measure of central tendency needs to be 'central' to the data in some sense, but as there is no perfect definition of a typical or central value, it is usual to look at several possible measures.

The most common measures of centre of a batch of data are:

1. the **mode**,
2. the **median**,
3. the **arithmetic mean**.

We will discuss the **mode** and the **median** by means of a definition only, but will concentrate on the measure that is generally used when further statistical analysis is required – namely, the **arithmetic mean.**

Important note:
You will find that, in most cases, statistical populations are far too large to warrant collecting all possible measurements. Therefore, although it is theoretically possible to do so, numerical summaries of populations (i.e. **parameters**) are seldom calculated.

So, although you will be expected to know the **symbols** denoting various population parameters, the only numerical summaries that you will be required to actually calculate will be those summarising a **sample** of data.

1. The mode

> The **mode** of a batch of data is the actual data value that occurs most frequently.

Since the most frequently occurring observation need not necessarily be located near the 'centre' of a batch, the mode is of limited use as a summary measure.

Another disadvantage of the mode is that it is not always well-defined.

For example:

(i) In some samples, all the data values occur with the **same** frequency. In this case we say that the sample has **no mode.**

(ii) Other situations arise when there are *two* data values occurring equally often and more frequently than the other values. Such a sample of data is said to be **bimodal** (i.e. having 2 modes).

EXAMPLE 8.10
Find the **mode** of the following samples of data:
(a) 18 19 18 20 18 18 20 21 37 18
(b) 27 25 29 36 14 24
(d) 19 18 20 35 18 18 19 46 20 19

Solution

(a) Mode = 18 (with frequency = 5).

(b) No mode (since all values have the same frequency of 1).

(c) The values of 18 and 19 both occur 3 times, so the sample is **bimodal**, with modes at 18 and 19.

2. The median

> The **median** of a batch of data is that value for which half the data are smaller and half are larger.

If the sample size, n, is **odd**, the median is the *middle* observation *after* the data have been ordered from smallest to largest.

If the sample size, n, is **even,** the median is the *average* of the *two* middle observations *after* the data have been ordered.

EXAMPLE 8.11
Find the **median** of the following samples of data:
(a) 7 3 2 8 11
(b) 18 19 18 20 18 18 20 21 37 18

Solution

(a) Ordered data: 2 3 7 8 11

Since $n = 5$ (odd), the median = 7, the middle measurement.

(b) Ordered data: 18 18 18 18 18 19 20 20 21 37

Since $n = 10$ (even), the median =

$\dfrac{18 + 19}{2} = 18.5$, the average of the two middle measurements.

3. The (arithmetic) mean

> The **mean** is the arithmetic average of all measurements in the batch, obtained by adding them all together and dividing this sum by the batch size.

Notation:
Let x refer to the measurement of interest, e.g. age, I.Q., height, income, etc.

Then, $\bar{x}$ denotes the **sample** mean

μ (Greek letter mu) denotes the **population** mean.

Since the method for calculating the mean depends upon the way the data are presented, we will look at
(i) data presented as a sample of **raw data**
(ii) sample data presented as a *G.F.D.*

(i) Measurements presented as a sample of raw data

The mean of a sample of raw data is defined to be

$$\bar{x} = \frac{\sum x}{n}$$

where n = sample size.

EXAMPLE 8.12

Find the **mean** of the following samples of data:

(a) 11 14 10 5

(b) 2 2 4 5 5 5 6 7 8 9 11

Solution

(a) By definition,

$$\bar{x} = \frac{\sum x}{n}$$

$$= \frac{11 + 14 + 10 + 5}{4}$$

$$= \frac{40}{4}$$

$$= 10$$

(b) By definition,

$$\bar{x} = \frac{\sum x}{n}$$

$$= \frac{64}{11}$$

$$= 5.818$$

(ii) Sample data presented as a G.F.D.

Because the individual values of the observations within a given interval of a G.F.D. are not known, it is usual to assume that each observation takes the value of the *midpoint* of that interval.

The midpoint of an interval is denoted by the letter *m*.

For a particular interval, if we multiply *m* by *f*, the frequency of values within that interval, we obtain the total of the measurements within that interval. Summing the interval totals will then give us the total of all measurements contained within the G.F.D., and $\bar{x}$ may be found in the usual manner, by dividing this sum by the total number of measurements, $n = \sum f$.

So, the mean of sample data presented as a G.F.D. may be found using the formula

$$\bar{x} = \frac{\sum fm}{\sum f} = \frac{\sum fm}{n}$$

where *m* = midpoint of each interval
  f = frequency of each interval
  fm = total of all measurements in each interval

$\sum f = n$ = sample size.

EXAMPLE 8.13

Find the mean of the following grouped frequency distribution:

Interval	Frequency
0 – 9	1
10 – 19	2
20 – 29	4
30 – 39	8
40 – 49	4
50 – 59	1

Solution

Step 1:

Add a column to the G.F.D. showing the midpoint of each interval. Remember, midpoint = $\frac{1}{2}$ (lower limit + upper limit).

Interval	Freq. (*f*)	Midpoint (*m*)
0 – 9	1	4.5
10 – 19	2	14.5
20 – 29	4	24.5
30 – 39	8	34.5
40 – 49	4	44.5
50 – 59	1	54.5

Step 2:

Add another column to the G.F.D., showing the product, *f* multiplied by *m* (i.e. *fm*), for each interval. Sum the values in the frequency, *f*, column and in the *fm* column.

Interval	*f*	*m*	*fm*
0 – 9	1	4.5	1 x 4.5 = 4.5
10 – 19	2	14.5	2 x 14.5 = 29.0
20 – 29	4	24.5	4 x 24.5 = 98.0
30 – 39	8	34.5	276.0
40 – 49	4	44.5	178.0
50 – 59	1	54.5	54.5
	$\sum f = n = 20$		$\sum fm = 640.0$

Step 3:
Calculate the mean of the *G.F.D.* by substituting the required values into the appropriate formula. So,

$$\bar{x} = \frac{\sum fm}{\sum f}$$

$$= \frac{640}{20}$$

$$= 32$$

Therefore, the mean of the given *G.F.D.* is $\bar{x} = 32$.

YOUR TURN

8.12 Find the mode(s) of the following samples of raw data:
  (a) 6 2 6 8 2 6 9 0
  (b) 15 25 32 9 14
  (c) 24 32 16 32 12 16 18 25

8.13 Find the median of the following samples of raw data:
  (a) 35 28 23 34 30
  (b) 11 3 7 15 9 17 5 13
  (c) 2.4 6.5 8.2 1.0 6.5 9.7

8.14 If someone claimed that the median of the data

      1 3 15 4 8

  is 15, you should immediately recognise that a blunder has been made. Explain.

8.15 Find the mean of the following samples of raw data:
  (a) 15 12 0 3 5 6 30
  (b) 34.3 52.8 38.3 39.9 42.5
      63.4 50.2 45.0

8.16 If someone claims that the mean of the data

      6 2 9 4 3

  is 10, you should immediately recognise that a blunder has been made. Explain.

8.17 Find the mean of each of the following grouped frequency distributions:
  (a)

Interval	f	m	fm
0 – 4	2	2	4
5 – 9	7		
10 – 14	15		
15 – 19	9		
20 – 24	7		

(b)

Interval	f
0 – 9	2
10 – 19	2
20 – 29	3
30 – 39	4
40 – 49	2
50 – 59	1

(c)

Interval	f
10 – 11	2
12 – 13	0
14 – 15	3
16 – 17	4
18 – 19	1

8.4.2 Measures of variability

In 8.4.1 we discussed the use of a **single** number, namely a measure of *central tendency* to condense and describe a batch of data. Clearly, however, it is possible to go too far trying to simplify masses of information, and vital aspects may be lost if only a single summary item is used.

This situation is very clearly demonstrated by the following fireside story, often told on cold winter nights.

> A Chinese warlord, off to do battle against his hated enemy, sets out one fine morning with his 600 able-bodied soldiers. Several hours later he comes to a river that must be crossed. But, being in a hurry to meet his enemy, the old warlord does not wish to spend the extra time travelling up-river to the bridge crossing. However, while standing on the bank scratching his head, knowing that most of his men cannot swim, he suddenly remembers having read somewhere that, at this time of year, all parts of river are only 1 metre deep **on the average.** So, he immediately orders all his soldiers to hold their weapons high and to wade across the river as quickly as possible.

Sometime later, the old warlord looks back at his troops and is amazed to see that only about a quarter of his men are following. He quickly enquires as to the reason for this and learns that $\frac{3}{4}$ of his men were drowned while crossing the river!

No one had mentioned that, although the river was only 1 metre deep **on the average**; in some places it was 3 centimetres while in others it was 2 metres.

(Needless to say the Chinese warlord lost his battle against his hated enemy!).

From the above story it is clear that we need some additional measure to supplement our measure of centre – namely, a measure of the **internal variation** of the data.

As we saw with measures of centre, there are several quantities that may be used as measures of variability.

However, we will limit ourselves here to a full discussion of the most common measure of variability (especially when further statistical analysis is required) – namely, the **standard deviation**, although we will mention just briefly, at the outset, another very simple measure of variability called the **range**.

The **range** of a sample of data is defined to be the difference between the largest and the smallest value in the sample.

Although widely used in summarising data made available to the general public, the range is rarely used in statistical analysis because it is too easily influenced by extreme values.

Statisticians prefer a measure of variability that involves all the items in a data set if possible. Because the range is seldom used in statistical analysis, it has no special symbol.

EXAMPLE 8.14
Consider the following sample of data:

18 19 18 20 18 37 21 20 18 18

The **range** is (37 − 18) = 19, suggesting a fairly wide spread or high variability in the data. However, if the value 37 is omitted from the sample, the range is reduced to 3, suggesting a very small variability in the data. Clearly, we need a measure of variability that is not unduly influenced by a few extreme values.

Let us turn now to our discussion of the **standard deviation**.

As the name implies, the standard deviation is a single value intended to measure the average or typical deviation of the data values from their mean. It is defined as follows:

The **standard deviation** of a sample of data is a single measure of variability obtained by summing the **squares** of the deviations from the mean, $(x - \bar{x})^2$, dividing this sum by $(n-1)$, where n is the sample size, and finally taking the square root of the result.

Notes:

(i) The reason for using the **squared** deviations from the mean when calculating the standard deviation is that summing the actual deviations, $(x - \bar{x})$, will *always* result in **zero**, i.e. $\sum (x - \bar{x}) = 0$ *always*, due to the cancellation effect of positive and negative deviations.

(ii) The reason for dividing by $(n-1)$ rather than n is too complicated to explain here – you should just accept that it is necessary for mathematical reasons.

(iii) The reason for taking the square root at the last step is to counteract the effect of squaring the residuals (deviations).

Notation:
Let x refer to the measurements of interest, e.g. age, I.Q., height, weight, etc.

Then: s (small s) denotes the **sample** standard deviation

σ (Greek letter lower case sigma) denotes the **population** standard deviation

As with the mean, the calculation of the standard deviation depends upon the way the data are presented. So again, we will look at:

(i) data presented as a sample of **raw data**

(ii) sample data presented as a *G.F.D.*

(i) Calculation of the sample standard deviation, s, for a sample of raw data
Following the definition given above, the **theoretical** formula for the standard deviation of a sample of raw data is

$$s = \sqrt{\frac{\sum (x - \bar{x})^2}{n-1}}$$

where $\bar{x}$ = sample mean

$(x - \bar{x})^2$ = the square of the deviation, $(x - \bar{x})$

n = sample size.

However, the above formula for s, although theoretically accurate according to the definition, is rather difficult and often tedious to use in practice, especially when $\bar{x}$ is a decimal number.

Fortunately, however, using basic algebra, it can be shown that the above **theoretical** formula for s reduces to

$$s = \sqrt{\frac{\sum x^2 - n\bar{x}^2}{n-1}}.$$

The formula above should be considered as the **computational** formula as it allows s to be calculated more easily and efficiently.

EXAMPLE 8.15

Find the standard deviation of the following sample data:

2 3 5 5 9

Solution

Let x: 2   3   5   5   9

Then x^2:   4   9   25   25   81

Now, $\bar{x} = \dfrac{\sum x}{n} = \dfrac{24}{5} = 4.8$,

and $\sum x^2 = 4 + 9 + 25 + 25 + 81 = 144$.

So,

$$s = \sqrt{\frac{\sum x^2 - n\bar{x}^2}{n-1}}$$

$$= \sqrt{\frac{144 - (5 \times 4.8^2)}{4}}$$

$$= \sqrt{\frac{144 - 115.2}{4}}$$

$$= \sqrt{7.2}$$

$$= 2.6832816$$

Therefore, the standard deviation of the given sample of raw data is 2.68.

Note:
A logical interpretation of this figure would be to say that '**on the average**' the data in the sample lie a distance 2.68 from the mean.

(ii) Calculation of s for a sample of data presented as a G.F.D.

As we saw when finding the mean of a sample of grouped data the individual values within a given interval of the G.F.D. are not known. So again we will assume that, in order to find the deviations from the mean, each observation takes the value of the midpoint, m, of that interval.

Hence, following the definition for the standard deviation, we would first subtract the mean, $\bar{x}$, from the midpoint of each interval, square the result and then multiply this squared deviation by the frequency of observations in that interval.

Finally, we would sum these results, divide the sum by $(n-1)$, where $n = \sum f$, and then take the square root.

The above procedure is summarised by the following **theoretical** formula:

$$s = \sqrt{\frac{\sum f(m - \bar{x})^2}{(\sum f) - 1}} = \sqrt{\frac{\sum f(m - \bar{x})^2}{n-1}}$$

where m = midpoint of each interval

f = frequency of each interval

$\bar{x} = \dfrac{\sum fm}{\sum f}$ = mean of sample data

$f(m - \bar{x})^2$ = total of all squared deviations,

$(m - \bar{x})^2$, in each interval

$\sum f = n$ = sample size.

However, as we found when calculating the standard deviation of a sample of raw data, the above **theoretical** formula is *not* an efficient method for determining s.

Nevertheless, using basic algebra, it can be shown once again that the above **theoretical** formula reduces to the following **computational** formula:

$$s = \sqrt{\frac{\sum fm^2 - n\bar{x}^2}{n-1}}$$

where $\bar{x} = \dfrac{\sum fm}{\sum f}$

$n = \sum f$.

Note:

There is a big difference between $\sum fm^2$, $\sum (fm)^2$ and $(\sum fm)^2$ – so beware that you do **not** fall into the trap!

fm^2 means square m, then multiply by f.

$(fm)^2$ means multiply m by f, then square the result.

EXAMPLE 8.16

Find the standard deviation of the following sample of grouped data.

Cost ($)	f
10 – 14	4
15 – 19	7
20 – 24	5
25 – 29	3
30 – 34	1

Solution

Step 1:

Make a table with 5 columns headed: interval, m (midpoint), f (frequency), fm, fm^2.

Interval	m	f	fm	fm^2
10 – 14	12	4	12 x 4 = 48	4 x 144 = 576
15 – 19	17	7	17 x 7 = 119	7 x 289 = 2023
20 – 24	22	5	110	2420
25 – 29	27	3	81	2187
30 – 34	32	1	32	1024

Step 2:

Find the sum of the values in the 'fm' column and in the 'f' column and hence find the mean $\bar{x}$. Also sum the fm^2 column.

f	fm	fm^2
4	48	576
7	119	2023
5	110	2420
3	81	2187
1	32	1024
$\sum f = 20$	$\sum fm = 390$	$\sum fm^2 = 8230$

So, $\bar{x} = \dfrac{\sum fm}{\sum f} = \dfrac{390}{20} = 19.5$.

Step 3:

Substitute the values from Step 2 into the computational formula below and hence calculate s.

$s = \sqrt{\dfrac{\sum fm^2 - n\bar{x}^2}{n-1}}$, where $n = \sum f$

$= \sqrt{\dfrac{8230 - (20 \times 19.5^2)}{19}}$

$= \sqrt{\dfrac{8230 - 7605}{19}}$

$= \sqrt{32.894737}$

$= 5.7353933$

Hence, the standard deviation of the given sample of grouped data is $s = 5.735$.

Comments:

(i) The **variance** of a batch of data is defined to be the square of the standard deviation.

Thus, s^2 = the variance of the **sample** data.

σ^2 = the **population** variance.

Note:

Because the variance is in square units it is sometimes difficult to interpret. *For example*, the variance of a sample of times taken to complete a 100-metre swim would be in 'square seconds'!!

(ii) If $n = 1$, then clearly $s = 0$!

If $n > 1$, then $s = 0$ only if *all* observations are of *equal* value.

(iii) If a constant k is *added to* (or *subtracted from*) each observation in a sample of data, then the standard deviation of the new sample has the **same value** as the standard deviation of the original data. **Why?** (Experiment with a simple set of data).

This is **not** true for the mean. The mean of the new sample will be the value of the mean of the original sample plus (or minus) k.

So, if $y = x + 4$ for all x and y, then

a) $\bar{y} = \bar{x} + 4$, and

b) $s_y = s_x$

(iv) If each observation in a sample of data is *multiplied* by a constant, k, both the mean and the standard deviation of the new sample of data will be k times the value of the mean and standard deviation in the original sample.

Similarly, if *division* by a constant, k, occurs, then the new mean and standard deviation will be the value of the original mean and standard deviation divided by k.

So,

a) if $y = 4x$ for all x and y, then

$$\bar{y} = 4\bar{x}, \text{ and}$$

$$s_y = 4s_x$$

b) if $y = \dfrac{x}{2} - 3$ for all x and y, then

$$\bar{y} = \dfrac{\bar{x}}{2} - 3, \text{ and}$$

$$s_y = \dfrac{s_x}{2}$$

Reasonableness checks:

It is very important to *always* check that any value you calculate is at least a reasonable one (keeping in mind that it is very easy to make a 'silly' error that can have disastrous results).

Here are some quick checks for both the mean and the standard deviation of a sample of data.

(i) The **mean** of a sample of data must have a value somewhere between the largest data value and the smallest data value. **Why?**

(ii) The **standard deviation**, s, can **never** be negative, since it is defined as the positive square root of a sum of squares.

(iii) The **standard deviation** can **never** exceed the **range** of the sample data. In fact, you will generally find that s is somewhere between $\dfrac{1}{3}$ and $\dfrac{1}{4}$ of the range.

8.18 Calculate
 (i) the range
 (ii) the variance
 (iii) the standard deviation

 of the following samples of raw data. Be sure to check that your answer to (iii) is 'reasonable'.
 a) 15 12 0 3 5 10 6 9
 b) 34.3 52.8 38.3 39.9 42.5 63.4 50.2 45.0

8.19 Find
 (i) the variance
 (ii) the standard deviation

 of the samples of grouped data represented by the following *GFD's*:

(a)

Interval	f
0 – 4	2
5 – 9	7
10 – 14	15
15 – 19	9
20 – 24	7

(b)

Interval	f
0 – 9	2
10 – 19	2
20 – 29	3
30 – 39	4
40 – 49	2
50 – 59	1

8.20 A certain sample of data has a mean of 6 and a standard deviation of 1.9. If all the sample values are now multiplied by 5 and then have 2 added, write down the value of the mean and standard deviation of the new sample.

8.4.3 Using your calculator to determine $\bar{x}$ and s

Most scientific calculators include a facility for calculating various statistical quantities such as $\bar{x}$, s, $\sum x$, $\sum x^2$, n etc.

In order to utilise these features, it is first necessary to put your calculator into standard deviation mode before entering the given data. Then you will need to refer to your instruction booklet to determine how to access the statistical functions and quantities you require.

(i) **Determining $\bar{x}$ and s of raw data using your calculator**

Step 1: Make sure that the calculator is in standard deviation mode by pressing, on our Casio *fx-82super*,

[mode] [.]   DEG   0. SD

Step 2: Enter the data values after clearing all memories.

Here, data [M+] data [M+] etc.

Step 3: Check that all data have been entered by pressing the sample size button.

$\boxed{n}$ $\boxed{\text{SAMPLE SIZE}}$

Step 4: Find the mean, $\bar{x}$.

$\boxed{\bar{x}}$ $\boxed{\text{MEAN X}}$

Step 5: Find the standard deviation, s.

$\boxed{\sigma_{n-1}}$ $\boxed{\text{ST. DEV..S}}$

Note:
There should be two standard deviation buttons on your calculator, *for example,*

$$\boxed{\sigma_n} \text{ and } \boxed{\sigma_{n-1}} \text{ or } \boxed{\sigma} \text{ and } \boxed{s}$$

The button $\boxed{\sigma_{n-1}}$ (or $\boxed{s}$) calculates the **sample** standard deviation, s, while the button $\boxed{\sigma_n}$ calculates the **population** standard deviation, σ.
Be *very* careful to use the correct button in a given situation.

EXAMPLE 8.17
Refer back to example 8.15.
Use the statistical buttons on your calculator to find the mean and standard deviation of the sample data:

2 3 5 5 9

Solution

Step 1: Standard deviation mode.
Press $\boxed{\text{mode}}$ $\boxed{.}$

Step 2: Enter data, after clearing memories.
Here, press 2 $\boxed{\text{M+}}$ 3 $\boxed{\text{M+}}$ 5 $\boxed{\text{M+}}$ 5 $\boxed{\text{M+}}$ 9 $\boxed{\text{M+}}$

Step 3: Check sample size.
$\boxed{n}$ $\boxed{5}$ As expected $n = 5$.

Step 4: Find the mean, $\bar{x}$.
$\boxed{\bar{x}}$ $\boxed{4.8}$

Step 5: Find the standard deviation, s.
$\boxed{\sigma_{n-1}}$ $\boxed{2.6832816}$

Therefore, using the calculator we find that

$\bar{x} = 4.8$ and $s = 2.6832816$

are the mean and standard deviation of the given sample data.

Note: As expected, this is the same result as that obtained in example 8.15.

(ii) Determining $\bar{x}$ and s of a *G.F.D.* using your calculator

Note:
Not all calculators have a facility for calculating the mean and standard deviation of a *G.F.D.*
We will go through the steps using an example.

EXAMPLE 8.18
Find the mean and standard deviation of the sample data summarised in the *G.F.D.* in example 8.16.

Cost ($)	f
10 – 14	4
15 – 19	7
20 – 24	5
25 – 29	3
30 – 34	1

Solution

Step 1: Standard deviation mode.
Here, press $\boxed{\text{mode}}$ $\boxed{.}$

Step 2: Clear all memories.
Here, $\boxed{\text{shift}}$ $\boxed{\text{AC}}$

Step 3: Enter data: midpoint times frequency, input.

12 $\boxed{\times}$ 4 $\boxed{\text{M+}}$ 17 $\boxed{\times}$ 7 $\boxed{\text{M+}}$ 22 $\boxed{\times}$ 5 $\boxed{\text{M+}}$
27 $\boxed{\times}$ 3 $\boxed{\text{M+}}$ 32 $\boxed{\times}$ 1 $\boxed{\text{M+}}$

Step 4: Check sample size.
$\boxed{n}$ $\boxed{20}$

Step 5: Find the mean, $\bar{x}$.
$\boxed{\bar{x}}$ $\boxed{19.5^{50}}$

Step 6: Find the standard deviation, s.
$\boxed{\sigma_{n-1}}$ $\boxed{5.7353933}$

Therefore, using the calculator we find that

$\bar{x} = 19.5$ and $s = 5.7353933$

are the mean and standard deviation of the data in the given *G.F.D.*

Note: As expected, this is once again the same result as that obtained in example 8.18.

YOUR TURN

8.21 Refer back to example 8.18.

Now find the mean and standard deviation of the given sample data using the statistical buttons on your calculator. Compare your answers.

a) 15 12 0 3 5 10 6 9

b) 34.3 52.8 38.3 39.9 42.5 63.4 50.2 45.0

8.22 Refer back to example 8.19.

Now find the mean and standard deviation of the data in the *G.F.D.'s* given in **(a)** and **(b)**. Again compare your answers.

(a)

Interval	f
0 – 4	2
5 – 9	7
10 – 14	15
15 – 19	9
20 – 24	7

(b)

Interval	f
0 – 9	2
10 – 19	2
20 – 29	3
30 – 39	4
40 – 49	2
50 – 59	1

8.5 Sets and events

> A **set** is the term given to any well-defined collection of distinct objects, each of which is said to be a member or **element** of that set.

In the above definition, 'well-defined' means that any objects may be easily classified as either a member or non-member of a particular set.

Notation: In general,

(i) **sets** are denoted by capital letters,

(ii) **elements** are denoted by lower case letters,

(iii) curly brackets, {…}, are used to enclose the contents of a given set.

The contents of any set may be described by

(i) listing the elements, **or**

(ii) stating the rules or conditions that determine the members.

EXAMPLE 8.19

Let sets A, B, and C be defined as follows:

(i) Set A consists of the numbers 6, 12, 13, 24, 50.

(ii) Set B consists of the prime numbers between 1 and 15 inclusive.

(iii) Set C consists of all the vowels in the English alphabet.

Re-define sets A, B and C using notation conventional to set theory.

Solution

(i) The contents of set A could be written

$$A = \{6, 12, 13, 24, 50\}.$$

(ii) The contents of set B could be written

$$B = \{b : b \text{ is a prime number between 0 and 15}\}.$$

(iii) The contents of set C could be written

$$C = \{c : c \text{ is a vowel in the English alphabet}\}. \text{ i.e. } C = \{a, e, i, o, u\}.$$

8.5.1 Types of sets

(i) Subsets

> If *every* element of a given set A also belongs to a certain set B, then A is said to be a **subset** of B. This is denoted $A \subset B$.

EXAMPLE 8.20

Suppose a single die is tossed once. Let

(i) set B be all possible numbers on the uppermost face, and

(ii) set A be only the *odd* numbers appearing on the uppermost face.

Using set notation, we could write the above as follows:

$$\text{Let } B = \{1, 2, 3, 4, 5, 6\}$$

$$\text{and } A = \{1, 3, 5\}.$$

Clearly, $A \subset B$ since all the elements in A are included in B.

(ii) The empty set and the universal set

> Any set that contains **no** elements at all is called the **empty** or **null** set and is denoted by the symbol $\varnothing$.

Note: The empty set, $\varnothing$, is itself a subset of any set.

> In some situations it is useful to consider an underlying reference set that consists of all possible relevant elements and of which all sets under discussion are subsets. This set is usually referred to as the **universal** set and is denoted U.

Note: Clearly, the universal set will change as our underlying frame of reference changes.

EXAMPLE 8.21

(a) Let set A consist of all the prime numbers between 24 and 28, inclusive.

Using set notation, this may be written as $A = \{x : 24 \le x \le 28 \text{ and } x \text{ is prime}\}$.

Then clearly, $A = \varnothing$ since there are no prime numbers between 24 and 28.

(b) Suppose that we are interested only in the possible outcomes from the toss of a single die.

Then the set $\{1, 2, 3, 4, 5, 6\}$ could be thought of as the universal set, U.

(iii) Mutually exclusive or disjoint sets

> If two sets A and B have **no** elements in common they are said to be **mutually exclusive** or **disjoint** sets.

EXAMPLE 8.22

Suppose that A is the set of *even* numbers from 6 to 16 and B is the set of *prime* numbers from 6 to 16, then A and B are mutually exclusive or disjoint sets. **Why?**

$$A = \{6, 8, 10, 12, 14, 16\}$$
$$B = \{7, 11, 13\}$$

Since A and B have no elements in common, they are said to be mutually exclusive sets.

8.5.2 Venn diagrams

> **Venn diagrams**, named after the English logician John Venn (1934 – 83), are simple methods of representing sets and set operations pictorially.

Note:
A rectangle is used to represent the universal set, U, and the subsets of U are shown as circles in appropriate regions within the rectangle.

EXAMPLE 8.23

(a) Let U = $\{x : x$ is a number between 1 and 20, inclusive$\}$

A = $\{a : a$ is an even number between 4 and 12, inclusive$\}$

B = $\{b : b$ is an odd number between 10 and 16, inclusive$\}$

Diagrammatically, the above sets may be represented as

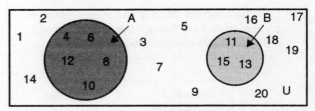

(b) Suppose that U = $\{a, e, i, o, u\}$

with A = $\{a, e, i\}$

and B = $\{a\}$,

then diagrammatically, we have

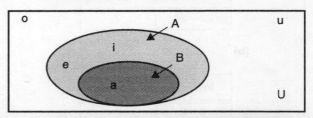

YOUR TURN

8.23 **(a)** Let A be the set of all real numbers, x, such that $x^2 = -4$.
That is, let A = $\{x : x^2 = -4\}$.
Write down all the possible elements of A.

(b) Let B be the set of all real numbers between 20 and 30 inclusive.
Let C be the set of odd numbers between 20 and 30 inclusive.
List the elements of B and C and hence show that C is a subset of B.

8.24 Let U be the set of all real numbers between 0 and 20.
Let A be the set of odd numbers between 0 and 20.
Let B be the set of all prime numbers between 0 and 20.
Show the relationship between A, B and U using a Venn diagram.

8.25 Let U be the set of all letters in the English language alphabet.

Let A be the set of vowels, B be the letters d to j and C be the letters n to v.

Show the relationship between A, B, C and U using a Venn diagram.

8.5.3 Set operations

1. Union

> The **union** of two sets A and B is defined to be the set of all elements that belong to A *or* B *or* both. This union is denoted $A \cup B$.

In the following Venn diagram, the shaded region reflects $A \cup B$ (i.e. the union of A and B).

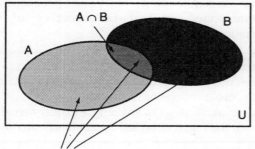

Here $(A \cup B)$ = elements of A or B or both.

2. Intersection

> The **intersection** of two sets A and B is defined to be the set of all elements that belong to *both* A *and* B (i.e. common to both sets). This intersection is denoted $A \cap B$.

In the following Venn diagram the shaded region represents $A \cap B$ (i.e. the intersection of A and B).

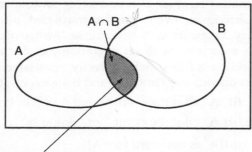

Here $(A \cap B)$ = elements common to both A and B

Note:
From the diagram above it should be clear that if A and B are mutually exclusive or disjoint sets then $A \cap B = \varnothing$ (the empty set).

Why?
If A and B are mutually exclusive they have **no** elements in common and thus do not overlap. Therefore, since $A \cap B$ contains no elements it is, by definition, an empty set.

3. Complement

> If A is a subset of some universal set U, then the **complement** of A is defined to be the set of all elements in U that are *not* in A. In this text, the complement is denoted $A^{|}$, (alt. notation, $\overline{A}$).

In the following Venn diagram the shaded region represents $A^{|}$ (i.e. the complement of A).

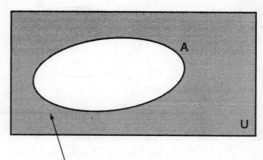

Here $A^{|}$ = all elements *not* in A.

4. Difference

> The **difference** between two sets A and B is defined to be the set of all elements in A but *not* in B. This difference is denoted $A - B$.

In the following Venn diagram the shaded region represents $A - B$ (i.e. the difference between A and B).

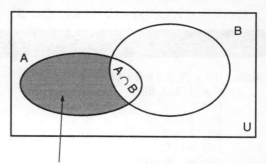

Here $(A - B)$ = all elements in A but *not* in B.

Note:

(i) $B - A$ is the set of all elements in B but *not* in A. Clearly $A - B$ and $B - A$ are quite different and are, in fact, mutually exclusive sets.

(ii) If A and B are themselves mutually exclusive, then $A - B = A$ and $B - A = B$. (Check these results on a Venn diagram if you are not convinced.)

EXAMPLE 8.24

Let $U = \{8, 9, 10, 11, 12, 13, 14, 15, 16, 17, 18\}$
$A = \{a : a \text{ is even}\} = \{8, 10, 12, 14, 16, 18\}$
$B = \{b : b \text{ is divisible by } 3\} = \{9, 12, 15, 18\}$
$C = \{c : c \text{ is prime}\} = \{11, 13, 17\}$

(a) Draw a Venn diagram to represent the relationship among the sets given.

(b) List the elements in the following sets:
$A \cup B, A \cup C, A \cap B, A \cap C, A \cup B \cup C,$
$A - B, A^{\mathrm{I}}, B \cap A^{\mathrm{I}}, B^{\mathrm{I}}, C \cap B^{\mathrm{I}}$

Solution

(a)

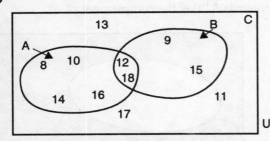

(b) $A \cup B = \{8, 9, 10, 12, 14, 15, 16, 18\}$
$A \cup C = \{8, 10, 11, 12, 13, 14, 16, 17, 18\}$
$A \cap B = \{12, 18\}$
$A \cap C = \varnothing$. Hence, A and C are mutually exclusive (as are B and C).
$A \cup B \cup C = \{8, 9, 10, 11, 12, 13, 14, 15, 16, 17, 18\} = U$
$A - B = \{8, 10, 14, 16\}$
$A^{\mathrm{I}} = \{9, 11, 13, 15, 17\}$
$B \cap A^{\mathrm{I}} = \{9, 15\}$
$B^{\mathrm{I}} = \{8, 10, 11, 13, 14, 16, 17\}$
$C \cap B^{\mathrm{I}} = \{11, 13, 17\} = C$. Hence, $C \subset B^{\mathrm{I}}$.

8.6 Probability

Because statisticians are concerned with the collection and analysis of data leading to the drawing of inferences about unknown situations, some understanding of basic probability is recommended so that the accuracy of such inferences may be explored.

> In statistics, the word **experiment** is used to describe any process that generates raw data (i.e. observations or numerical measurements). On any single experimental trial, the result obtained is called an **outcome.**

For example: On a single coin toss (experiment) there are two possible outcomes – namely, a head or a tail.

> Given any statistical experiment, the set of *all* possible outcomes is called the **sample space** and is denoted U (i.e. the *universal set*). We can now define an **event** as any subset of the sample space.

Note:

An *event*, in the probabilistic context, is just an alternative term for a *set* and hence any statement concerning an event can be translated into set theory language and vice versa. Moreover, the rules associated with set operations, as discussed in section 8.5.2, can be directly translated into event operations. Thus, if A and B are events, then

(i) $A \cup B$ is the event {either A or B or both}.
(ii) $A \cap B$ is the event {both A and B}.
(iii) A^{I} is the event {not A}.
(iv) $A - B$ is the event {A but not B}.
(v) If $A \cap B = \varnothing$, then A and B are disjoint or mutually exclusive events.

Now let us define what we mean by the probability or likelihood of occurrence of an event

A, denoted $P(A)$, resulting from a statistical experiment.

EXAMPLE 8.25

If a single card is drawn from a standard pack, find the probability that it is a spade.

Solution

If A is the event 'a spade is drawn',
then $P(A)$ = Prob. {a spade is drawn}.
From the definition, we know that

$$P(A) = \frac{n}{N}.$$

Here, N = no. of possible outcomes (equally likely), and

n = no. of outcomes constituting A
   = 13 (since there are 13 spades in a standard pack).

Therefore, the probability that a spade is drawn may be written

$$P(A) = \frac{13}{52} = \frac{1}{4}.$$

EXAMPLE 8.26

A bag contains 7 red balls, 5 blue balls and 4 green balls. Find the probability that if one ball only is drawn at random it will be green.

Solution

Let A be the event 'a green ball is drawn'.

Then $P(A) = \frac{4}{16} = \frac{1}{4}.$ **Why?**

Because, N = total possible outcomes
      = 16 balls of any colour,

n = total outcomes comprising event A
   = 4 green balls

Therefore, $P(A) = P(\text{Green}) = \frac{4}{16} = \frac{1}{4}.$

8.29 Consider an experiment where a single ticket is drawn from a box containing 10 tickets numbered 1 to 10.
  (a) List the sample space, U, for this experiment.
  (b) List the subsets representing the following events:
    (i) A: a number greater than 5
    (ii) B: an even number
    (iii) B^{I}
    (iv) $A^{\text{I}} \cap B$
    (v) $A \cup B^{\text{I}}$

8.30 Three girls, Joy, Pam and Sandra, are to race against each other in a 100 metre dash. One of them comes first, one second and one third (no ties):
  (a) List the sample space of all possible placings.
  (b) List the subsets corresponding to the following events:
    (i) Joy wins
    (ii) Sandra comes last
    (ii) Pam does not win
    (iv) Joy does not come last.

8.31 Once again, consider a single draw from a standard pack of cards. Find the probability that the card drawn is
  (a) a king
  (b) a heart
  (c) the ace of clubs
  (d) a black card
  (e) not the queen of spades.

8.32 A number is selected at random from the numbers 1 to 15 inclusive. Find the probability that the number selected is
  (a) less than 10
  (b) a prime number
  (c) divisible evenly by 5
  (d) a perfect square
  (e) not divisible evenly by 2.

8.33 A letter is chosen at random from the word **UNIVERSITY**. Find the probability that the letter is
  (a) a T
  (b) a vowel
  (c) an I
  (d) a consonant.

8.6.1 Axioms and rules of probability

(a) Axioms

The following axioms arise naturally from the definition of probability and also from the knowledge that the occurrence of any event A is either impossible, certain or somewhere in between.

> **Axiom 1**
> $0 \le P(A) \le 1$, for any event A.

In words, this axiom states the probability that event A occurs is some number between 0 and 1 (inclusive). By convention, it is usually expressed as a decimal.

> **Axiom 2**
> $P(A) = 0$, whenever $A = \varnothing$.

In words, this axiom states that whenever the occurrence of event A is impossible, then A is an empty set implying that $P(A) = 0$.

> **Axiom 3**
> $P(A) = 1$, whenever $A = U$.

In words, this axiom states that whenever event A is certain to occur, then A is the universal set implying that $P(A) = 1$.

(b) Rules

The following reasonably simple rules of probability are consequences of both the set theory discussed in the previous section and the axioms of probability given above.

(i)
> If $A \subset B$, then $P(A) \le P(B)$

In words, this rule states that if event A is a subset of event B, then the probability that A occurs must be less than or equal to the probability that B occurs.

Note: The equality is true only when $A = B$.

EXAMPLE 8.27
Consider a single draw from a standard pack of cards. Let events A and B be defined as follows:

    A: the 6 of clubs is drawn.

    B: any club is drawn.

Clearly, $A \subset B$ since A is one member of B.

Furthermore, since $P(B) = \dfrac{13}{52} = \dfrac{1}{4}$ and $P(A) = \dfrac{1}{52}$, we see that $P(A) < P(B)$.

Note:

$P(B) = \dfrac{13}{52}$, since 13 out of the 52 cards in the standard pack are clubs. Similarly, $P(A) = \dfrac{1}{52}$ since there is only *one* six of clubs in a standard pack, i.e. 1 out of 52.

(ii)
> $P(A^{|}) = 1 - P(A)$

In words, the above rule states that if A is any event in the sample space U, then the probability that the complement of A occurs (i.e. P (not A)) is just one minus the probability that event A occurs.

Note: The above result is true because $A + A^{|} = U$ and $P(U) = 1$.

EXAMPLE 8.28
Again consider a single draw from a standard pack of cards. Let event A be defined as 'a heart is drawn'. Then it follows that event $A^{|}$ must be 'a heart is **not** drawn'.

It should now be clear that

$$P(A^{|}) = P(\text{not a heart})$$
$$= P(\text{a club or a spade or a diamond})$$
$$= \frac{39}{52} = \frac{3}{4}$$

Hence, since
$$P(A) = P(\text{heart})$$
$$= \frac{13}{52} = \frac{1}{4}, \text{ we have that}$$
$$P(A^{|}) = 1 - P(A).$$

(iii) The additive rule of probability
This rule applies to the *union* of two events A and B, and states that

> $$P(A \cup B) = P(A) + P(B) - P(A \cap B)$$

In words, this rule states that the probability of the occurrence of either of events A or B (or both) is equal to the sum of their individual probabilities minus the probability of their joint occurrence (to eliminate the double counting of the overlap region).

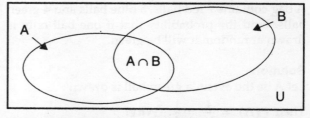

Note: Although $P(A)$ and $P(B)$ *both* include $(A \cap B)$, when determining $P(A \cup B)$, the intersection $(A \cap B)$ is allowed to contribute only once.

EXAMPLE 8.29

Suppose that there are two events, S and T, such that $P(S) = 0.4$, $P(T) = 0.5$ and $P(S \cap T) = 0.15$. Find $P(S \cup T)$ using the *additive rule*.

Solution

From the rule, we know that

$$P(S \cup T) = P(S) + P(T) - P(S \cap T)$$
$$= 0.4 + 0.5 - 0.15$$
$$= 0.75$$

Important:

If two events A and B are *mutually exclusive* (i.e. have no elements in common), then

$$P(A \cup B) = P(A) + P(B).$$

Why?

If two events are mutually exclusive they cannot occur together. Hence, their intersection $A \cap B$ is $\varnothing$ (the empty set) and clearly $P(A \cap B) = 0$. Therefore, the true additive rule

$$P(A \cup B) = P(A) + P(B) - P(A \cap B)$$

reduces to

$$P(A \cup B) = P(A) + P(B)$$

whenever events A and B are *mutually exclusive*.

Note:

The above observation may be readily extended to cover several events.

Hence, if A_1, A_2, A_3, ..., A_n are mutually exclusive events such that $A_i \cap A_j = \varnothing$ for any two events, then

$$P(A_1 \cup A_2 \cup A_3 \cup ... \cup A_n) = P(A_1) + ... + P(A_n)$$

$$= \sum_{i=1}^{n} P(A_i),$$

since $P(A_i \cap A_j) = 0$ for any i, j.

EXAMPLE 8.30

If a single draw is made from a standard pack of cards, find the probability of drawing a jack or a queen or a king.

Solution

Let events A, B and C be defined as follows:

A: a jack is drawn
B: a queen is drawn
C: a king is drawn

Now, since each event is composed of 4 of the 52 possible outcomes, we have

$$P(A) = P(\text{jack}) = \frac{4}{52} = \frac{1}{13}$$

$$P(B) = P(\text{queen}) = \frac{4}{52} = \frac{1}{13}$$

$$P(C) = P(\text{king}) = \frac{4}{52} = \frac{1}{13}$$

Furthermore, since events A, B and C are mutually exclusive (as they *cannot* occur at the same time), we know from the *additive rule*, that

$$P(A \cup B \cup C) = P(A) + P(B) + P(C)$$

$$= \frac{1}{13} + \frac{1}{13} + \frac{1}{13}$$

$$= \frac{3}{13}$$

$P(A \cup B \cup C) = P(A) + P(B) + P(C) + P(A \cup B) + P(A \cup C) + P(B \cup C) - P(A \cap B \cap C)$

YOUR TURN

8.34 Suppose that the sample space for an experiment is U = {A, B, C}. Determine which, if any, of the following probabilities are permissible and give a brief reason for your answer.

(a) $P(A) = \frac{1}{3}$, $P(B) = \frac{1}{6}$ and $P(C) = \frac{1}{2}$

(b) $P(A) = \frac{1}{2}$, $P(B) = \frac{1}{4}$ and $P(C) = \frac{1}{3}$

(c) $P(A) = -\frac{1}{4}$, $P(B) = \frac{3}{4}$ and $P(C) = 0$

(d) $P(A) = \frac{3}{4}$, $P(B) = 0$ and $P(C) = \frac{1}{4}$

8.35 A pet shop has an aquarium full of tropical fish containing 5 guppies, 10 neons, 8 angel fish, 4 swordtails and 3 cat fish. If a young girl comes into the shop and buys one fish at random, find the probability that she selects
(a) a catfish
(b) either a guppy or a swordtail
(c) not an angel fish
(d) a neon, an angel fish or a guppy
(e) neither a cat fish nor a guppy.

8.36 Let A and B be two events such that $P(A) = 0.36$, $P(B) = 0.52$ and $P(A \cap B) = 0.15$.
Using a Venn diagram, or otherwise, find the following probabilities.
(a) $P(A^|)$ **(b)** $P(B^|)$ **(c)** $P(A \cup B)$

8.37 Determine which, if any, of the following probability statements are permissible and give brief reasons for your answer.
(a) The probability that Mary Smith will fail her final Statistics exam is 0.24 and the probability that she will pass this same exam is 0.86.
(b) The probability that there will be a rise in the stock market share prices is 0.58 and the probability that there will be a fall is 0.32.
(c) The probability that Paul Jones will forget his raincoat is 0.25 and the probability that he will forget both his raincoat *and* also his umbrella is 0.4.

Summary

In this chapter we have had a brief look at two somewhat diverse topics – namely,

(i) descriptive statistics, and

(ii) set theory and classical probability.

In *section 1* we presented some common terminology and you should now be able to distinguish between
- categorical and numerical data
- discrete numerical and continuous numerical data
- a population and a sample
- a parameter and a statistic
- a measure of centre and a measure of variability.

Sections 2 to 4 examined methods of summarising data. These included

(a) **tabular displays**
- an **array** or ordered batch of data
- a **frequency distribution,** where the possible data values are listed together with their corresponding frequencies
- a **grouped frequency distribution** (*G.F.D.*), where the data values are classified into non-overlapping intervals of equal width with the frequency of values in each interval listed in an adjacent column.

(b) **pictorial displays**
- a **pie chart** or a **bar chart** for categorical data
- a **histogram** and a **frequency polygon** for continuous data.

(c) **numerical summaries** of a **single** sample of data
- measures of centre: mode, median, mean
- measures of variability: range, variance, standard deviation.

By definition:
The **mode** is the data value that occurs most often.
The **median** is the value that divides the **ordered** sample of data in half.
The **range** is the difference between the largest and the smallest value in the sample of data.
The **mean, variance** and **standard deviation** are best defined by the following table.

Note:
Although these numerical summaries exist for populations (called **parameters**) as well as for samples (called **statistics**), actual calculations are performed only on **sample** data in this chapter.

Summary measure	Population notation	Sample notation	Raw data	Grouped data G.F.D.
Mean	μ (mu)	$\bar{x}$	$\bar{x} = \dfrac{\sum x}{n}$ where n = sample size	$\bar{x} = \dfrac{\sum fm}{\sum f}$ $= \dfrac{\sum fm}{n}$ where m = interval midpoint f = interval frequency
Variance	σ^2	s^2	$s^2 = \dfrac{\sum x^2 - n\bar{x}^2}{n-1}$	$s^2 = \dfrac{\sum fm^2 - n\bar{x}^2}{n-1}$ where m = interval midpoint f = interval frequency $n = \Sigma f$ = sample size $\bar{x} = \dfrac{\sum f}{n}$
Standard deviation	σ	s	$s = \sqrt{s^2} = \sqrt{\dfrac{\sum x^2 - n\bar{x}^2}{n-1}}$	$s = \sqrt{s^2} = \sqrt{\dfrac{\sum fm^2 - n\bar{x}^2}{n-1}}$

Sections 5 and 6 concentrated on the **theory of sets** and the concepts of **classical probability**.

1. Set theory

(a) Types of sets: the following terms should be familiar:

- subset, $A \subset B$: A is contained in B.
- the universal set, U: contains *all* relevant elements.
- the empty set, $\emptyset$: contains *no* elements.
- mutually exclusive sets: sets with no common elements.

(b) Set operations: these are best explained using Venn diagrams if possible.

- union, $A \cup B$: elements in A or B or both.
- intersection, $A \cap B$: elements in A and B together.
- complement, $A^{|}$: elements not in A.
- difference, $A - B$: elements in A but not in B.

2. Probability

The following terms should be familiar (refer back to *Section 6*).

- experiment
- outcome
- event
- sample space

(a) Definitions of probability

- **The empirical definition**

$$P(A) = \frac{\text{no. of times A occurred}}{\text{total no. of repetitions of experiment}}$$

- **The classical definition**

$$P(A) = \frac{n}{N}, \text{ where } n = \text{outcomes in event A}$$

$$N = \text{total outcomes.}$$

(b) Axioms of probability

- $0 \leq P(A) \leq 1$, for any event A
- $P(A) = 0$, whenever $A = \emptyset$
- $P(A) = 1$, whenever $A = U$.

(c) Rules of probability

- $A \subset B$ implies that $P(A) \leq P(B)$. **Why?**
- $P(A^{|}) = 1 - P(A)$
- The *additive rule*:
  $P(A \cup B) = P(A) + P(B) - P(A \cap B)$
- $P(A \cup B) = P(A) + P(B)$, only if A and B are mutually exclusive. **Why?**

8.1 The ages at death of the first 30 presidents of the U.S.A. are given below:

Washington	67	Polk	53	Arthur	56
Adams	90	Taylor	65	Cleveland	71
Jefferson	83	Fillmore	74	Harrison	67
Madison	85	Pierce	64	McKinley	58
Monroe	73	Buchanan	77	Roosevelt	60
Adams	80	Lincoln	56	Taft	72
Jackson	78	Johnson	66	Wilson	67
Van Burren	79	Grant	63	Harding	57
Harrison	68	Hayes	70	Gooldige	60
Tyler	71	Garfield	49	Hoover	90

(a) Construct a grouped frequency distribution for the data using intervals of width 10. (That is, 40 – 49, 50 – 59, 60 – 69 etc.)

(b) From your *GFD* in **(a)** draw
(i) a histogram
(ii) a frequency polygon
to display the given data.

(c) Write down the interval limits for the intervals 40 – 49 and 60 – 69.

(d) Write down the interval boundaries for the intervals 50 – 59 and 70 – 79.

(e) What is the midpoint of the interval 50 – 59?

8.2 The graph below summarises the scores obtained by 100 students on a questionnaire designed to measure aggressiveness. (The scores are integer values and range from 0 to 20. A high score indicates a high level of aggression.)

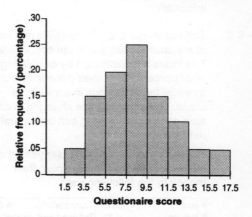

(a) Which interval contains the highest proportion of test scores?

(b) What proportion of scores lies between 3.5 and 5.5?

(c) What proportion of scores exceeds 11.5?

(d) How many students scored less than 5.5?

8.3 The data in the table below show the composition of the labour force in two factories.

Category of worker	Number of workers	
	Factory A	Factory B
Skilled manual	270	200
Unskilled manual	600	100
Managerial	50	40
Clerical	80	60
Total	1000	400

Compare the 2 sets of data above using
(i) bar charts
(ii) pie charts

(*Note* that the workforces differ in size).

8.4 An early investigation into the eggs of the cuckoo was reported in the first volume of the journal *Biometrika* in 1902 in the form of a grouped frequency distribution.

Breadth (mm)	Frequency
13.75 – 14.25	1
14.25 – 14.75	1
14.75 – 15.25	5
15.25 – 15.75	9
15.75 – 16.25	73
16.25 – 16.75	51
16.75 – 17.25	80
17.25 – 17.75	15
17.75 – 18.25	7
18.25 – 18.75	0
18.75 – 19.25	1

Draw a histogram to display the above information.

8.5 The consumer price index (*CPI*) is a way of measuring changes in the cost of living. The index is constructed by calculating what percentage of their expenditure an average family spends in a variety of areas. The effect of price changes in each area on the cost of living can be accurately determined.

Suppose that in one particular year, the average Australian family spent their money in the following way:

Food	19.0%	Transportation	18.0%
Clothing	6.8%	Tobacco and alcohol	
Housing	14.2%		7.2%
Housing equipment and operation	18.4%	Health and personal care	5.6%
		Recreation and education	10.8%

Display the above information by drawing
(i) a bar chart
(ii) a pie chart

8.5 *Cont.*

Prepare another set of charts showing the breakdown of **your** current spending. How does it compare with the above information? What do you believe are the main reasons for the differences?

8.6 The scores for a statistics test are as follows:
89 89 76 96 77 94 92 88 85 66
69 79 95 50 91 83 88 82 58 18
(a) Compute the mean, median and mode of these data.
(b) Which of the 3 measures of central tendency do you think would best represent the achievement of the class?
(c) Eliminate the two lowest scores and again compute the mean, median and mode. Which measure of central tendency is most affected by extremely low scores?

8.7 Consider the following two samples:

Sample 1	10 0 1 9 10 0 8 1 1 9
Sample 2	0 5 10 5 5 5 6 5 6 5

(a) Examine both samples and state which one you believe has the greater variability.
(b) Calculate the range of each sample. Does the result agree with your answer to part (a)? Explain.
(c) Calculate the standard deviation of each sample. Does the result agree with your answer to part (a)? Explain.
(d) Which of the two, the range or the standard deviation provides a better measure of variability? Why?

8.8 The following rainfall registrations were recorded in millimetres at 21 weather stations over a weekly period:
10 6 13 8 16 13 24 5 1 19 6
9 2 12 14 20 16 15 8 21 17
Calculate:
(a) the mean, and
(b) the standard deviation of the sample data.

8.9 A shopper compares the prices of a certain commodity at a number of different supermarkets. She finds the following prices (in cents):

63 56 65 48 73 59 72 63 65 60
44 79 63 61 66 69 64 64 57 63

For the sample data, determine:
(a) the mode

8.9 *Cont.*
  (b) the mean
  (c) the range
  (d) the variance
  (e) the standard deviation.

8.10 **(a)** Find the mean and standard deviation of the sample data:
  3 6 2 1 7 5
  (b) By adding 5 to each of the numbers in **(a)** above, we obtain the sample:
  8 11 7 6 12 10
  With minimum calculation, write down the mean and standard deviation of this new sample.
  (c) By multiplying each of the numbers in **(a)** above by 2 and then adding 5, we obtain the sample
  11 17 9 7 19 15
  Again with minimum calculation, write down the mean and standard deviation of this new sample.

8.11 The following table classifies the closing prices of shares of HB Mines Limited for 150 trading days on the Stock Exchange:

Interval midpoint, x	Frequency, f
93	19
100	32
107	28
114	34
121	18
128	11
135	8

Calculate the mean and standard deviation of the above grouped sample data.

*****8.12** A random sample of 91 bales of wool has the mass (in kilograms) of each recorded. The resulting data are summarised in the following *GFD*:

Bale mass (*kg*)	Frequency, *f*
150 – 159	3
160 – 169	7
170 – 179	11
180 – 189	13
190 – 199	24
200 – 209	20
210 – 219	7
220 – 229	5
230 – 239	1

For this distribution, calculate the mean and standard deviation and hence find (approximately) the proportion of bales whose masses lie within one standard deviation of the mean (that is, between $\bar{x} - s$ and $\bar{x} + s$).

8.13 A certain sample of data has a mean of 40 and a standard deviation of 8. What would the mean and standard deviation become if:
  (a) all data values were doubled?
  (b) 25 was subtracted from all values?
  (c) all data values were divided by 10?

8.14 The following sample of data shows the duration of 20 randomly selected long distance telephone calls:

  8 12 13 15 16 4 7 21 10 14
  18 4 11 25 13 6 13 8 15 9

  (a) Calculate the mean, variance and standard deviation of the above sample data.
  (b) Summarise the given raw data into a *GFD* with intervals of width 4 (that is, 4 – 7, 8 – 11, 12 – 15 etc).
  (c) Calculate the mean, variance and standard deviation of your *GFD* in **(b)**.
  (d) Compare your sample statistics in **(a)** with those calculated in **(c)**. Comment on any differences. Which are more accurate? Why?

8.15 Consider the sample space
  S = { copper, sodium, nitrogen, potassium, uranium, oxygen, zinc}, and the events
  A = {copper, sodium, zinc}
  B = {sodium, nitrogen, potassium}
  C = {oxygen, uranium, zinc}
  List the elements of the sets corresponding to the following events
  (a) $(A \cup C)$ **(b)** $(A \cap B)$
  (c) $(A \cap C) \cup B$

8.16 A coin is tossed 3 times. Write down each of the possible 'head/tail' sequences. (*For example*, 'head' followed by 'tail' followed by 'head' = HTH). Also write down the number of heads in each sequence.

  Using the above information, now calculate the probability of
  (a) 3 heads
  (b) 3 tails
  (c) 2 heads and 1 tail
  (d) at least 2 heads
  (e) 1 head and 2 tails.

8.17 Two snapdragon seeds are planted in a pot. The seeds grow into flowers that are white (w), pink (p) or red (r). The sample space can be written as
  S = {ww, wp, wr, pp, pr, rr}
  (a) Show on a Venn diagram the following events:

8.17 **(a)** *Cont.*

　　J:　where J = both plants had flowers of the same colour

　　K:　where K = you obtained at least one white flower

　　L:　where L = you obtained a red-flowered plant and one of another colour.

(b) Write down all the points of the sample space that are in each of the following events:

(i) $J^|$　　　　**(ii)** $J \cap K$

(iii) $K^|$　　　**(iv)** $J \cup L$

(v) $K \cap L$

(c) Describe in words each of the events in **(b)**.

8.18 At a certain university, 80 students are selected at random. Of the 20 females, 5 will major in statistics, while of the 60 males, 15 will major in statistics.

If *one* of the 80 students is selected at random, find the probability that the student selected will

(a) be a female

(b) **not** major in statistics

(c) either be a male **or** major in statistics

(d) be a male **and** major in statistics

(e) **not** major in statistics and **not** be a male

(*Hint:* use Venn diagrams wherever possible.)

8.19 A card is drawn from a standard pack.

Let A be the event: a red card is drawn,

　　B be the event: the card is greater than 2 but less than 9.

Find

(a) $P(A)$　　　　**(b)** $P(B)$

(c) $P(A \cap B)$　**(d)** $P(A \cup B)$

8.20 Of a total of 50 sales made at a small women's fashion shop, 20 were made on Monday and 30 on Tuesday.

Of the Monday sales, 5 were made by Visa Card.

Of the Tuesday sales, 10 were made by Visa Card.

If *one* sale is taken at random, find the probability that

(a) it was made by a Visa Card

(b) it was made by a Visa Card **and** it occurred on Monday

(c) it was made on Tuesday

(d) it was made on Monday **and** it was **not** by Visa Card.

8.21 The Venn diagram below depicts an experiment with six simple events. The events A and B are also shown.

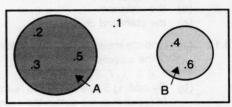

The probabilities of the simple events are as follows:

$$P(1) = P(2) = P(4) = \frac{2}{9};$$

$$P(3) = P(5) = P(6) = \frac{1}{9}.$$

(a) Find $P(A)$.

(b) Find $P(B)$.

(c) Find $P(A \cup B)$; that is, the probability of A or B (or both).

(d) Find $P(A \cap B)$; that is, the probability that the events *A* and *B* occur simultaneously.

*8.22 A hospital reports that two patients have been admitted who have contracted Legionnaire's disease. Suppose our experiment consists of observing whether patients survive or die as a result of the disease. The simple events and the probabilities of their occurrence are shown in the following table (where S in the first position means that patient 1 survives, D in the first position means that patient 1 dies, etc.).

Sample events	Probabilities
SS	.81
SD	.09
DS	.09
DD	.01

Find the probability of each of the following events:

(a) A: {both patients survive the disease}

(b) B: {at least one patient dies}

(c) C: {exactly one patient survives the disease}.

*8.23 Telephone enquiries to the office of a Sydney importing company are directed to one of the staff members who is available that day. On a particular day, eight staff members are present. The following table shows information about each one.

*8.23 *Cont.*

Name	Sex	Marital status	Department
Mr. Al–Hassan	Male	Married	Sales
Ms Dennis	Female	Married	Administration
Ms Chang	Female	Single	Sales
Mr Bennito	Male	Married	Sales
Ms Brown	Female	Married	Sales
Mr Lee	Male	Single	Administration
Ms Giorgiano	Female	Single	Administration
Ms Ng	Female	Single	Sales

A telephone enquiry arrives and is directed at random to one of the available staff.

(a) Write down the sample space for this experiment.

(b) What points of the sample space are in the following events?

  (i) A = the call is directed to a male staff member

  (ii) B = the call is directed to a married person

  (iii) C = the call is directed to a member of the sales staff.

(c) Show these events A, B and C on a Venn diagram.

(d) Find the probability of each of the events A, B and C.

(e) List the points of the sample space that are in the following events:

  (i) A^I **(ii)** $A \cap C$ **(iii)** $B \cup C$

  (iv) $A \cup A^I$ **(v)** $A \cap B$

(f) Find the probability of each of the events **(i)** to **(v)**, in **(e)**.

*8.24 Of the 100 people who applied for a position with a firm during the past year, 40 had previous work experience, 30 had an appropriate technical certificate and 20 had both work experience and a technical certificate. If an applicant is chosen at random, find the probability that he/she had

(a) either previous experience or a certificate or both

(b) either previous experience or a certificate but **not** both.

9 Extras – Geometry

Introduction

This is the first of six chapters covering the more specialised topics considered to be particularly useful for specified discipline areas. **Elementary geometry**, the topic for this chapter, is, in most tertiary institutions, assumed knowledge for those students intending to study further in the fields of chemistry, physics and mathematics.

While geometry itself covers a very broad range of subject areas, both theoretical and practical, we will confine ourselves here to just *three* important branches – namely, plane geometry, solid geometry and coordinate geometry.

After defining some important terms in *section 1*, we then examine the concept of an angle in *section 2* and present, by means of diagrams, the main types of angles.

Section 3 discusses the branch of geometry dealing with *plane figures*, where

> **a plane figure** is defined as a 2-dimensional or flat closed figure.

Note: In this context, a closed figure is one that has the same starting and finishing point.

Section 4 examines the *solid geometry* branch, where

> **a solid figure** is defined as a 3-dimensional closed figure – that is, one that takes up space.

Finally, in *section 5* we present, very briefly, the main aspects of *coordinate geometry*, a branch of geometry where all discussions are restricted to the **Cartesian or (x,y) plane.**

Important:

It should be noted that it is *not* the function of this textbook to present theoretical proofs of various results, but rather to discuss the results themselves together with their implications for practical problem solving.

Also, since each section of the chapter involves, for the most part, a presentation of the different types of figures together with their associated properties and formulae, there is little need for examples and exercises throughout. However, several examples and a number of practical exercises using the information given, will be included at the end.

9.1 Some useful terminology

In geometry, there are several terms and symbols that are frequently used, and in all future sections of this chapter it will be assumed that these terms are familiar and understood.

(i) line

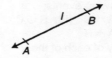

This is denoted by l or sometimes called the line *AB*.

(ii) line segment (or interval)

This is a *portion* of a line and is called the line segment (interval) *AB*.

(iii) parallel

This is a term that usually refers to lines or planes lying in the same direction and remaining the same distance apart. It is denoted //.

(iv) perpendicular

This is a term that again refers to lines and/or planes where the angle between them is 90° (a right angle). It is denoted ⊥.

Note: The concept of an angle between two lines will be discussed in *section 2*.

(v) perimeter

This is a measurement of the *boundary* of a closed 2-dimensional flat figure.

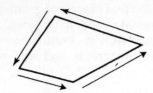

That is, perimeter = the sum of the length of each side.

(vi) area

This measures the space *inside* the boundary of a closed 2-dimensional flat figure.

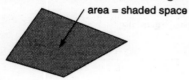

area = shaded space

(vii) surface area

This is sum of the area of each side of a 3-dimensional solid figure.

(viii) volume

This measures the space enclosed by a 3-dimensional solid figure.

Note:

Units of measure for (v) to (viii) above:

(v) perimeter: original units; e.g. cm.

(vi) area: squared units; e.g. cm^2.

(vii) surface area: squared units; e.g. cm^2.

(viii) volume: cubed units; e.g. cm^3.

The reason for these units of measure will become clearer when you see the appropriate formula for calculating the required measurement in a given practical situation.

9.2 Angles

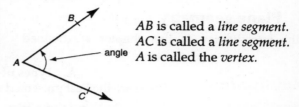

AB is called a *line segment*.
AC is called a *line segment*.
A is called the *vertex*.

The angle between AB and AC, shown on the diagram as , is denoted by the symbol $\angle$ and is commonly called $\angle A$, or $\angle BAC$ or $\angle CAB$. The units of measure for all angles is degrees, denoted °, and the actual size of a given angle may be determined using an instrument called a *protractor*,

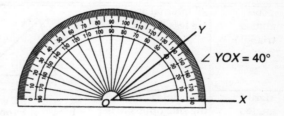

$\angle YOX = 40°$

9.2.1 Types of angles

Table 9.1: Types of angles

Name	Brief description	Diagram
1. acute angle	an angle less than 90°	
2. right angle	an angle of exactly 90°	
3. obtuse angle	an angle between 90° and 180°	
4. straight angle	an angle of exactly 180°	
5. reflex angle	an angle greater than 180° (but less than 360°)	or
6. one revolution	the angle included in one full turn is 360°	

Note:

(i) 360° = 1 revolution (full turn).

(ii) 1° may be divided into 60 equal parts called *minutes*, denoted '; that is, 60' = 1°.

(iii) 1' may be divided into 60 equal parts called *seconds*, denoted "; that is, 60" = 1'.

9.3 Plane geometry

In the introduction to this chapter we defined plane geometry to be that branch of geometry that includes 2-dimensional closed, flat figures.

The following sub-sections examine several of these by presenting diagrams and any relevant formulae.

9.3.1 The triangle

> A **triangle** is a 3-sided closed plane figure, where the sum of the three internal angles is 180°. It is denoted by Δ.

A. Types of triangles

These are the best presented using the following table.

Table 9.2: Types of triangles

Name	Brief description	Diagram where b = length of base of the triangle, h = perpendicular height of the triangle	Perimeter, P (units)	Area, A (units^2)
1. acute angled triangle	all internal angles are less than 90°		$P = b + a + c$ where b = length of AC c = length of AB a = length of BC	$A = \frac{1}{2}bh$ where b = length of the base, AC h = perpendicular height from the base
2. right angled triangle	one angle is 90° and the side opposite is called the **hypotenuse**	BC is called the **hypotenuse**	$P = b + a + c$	$A = \frac{1}{2}bh = \frac{1}{2}bc$ (since here $c = h$)
3. obtuse angled triangle	one internal angle is between 90° and 180°		$P = b + a + c$	$A = \frac{1}{2}bh$
4. scalene triangle	all 3 sides are of different length (i.e. 3 unequal sides)		$P = b + a + c$	$A = \frac{1}{2}bh$
5. isosceles triangle	any 2 sides are of equal length		$P = b + 2a$ (since $AB = BC$)	$A = \frac{1}{2}bh$
6. equilateral triangle	all 3 sides are the same length		$P = 3b$ (since $AB = BC = AC$)	$A = \frac{1}{2}bh$

B. Similar triangles

> Two triangles are said to be **similar** if one is an enlargement of the other. That is,
>
> **(i)** if corresponding angles are *equal*, and
>
> **(ii)** if corresponding sides are in the *same ratio*.

For example:

The two triangles following are *similar* because they are equiangular. Hence, their corresponding sides must be in the same ratio.

Note: In similar triangles, 'corresponding' sides are those *opposite* equal angles.

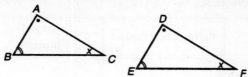

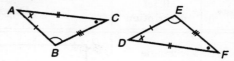

their corresponding angles and corresponding sides are *equal*.

Here:

(i) $\angle B = \angle E$; $\angle A = \angle D$; $\angle C = \angle F$.

(ii) *AB* and *DE*; *AC* and *DF*; *BC* and *EF* are all pairs of corresponding sides.

C. Congruent triangles

> Two triangles are said to be **congruent** if they are exactly the same size and shape. That is, each angle in one triangle has an equal angle in the other, and each side in one has an equal side in the other.

For example:
The two triangles following are *congruent* because

Note:
The sides that 'correspond' in congruent triangles are those *opposite* the angles that correspond. Therefore, if $\angle B = \angle E$, then $AC = DF$. Similarly, since $\angle A = \angle D$, then $BC = EF$ and since $\angle C = \angle F$, then $AB = DE$.

9.3.2   The quadrilateral

> A **quadrilateral** is a 4-sided closed plane figure, where the sum of the four interior angles is 360°.

Types of quadrilaterals
As with triangles, these are figures are best presented using the following table.

Table 9.3: Types of quadrilaterals

Name	Brief description	Diagram	Perimeter, P (units)	Area, A (units²)
1. a trapezium	*one* pair of opposite sides parallel	Here $AB \,//\, DC$ h = perpendicular height a and b = lengths of the parallel sides c and d = lengths of remaining sides	P = distance around the boundary $= a + d + b + c$	A = area of $\triangle ABC$ + area of $\triangle ADC$. $= \frac{1}{2}ah + \frac{1}{2}bh$ $= \frac{1}{2}h(a+b)$
2. a parallelogram	*both* pairs of opposite sides parallel	$AB \,//\, DC$ and $AD \,//\, BC$	$P = 2a + 2b$ $= 2(a+b)$	$A = \frac{1}{2}bh + \frac{1}{2}bh = bh$
3. a rectangle	a parallelogram with one angle a right angle (hence all angles must be 90°)	$WX \,//\, YZ$ and $WZ \,//\, XY$ Also $\angle W = \angle Z = \angle Y = \angle Z = 90°$	$P = 2a + 2b$ $= 2(ab)$	$A = ab$

Table 9.3: Types of quadrilaterals

Name	Brief description	Diagram	Perimeter, P (units)	Area, A (units^2)
4. a square	a rectangle with four equal sides		$P = x + x + x + x$ $= 4x$	$A = x \times x$ $= x^2$

9.3.3 The circle

Table 9.4: The circle

Description	Diagram	Circumference, C (units)	Area, A (units^2)
all points are equidistant (an equal distance) from a common centre point	d = diameter $r = \frac{1}{2}d$ = radius (plural: radii) C = circumference = distance around the boundary	$C = 2\pi r$ or $C = \pi d$ *Note:* the distance around the boundary (perimeter) of a circle is called the **circumference**	$A = \pi r^2$ where r = radius

Note:

Calculation of the circumference and area of a circle involves multiplication by π (pi). Although the value of π may be approximated by $\frac{22}{7}$, a more accurate result will be obtained from your calculator. Even though π is located above the button $\boxed{\text{EXP}}$, on our calculator, it is accessed by simply pressing this button $\boxed{\text{EXP}}$ *without* pressing $\boxed{\text{Shift}}$.

Table 9.5: Parts of a circle

1. A *semicircle* is half a circle.		2. A *segment* is a piece of a circle cut off by a straight line.	
3. A *sector* is a part of a circle bounded by two radii.		4. An *arc* is any part of the circumference.	
5. A *chord* is a straight line joining two points on the circumference.		6. A *tangent* is a straight line drawn outside the circle that touches the circumference at one point only.	

EXAMPLE 9.1

For each of the following figures, determine:

(i) the perimeter (or circumference, if appropriate), and

(ii) the area.

(a)

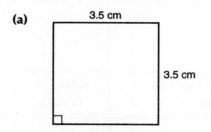

(b)

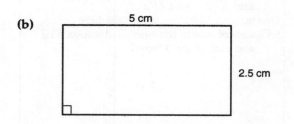

(c)

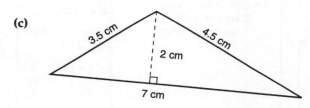

(d)

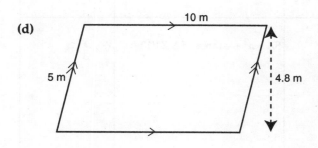

(e)

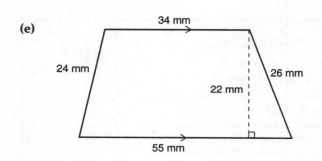

(f)

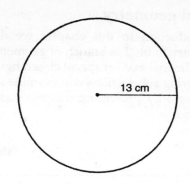

Solution

Let P = perimeter, C = circumference and A = area.

(a) **(i)** $P = 4(3.5) = 14$ cm

(ii) $A = (3.5)^2 = 12.25$ square cm

(b) **(i)** $P = 2(5) + 2(2.5)$
$= 10 + 5 = 15$ cm

(ii) $A = 5 \times 2.5 = 12.5$ square cm

(c) **(i)** $P = 3.5 + 4.5 + 7 = 15$ cm

(ii) $A = \frac{1}{2}(7) \times 2 = 7$ square cm

(d) **(i)** $P = 2(10) + 2(5) = 30$ m

(ii) $A = 10 \times 4.8 = 48$ square m

(e) **(i)** $P = 34 + 26 + 55 + 24 = 139$ mm

(ii) $A = \frac{1}{2} \times 22 \times (34 + 55) = 979$ square mm

(f) **(i)** $C = 2 \times \pi \times 13 = 81.68$ cm

(ii) $A = \pi \times 13^2 = 530.93$ square cm

EXAMPLE 9.2

Mr Smith's backyard is big enough to build a tennis court. If the length and breadth is 100 m by 50 m,

(i) what is the required perimeter of the court?

(ii) how much will it cost Mr Smith to fence the court at $5 per metre?

(iii) how much space will the tennis court cover?

Solution

(i) $P = 100 + 100 + 50 + 50 = 300$ metres.

(ii) Cost = $300 \times 5 = \$1500$.

(iii) $A = 100 \times 50 = 5000$ square metres.

9.4 Solid geometry

In the introduction to this chapter we defined **solid** geometry to be that branch of geometry that includes 3-dimensional (or space) closed figures.

The following sub-sections will examine several of these figures by presenting diagrams and any relevant formulae.

9.4.1 The pyramid

A **pyramid** is a solid figure with triangular sides (faces) meeting at a common point (called the *vertex*). The type of pyramid is named according to the shape of its base.

Table 9.6: Types of pyramids

Name	Diagram	Surface area, SA (units^2)	Volume, V (units^3)
1. triangular pyramid	A = area of base XYZ h = perpendicular height from vertex to base	SA = area XYZ + area XVZ + area XVY + area ZVY That is, SA = sum of area of the base and each of the 3 faces	$V = \frac{1}{3}Ah$ where A = area XYZ
2. rectangular pyramid	A = area of base $h = \perp$ height	SA = area of base (ab) + 2(area of $\Delta\,XVY$) + 2 (area of $\Delta\,ZVY$)	$V = \frac{1}{3}Ah$ where $A = ab$ $\therefore V = \frac{1}{3}abh$
3. square pyramid		$SA = a^2 + 4$ (area of $\Delta\,XVY$)	$V = \frac{1}{3}a^2h$
4. cone		$SA = \pi ra +$ base area $= \pi ra + \pi r^2$ $= \pi r(a + r)$	$V = \frac{1}{3}\pi r^2 h$

9.4.2   The prism

A **prism** is a solid figure whose base and top
have the same plane shape and size. The type
of prism is named according to the shape of its
base.

Table 9.7: Types of prisms

Name	Diagram	Surface area, *SA* (units²)	Volume, *V* (units³)
1. triangular prism	A = area of base $h = \perp$ height	$SA = 2$ (area of base) $+ dh + bh + ch$ $= 2$(area BCD) $+ h(d + b + c)$	$V = Ah$ where A = area of base
2. rectangular prism		$SA = 2xy + 2yh + 2xh$ $= 2\,(xy + xh + yh)$	$V = xyh$
3. square prism (cube)		$SA = 6a^2$	$V = a^3$
4. cylinder (or circular prism)	r = radius $h = \perp$ height	$SA = 2\pi rh + 2$(base area) $= 2\pi rh + 2\pi r^2$ $= 2\pi r(h + r)$	$V = \pi r^2 h$

9.4.3 The sphere

A **sphere** is a solid figure whose surface consists of points that are all the same distance from a fixed point, called the *centre*.

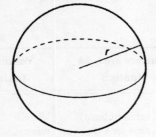

Surface area:
$SA = 4\pi r^2$

Volume:
$V = \dfrac{4}{3}\pi r^3$

EXAMPLE 9.3
Find

(i)   the total surface area, and

(ii)   the volume

of

(a)   a rectangular prism with base 85 mm by 60 mm and height 4 mm

(b)   a square pyramid with a base edge of 28.4 metres, a height of 22.8 metres and a triangular face area of 355 square metres

(c)   a cone with radius 6 cm, edge 10 cm and height 8 cm.

Solution

(a)   (i) $SA = 2(85 \times 60) + 2(85 \times 4) + 2(60 \times 4)$

$= 10200 + 680 + 480$

$= 11360$ square mm

(ii) $V = 85 \times 60 \times 4 = 20400$ cubic mm

(b)   (i) $SA = (28.4)^2 + 4(355)$

$= 806.56 + 1420$

$= 2226.56$ square metres

(ii) $V = \dfrac{1}{3} \times 28.4^2 \times 22.8$

$= 6129.856$ cubic metres

(c)   (i) $SA = (\pi \times 6 \times 10) + (\pi \times 6^2)$

$= 301.5929$ square cm

(ii) $V = \dfrac{1}{3} \times \pi \times 6^2 \times 8$

$= 301.593$ cubic cm

9.5   Coordinate geometry

Keeping in mind that in this branch of geometry our discussions are restricted to the cartesian or (x,y) plane, the following definitions are very useful.

Let (x_1, y_1) and (x_2, y_2) be any two points in the cartesian (x,y) plane.
Then,

1.   the **distance**, d, between these two points is given by

$$d = \sqrt{(x_2 - x_1)^2 + (y_2 - y_1)^2}$$

2.   the **midpoint** of the interval joining these two points is the point given by

$$\text{midpoint} = \left(\frac{x_1 + x_2}{2}, \frac{y_1 + y_2}{2} \right)$$

3.   the **gradient** (or slope) of the interval joining these two points is given by

$$m = \frac{y_2 - y_1}{x_2 - x_1}.$$

EXAMPLE 9.4
Find

(i)   the distance

(ii)   the midpoint

(iii) the gradient

of the interval (line segment) joining $A\,(0,1)$ and $B\,(-3,5)$.

Solution
Here $x_1 = 0$, $x_2 = -3$, $y_1 = 1$, $y_2 = 5$. Therefore,

(i)   distance, $d = \sqrt{(x_2 - x_1)^2 + (y_2 - y_1)^2}$

$= \sqrt{(-3 - 0)^2 + (5 - 1)^2}$

$= \sqrt{9 + 16}$

$= \sqrt{25}$

$= 5$

(ii)   midpoint of the interval:

$(\dfrac{x_1 + x_2}{2}, \dfrac{y_1 + y_2}{2})$ or $(\dfrac{-3}{2}, \dfrac{6}{2})$ or $(-1.5, 3)$

(iii) gradient, $m = \dfrac{y_2 - y_1}{x_2 - x_1}$

$= \dfrac{5 - 1}{-3 - 0}$

$= \dfrac{4}{-3}$

$= -1.33$

9.1

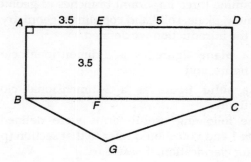

From the diagram above, name the following plane figures:

(i) *ABFE*
(ii) *BGC*
(iii) *EFCD*
(iv) *EBCD*

9.2 Determine

(i) the perimeter, (or circumference, as appropriate), and

(ii) the area

of the following plane figures:

(a)

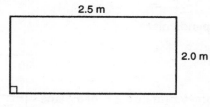

(b)

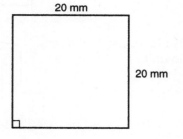

(c)

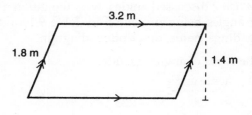

(d)

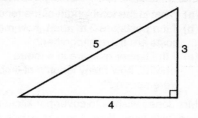

(e)

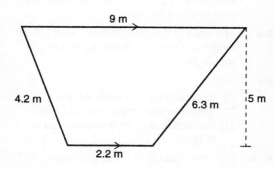

(f)

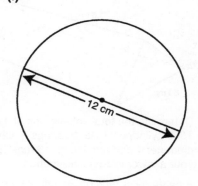

9.3 Calculate

(a) the perimeter of a square with side lengths 8.4 cm.

(b) the area of a rectangle, 8 cm long and 6 cm wide.

(c) the area of a triangle, base length 5 cm and vertical height 8 cm.

(d) the perimeter of a parallelogram, 12 m long and 10 m wide.

9.4 A rectangular prism is 3 cm wide, 2 cm deep and 4 cm high. Find

(i) the area of each side (face)

(ii) the total surface area of the figure.

9.5 The base of a triangular pyramid is an equilateral triangle with an area of 6 cm². If the perpendicular height of the pyramid is 10 cm, find its volume.

9.6 A soccer ground has a rectangular fence that measures 120 metres by 78 metres.

(a) What is the total length of the fence?

(b) If the posts are 3 m apart, how many posts are there altogether?

(c) If the timber rails are in 4 metre lengths, how many lengths of timber were needed?

9.7 Mrs Jones wishes to cover her kitchen floor with vinyl. If her floor measures 4 metres by 3 metres and if vinyl costs $21.25 per square metre, how much will it cost to cover the floor?

9.8 The area of a rectangle is 27.8 cm². If the length is 4 cm, find the width of the rectangle.

9.9 Find

(i) the diameter of a circle whose circumference measures 517 mm

(ii) the radius of a circle whose area is 6.08 m².

9.10 Find the volume, in cubic millimetres, of the triangular prism shown below.

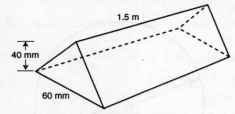

9.11 Determine the total volume of metal in 1500 cylindrical roller bearings, each of diameter 4 mm and length 15 mm. Give your answer in cubic millimetres.

9.12 A rectangular tank has length 2.65 m, breadth 480 mm and depth 3.25 mm. Find the volume of the tank

(i) in cubic millimetres

(ii) in cubic metres.

9.13 What length of copper rod of diameter 12 mm can be made from 1 cubic metre of copper? Answer in kilometres.

9.14 What is the surface area of the solid figure in

(i) exercise 9.5

(ii) exercise 9.10

(iii) exercise 9.11 (*SA* of 1 bearing only)

(iv) exercise 9.12.

9.15 Find

(i) the surface area

(ii) the volume

of each of the following spheres:

(a) radius = 56.0 mm

(b) diameter = 587 mm.

Summary

The main purpose of this chapter, the first of the extra chapters covering more specialised topics, is to examine three important branches of geometry – namely, plane, solid and coordinate geometry.

In the introduction we defined:

(i) a **plane** figure as a 2-dimensional closed figure, and

(ii) a **solid** figure as a 3-dimensional closed figure.

The following useful terms were defined in *section 1* and you should refer to that section (page 120) for clarification, if necessary.

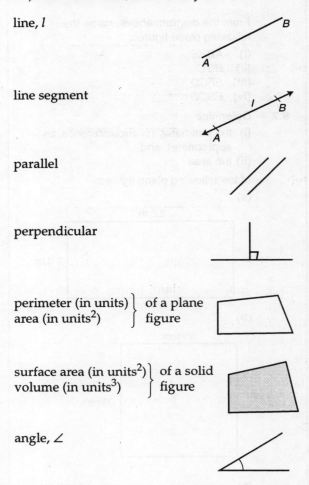

line, *l*

line segment

parallel

perpendicular

perimeter (in units) ⎫ of a plane
area (in units²) ⎭ figure

surface area (in units²) ⎫ of a solid
volume (in units³) ⎭ figure

angle, ∠

Section 2 discussed **angles**, with the different types of angles being presented in Table 9.1 under the headings: name, description, diagram.

The types of angles include:

(i) acute angle

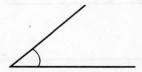

(ii) right angle

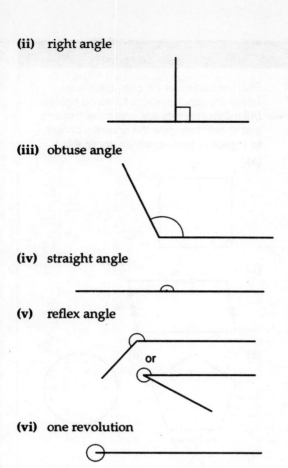

(iii) obtuse angle

(iv) straight angle

(v) reflex angle

or

(vi) one revolution

Note:
You should refer to Table 9.1 for a full description.

Section 3 examined **plane** figures and is divided into three sub-sections, covering

(i) the triangle

(ii) the quadrilateral

(iii) the circle.

Table 9.2 on page 122 presents the different types of triangles under the headings: name, description, perimeter formula, area formula.
The types of triangles include:

(i) acute angled

(ii) right angled

(iii) obtuse angled

(iv) scalene

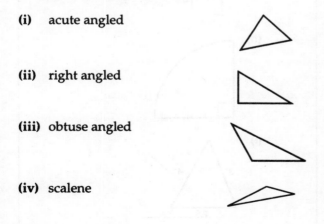

(v) isosceles

(vi) equilateral

The terms *similar triangles* and *congruent triangles* are also defined (see page 122).

Table 9.3 on page 123 presents different types of quadrilaterals together with the formulae for calculating the perimeter and the area. These figures include

(i) a trapezium

(ii) a parallelogram

(iii) a rectangle

(iv) a square.

The last of the plane figures – namely, the circle, is presented in Table 9.4 on page 124, and Table 9.5 on page 124, defining briefly the various parts of a circle.

Section 4 examines **solid** figures and is divided into three sub-sections covering

(i) the pyramid

(ii) the prism

(iii) the sphere.

Tables 9.6 and 9.7 include the different types of pyramids and prisms, together with appropriate formulae for calculating both the surface area and the volume.
The following figures are presented:

(i) triangular ⎫
(ii) rectangular ⎪ pyramid
(iii) square ⎬
(iv) cone ⎭

(v) triangular ⎫
(vi) rectangular ⎪ prism
(vii) square (cube) ⎬
(viii) cylinder ⎭

The last of the solid figures – namely, the sphere – is then defined, giving also the formulae for calculating the surface area and the volume.

Finally, *section 5* covers briefly the branch of coordinate geometry where all definitions and calculations are restricted to the Cartesian (x,y) plane only.

The following definitions are very important:

If (x_1, y_1) and (x_2, y_2) are two points in the (x, y) plane, then

(i) the *distance* between the points is

$$d = \sqrt{(x_2 - x_1)^2 + (y_2 - y_1)^2}$$

(ii) the *midpoint* of the line is

$$(\frac{x_1 + x_2}{2}, \frac{y_1 + y_2}{2})$$

(iii) the *gradient* of the line is

$$m = \frac{y_2 - y_1}{x_2 - x_1}.$$

9.1 Find the perimeter (or circumference, where appropriate) of the following figures. (All measurements are in centimetres and you should calculate the answers correct to 2 decimal places, where applicable.)

(a)

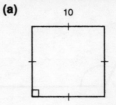

(b)

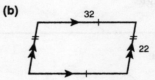

(c)   **(d)**

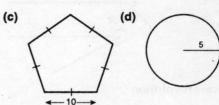

(e)

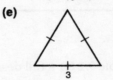

(f)

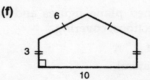

(g)

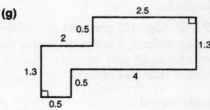

(h)

(i)

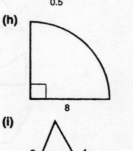

9.2 Find the area of each of the following figures. (Calculate your answers correct to 2 decimal places, where appropriate.)

(a)

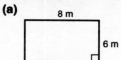

8 m
6 m

(b)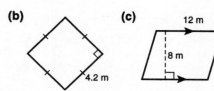

4.2 m

(c)

12 m
8 m

(d)

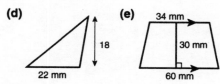

18
22 mm

(e)

34 mm
30 mm
60 mm

(f)

7 cm

(g)

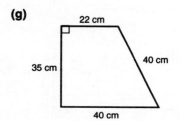

22 cm
35 cm
40 cm
40 cm

***(h)**

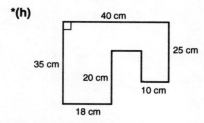

40 cm
35 cm
25 cm
20 cm
10 cm
18 cm

9.3 Find **(i)** the volume, and

(ii) the surface area

of each of the following figures, (correct to 3 decimal places, where appropriate).

(a)

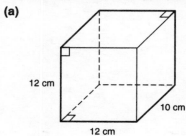

12 cm
12 cm
10 cm

9.3 *Cont.*

(b)

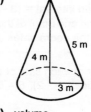

12 cm
6 cm

(c)

5 m
4 m
3 m

(d) volume only

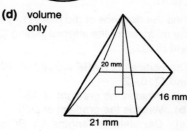

20 mm
16 mm
21 mm

(e)

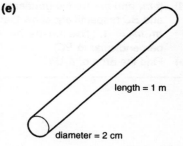

length = 1 m
diameter = 2 cm

***(f)** volume only

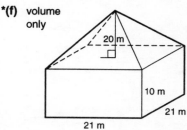

20 m
10 m
21 m
21 m

***9.4** **(a)** Calculate the total surface area of a 5 cm cube.

(b) A steel cylinder with a 12 metre radius and standing 13 metres high sits flat on the ground and is to be painted on the outside. Calculate the surface area to be painted if the top and base cannot be touched. Calculate the total cost if paint sells at $20 per 4-litre tin. (A 4-litre tin of paint covers an area of 15 square metres.)

9.5 Find the **(i)** distance

(ii) midpoint

(iii) gradient

of the line segment joining

(a) (4,2) and (7,6)

(b) (2,0) and (−1,3)

(c) (−5,−1) and (3,−2)

9.6 Show that $\triangle ABC$ is equilateral, given that A, B and C are the points (0,0) ($\sqrt{3}$,1) and (0,2).

9.7 The midpoint of an interval is (3,−1). If the coordinates of one end point of the interval are (−1,−2), find the coordinates of the other end point.

9.8 Find the gradient of each side of the triangle with vertices A(2,2), B(4,8) and C(5,4).

*9.9 Find the distance of the point (1,4) from the midpoint of the interval joining (3,−2) and (5,2).

*9.10 The vertices of $\triangle ABC$ are A(0,0), B(6,4) and C(10,−2).

(a) What is the gradient of AB?

(b) What is the gradient of BC?

(c) Calculate the lengths AB, BC and AC.

(d) If m_1 and m_2 are the gradients of AB and BC respectively, show that $m_1.m_2 = -1$. (This means that AB is perpendicular to BC)

(e) Find the area of $\triangle ABC$.

10 Extras – Trigonometry

Introduction

This is the second of the *extra* chapters and introduces one of the most important branches of mathematics – namely, **trigonometry**. Broadly speaking, **trigonometry** is defined as the study of the properties of triangles and trigonometric functions and of their applications.

The word **trigonometry** actually originated from the Greek word: *trigonom* meaning 'triangle' and *metron* meaning 'measurement'. Although the first uses of trigonometry can be traced as far back as 1700 BC, it was Hipparchus (146 - 127 BC), one of the greatest of the ancient astronomers, who made the first major contributions to the subject by developing some trigonometric tables and by investigating many of the properties of triangles.

There were some further advances in this field from that point, but it was not until the 17th century that any rapid progress was made in developing the subject to its current form. Today, **simple** trigonometry is mainly concerned with the calculation of angle sizes and side lengths of right-angled triangles and is applied in such diverse fields as surveying, engineering, architecture, building, communications, chemistry, physics, astronomy and navigation.

As is true for all branches of mathematics and statistics, trigonometry has its own 'jargon' words as well as making frequent use of both Roman and Greek symbols. The jargon words, no doubt familiar to many readers, include:

tangent (abb. tan)

sine (abb. sin)

cosine (abb. cos)

cotangent (abb. cot)

secant (abb. sec)

cosecant (abb. cosec)

All of the above terms are clearly defined in the appropriate sections of this chapter. The relevant Greek symbols, commonly used to represent angles, include:

α (alpha), Φ (phi) and θ (theta).

In *Section 1* headed *terminology and notation*, we revise what we learned about right-angled triangles in the preceding chapter, labelling the various parts where appropriate. We also re-define the term **ratio**, previously discussed in chapter 1.

Section 2 presents the three basic trigonometric ratios – namely,

   (i) the sine ratio

   (ii) the cosine ratio

   (iii) the tangent ratio.

The appropriate formula for each is given, followed by an indication of when and why each is used.

In *Section 3* we briefly define another set of useful trigonometric ratios generally referred to as the **reciprocal ratios**.

Section 4 builds on the concepts presented in earlier sections and discusses the solving of problems associated with *non-right-angled triangles*. These problems include determining side lengths, internal angles and areas, and the solution methods involve using

   (i) the sine rule, and

   (ii) the cosine rule.

The last section, *Section 5*, headed *Pythagorean identities*, presents three additional trigonometric identities (originating from Pythagoras' theorem) that allow more complex trigonometric relationships to be manipulated and simplified.

Keep in mind that this is only the first of *two* chapters, each of which discusses a wide variety of topics that are included in that branch of mathematics called **trigonometry.**

10.1 Terminology and notation

A. Since simple trigonometry is concerned only with the calculation of side lengths and angle sizes of *right-angled* triangles, all the trigonometric ratios presented in the following sections are defined in terms of the sides of a right-angled triangle. Hence, complete knowledge of a right-angled triangle is essential for full understanding.

Recall that in a right-angled triangle, the longest side is called the **hypotenuse**, while the remaining two sides are labelled according to which acute angle is given.

For example:

(a) Given $\angle\Phi$ in $\triangle ABC$,

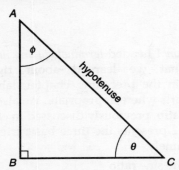

(i) BC is called the **opposite** side

(ii) AB is called the **adjacent** side.

(b) Given $\angle\theta$ in $\triangle ABC$,

(i) AB is called the **opposite** side

(ii) BC is called the **adjacent** side.

B. *Section 2* following examines the three trigonometric *'ratios'*. However, in this context, the term *'ratio'* is *not* defined $a : b$ (i.e. a is to b), but rather as the fraction (or proportion) $\frac{a}{b}$.

C. Labelling a triangle, especially in a trigonometric context, usually follows the convention that

(i) each of the three angles is denoted by a capital letter, and

(ii) each of the three sides is denoted by the small letter corresponding to the label attached to its opposite angle.

For example: In $\triangle ABC$,

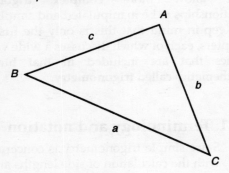

(i) side BC is denoted 'a'

(ii) side AC is denoted 'b'

(iii) side AB is denoted 'c'.

10.2 The trigonometric ratios

In order to understand fully the concept of trigonometric ratios, consider the following diagram presenting a series of **'similar'** triangles.

Remember that **similar** triangles were defined in Chapter 9 as two or more triangles whose corresponding angles are equal and whose corresponding sides are in the same ratio.

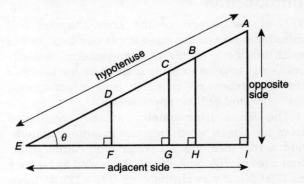

For example:

(i) In triangles AEI and CEG,

$$\frac{AI}{EI} = \frac{CG}{EG}.$$

(ii) In triangles DEF and BEH,

$$\frac{EF}{ED} = \frac{EH}{EB}.$$

(iii) In triangles CEG and BEH,

$$\frac{EG}{CG} = \frac{EB}{BH}.$$

Similarly, with all other triangles included in the diagram, the ratios of corresponding sides are equal.

10.2.1 The sine ratio

In any right-angled triangle, the ratio

$$\frac{\text{opposite side}}{\text{hypotenuse}}$$

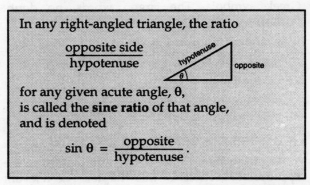

for any given acute angle, θ, is called the **sine ratio** of that angle, and is denoted

$$\sin\theta = \frac{\text{opposite}}{\text{hypotenuse}}.$$

For example: In the following diagram, the sine ratio of angle B is $\frac{3}{4}$.

That is, $\sin B = \dfrac{\text{opposite}}{\text{hypotenuse}} = \dfrac{3}{4}$.

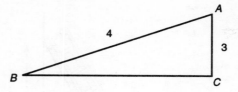

Note:

1. From the diagram at the beginning of this section it should be clear that, for any given angle, θ, the sine ratio *always* has the same value no matter how large the right-angled triangle.

2. The **sine ratio** can be used to
    (a) find the length of the opposite side or the hypotenuse of a right-angled triangle,
    (b) evaluate an angle given the opposite side and the hypotenuse.

    For **(a)** and **(b)** above, a scientific calculator is essential.

EXAMPLE 10.1: Determining a side length
Find the value of x in **(a)** and **(b)** below, correct to 2 decimal places.

(a)

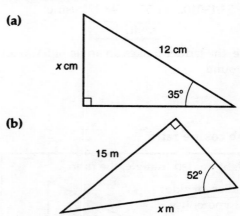

(b)

Solution

(a) From the sine ratio we know that

$$\sin \theta = \frac{\text{opposite}}{\text{hypotenuse}}.$$

Therefore, on substitution, we have

$$\sin 35° = \frac{x}{12}$$

or $x = 12 \sin 35°$ (by multiplying both sides by 12).

Using the calculator to evaluate $12 \sin 35°$, we

have

12 [x] 35 [sin] [=] [6.8829172]

Therefore, $x = 6.88$ cm (to 2 decimal places).

(b) From the sine ratio,

$$\sin 52° = \frac{15}{x}$$

or $x = \dfrac{15}{\sin 52°}$ (multiply both sides by x and divide both sides by $\sin 52°$ to make x the subject).

Using the calculator to evaluate $\dfrac{15}{\sin 52°}$, we have

15 [÷] 52 [sin] [=] [19.035273]

Therefore, $x = 19.04$ metres (to 2 decimal places).

EXAMPLE 10.2: Determining an angle
In the following diagram, find the angle marked θ, to the nearest degree.

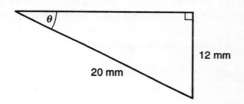

Solution
Since the lengths of both the opposite side and the hypotenuse are given, the sine ratio is appropriate for evaluating the angle θ.

We know that

$$\sin \theta = \frac{\text{opposite}}{\text{hypotenuse}}$$

$$= \frac{12}{20}$$

$$= \frac{3}{5}$$

Since $\sin \theta$ is known, we can now determine the value of θ by using $\sin^{-1}$. This is located on most calculators *above* the [sin] key and is usually accessed by pressing [shift] [sin] or [inv] [sin].

Therefore, to find θ we have

3 [÷] 5 [=] [sin⁻¹] [36.869898]

Therefore, $\theta = 37°$ (to the nearest whole degree).

Note:

The answer obtained above – namely, 36.869898 – is in decimal form. (That is, .869898 indicates the decimal part of one degree.)

However, from the previous chapter (Ch. 9) we know that $1° = 60$ minutes (denoted 60') and $1' = 60$ seconds (denoted 60").

Hence, if we wish to obtain the exact value of an angle in degrees (°), minutes (') and seconds ("), we can convert the degrees (in decimal form) to °,', " by pressing $\boxed{\text{shift}}$ $\boxed{\text{° ' "}}$ on our calculator.

In Example 10.2 above, we could find the exact value of θ in degrees, minutes and seconds by pressing

$$3 \boxed{÷} 5 \boxed{=} \boxed{\sin^{-1}} \boxed{\text{shift}} \boxed{\text{° ' "}} \quad \boxed{36°52°11.63}$$

Therefore, $θ = 36° \ 52' \ 11"$.

EXAMPLE 10.3: Word problems

(a) A girl is flying a kite on 150 metres of light line that makes an angle of 56° with the horizontal. What is the vertical height of the kite above the girl's hand? (Give the answer to the nearest metre.)

(b) A 25 metre ladder standing on level ground reaches 20 metres up a vertical wall. Find, to the nearest degree, the angle the ladder makes with the ground.

Solution

(a)

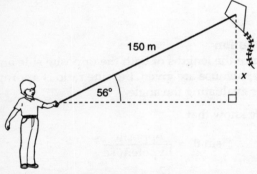

From the diagram, we require the length of x.

Using the sine ratio, we know that

$$\sin θ = \frac{\text{opposite}}{\text{hypotenuse}}.$$

Hence, $\sin 56° = \dfrac{x}{150}$

or $x = 150 \sin 56°$ (making x the subject).

Using the calculator to evaluate $150 \sin 56°$, we have

$$150 \boxed{×} 56 \boxed{\sin} \boxed{=} \boxed{124.35564}$$

∴ $x = 124.35564$ metres.

Thus, the kite is 124 metres higher than the girl's hand.

(b)

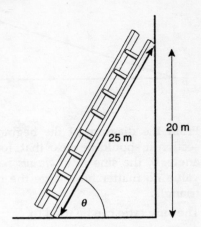

As shown in the above diagram, we are required to find θ, the angle the ladder makes with the ground.

Using the sine ratio, we have

$$\sin θ = \frac{20}{25} = \frac{4}{5}.$$

Using the calculator to find θ, we have

$$4 \boxed{÷} 5 \boxed{=} \boxed{\sin^{-1}} \boxed{53.130102}$$

∴ $θ = 53.130102°$ or $53° \ 7' \ 48.37"$ using $\boxed{\text{shift}}$ $\boxed{\text{° ' "}}$.

Hence, the ladder makes an angle of 53° with the ground.

10.2.2 The cosine ratio

> In any right-angled triangle, the ratio
>
> $$\frac{\text{adjacent side}}{\text{hypotenuse}}$$
>
> for any given acute angle, θ, is called the **cosine ratio** of that angle, and is denoted
>
> $$\cos θ = \frac{\text{adjacent}}{\text{hypotenuse}}.$$

For example: In the following diagram, the cosine ratio of angle B is $\frac{6}{8}$.

That is, $\cos B = \dfrac{\text{adjacent}}{\text{hypotenuse}} = \dfrac{6}{8}$.

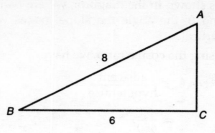

Note:

1. Once again, it should be clear from the diagram at the beginning of this section, that for any given angle, θ, the cosine ratio *always* has the same value no matter how large the right-angled triangle.

2. The **cosine ratio** can be used to

   (a) find the length of the adjacent side or the hypotenuse of a right-angled triangle,

   (b) evaluate an angle given the adjacent side and the hypotenuse.

   For (a) and (b) above, a scientific calculator is essential.

EXAMPLE 10.4: Determining a side length
Find the value of x in (a) and (b) below, correct to 2 decimal places.

(a)

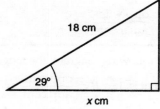

(b)

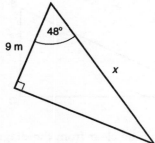

Solution

(a) From the cosine ratio we know that

$$\cos \theta = \frac{\text{adjacent}}{\text{hypotenuse}}.$$

Therefore, on substitution, we have

$$\cos 29° = \frac{x}{18}$$

or $x = 18 \cos 29°$ (multiplying both sides by 18).

Using the calculator to evaluate 18 cos 29°, we have

18 $\boxed{\text{x}}$ 29 $\boxed{\text{cos}}$ $\boxed{=}$ $\boxed{15.743155}$

∴ $x = 15.74$ cm (to 2 decimal places).

(b) From the cosine ratio,

$$\cos 48° = \frac{9}{x}$$

or $x = \frac{9}{\cos 48°}$ (making x the subject).

Using the calculator to evaluate $\frac{9}{\cos 48°}$, we have

9 $\boxed{÷}$ 48 $\boxed{\text{cos}}$ $\boxed{=}$ $\boxed{13.450289}$

∴ $x = 13.45$ metres (to 2 decimal places).

EXAMPLE 10.5: Determining an angle
In the following diagram, find the angle marked θ, to the nearest degree.

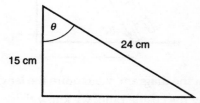

Solution
Since the lengths of both the adjacent side and the hypotenuse are given, the cosine ratio is appropriate for evaluating the angle θ.
We know that

$$\cos \theta = \frac{\text{adjacent}}{\text{hypotenuse}}$$

$$= \frac{15}{24}.$$

Since cos θ is known, we can now determine the value of θ by using $\cos^{-1}$. This is located on most calculators *above* the $\boxed{\text{cos}}$ button and is accessed by pressing $\boxed{\text{shift}}$ $\boxed{\text{cos}}$ or $\boxed{\text{inv}}$ $\boxed{\text{cos}}$.
Therefore, to find θ we have

15 $\boxed{÷}$ 24 $\boxed{=}$ $\boxed{\cos^{-1}}$ $\boxed{51.317813}$

Therefore, θ = 51° (to the nearest whole degree).

To find the *exact* size of θ (in degrees, minutes and seconds) we would press

15 $\boxed{÷}$ 24 $\boxed{=}$ $\boxed{\cos^{-1}}$ $\boxed{\text{shift}}$ $\boxed{° ' "}$ $\boxed{51° 19° 4.13}$
Therefore, more exactly, θ = 51° 19' 4.13".

EXAMPLE 10.6: Word problems

(a) A 25 metre ladder on level ground makes an angle of 62° with the horizontal as it leans against a wall. How far is the foot of the ladder from the base of the wall? (Answer correct to 1 decimal place.)

(b) A woman skis 450 metres straight down an even slope and in so doing loses 60 metres in height. What angle (to the nearest degree) does the slope make with the vertical?

Solution

(a)

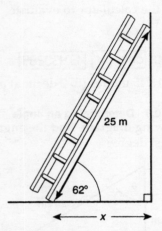

From the diagram, we require the length of x.

Using the cosine ratio, we know that

$$\cos \theta = \frac{\text{adjacent}}{\text{hypotenuse}}.$$

Hence, $\cos 62° = \dfrac{x}{25}$

or $x = 25 \cos 62°$ (making x the subject).

Using the calculator to evaluate $25 \cos 62°$, we have

25 ⨉ 62 |cos| | = | | 11.736789 |

Therefore, $x = 11.736789$ metres.

Thus, the foot of the ladder is 11.7 metres from the base of the wall.

(b)

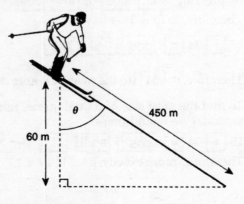

As shown in the diagram, we are required to find θ, the angle the slope makes with the vertical.

Using the cosine ratio, we have

$$\cos \theta = \frac{\text{adjacent}}{\text{hypotenuse}}$$

$$= \frac{60}{450}.$$

Using the calculator to find θ, we have

60 | ÷ | 450 | = | |cos⁻¹| | 82.337744 |

Therefore, $\theta = 82.337744°$ or 82° 20' 16" using

|shift| |° ' "| .

Hence, the slope of the mountain makes an angle of 82° with the vertical.

10.2.3 The tangent ratio

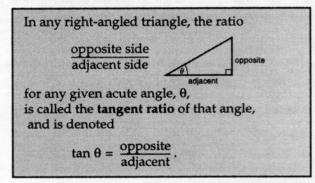

In any right-angled triangle, the ratio

$$\frac{\text{opposite side}}{\text{adjacent side}}$$

for any given acute angle, θ, is called the **tangent ratio** of that angle, and is denoted

$$\tan \theta = \frac{\text{opposite}}{\text{adjacent}}.$$

For example: In the following diagram, the tangent ratio of angle B is $\dfrac{9}{12}$. That is, $\tan B = \dfrac{\text{opposite}}{\text{adjacent}} = \dfrac{9}{12}$.

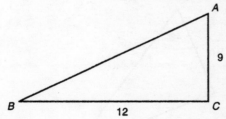

Note:

1. Again it should be clear from the diagram at the beginning of this section, that for any given angle, θ, the tangent ratio *always* has the same value regardless of the size of the right-angled triangle.

2. The **tangent ratio** can be used to

(a) find the length of the opposite or the adjacent side of a right-angled triangle

(b) evaluate an angle given the lengths of the opposite and adjacent sides.

For **(a)** and **(b)** above, a scientific calculator is essential.

EXAMPLE 10.7: Determining a side length

Find the value of x in **(a)** and **(b)** below, correct to 2 decimal places.

(a)

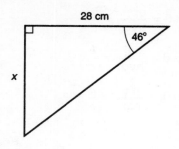

(b)

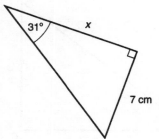

Solution

(a) From the tangent ratio we know that

$$\tan \theta = \frac{\text{opposite}}{\text{adjacent}}.$$

Therefore, on substitution, we have

$$\tan 46° = \frac{x}{28}$$

or $x = 28 \tan 46°$ (making x the subject).

Using the calculator to evaluate $28 \tan 46°$, we have

$$28 \boxed{\times} 46 \boxed{\tan} \boxed{=} \boxed{28.994849}$$

Therefore, $x = 28.99$ cm (to 2 decimal places).

(b) From the tangent ratio,

$$\tan 31° = \frac{7}{x}$$

or $x = \dfrac{7}{\tan 31°}$ (making x the subject).

Using the calculator to evaluate $\dfrac{7}{\tan 31°}$, we have

$$7 \boxed{\div} 31 \boxed{\tan} \boxed{=} \boxed{11.649956}$$

Therefore, $x = 11.65$ metres (to 2 decimal places).

EXAMPLE 10.8 : Determining an angle

In the following diagram, find the angle marked θ, to the nearest degree.

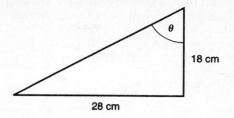

Solution

Since the lengths of both the adjacent and opposite sides are given, the tangent ratio is appropriate for evaluating the angle θ.

We know that, $\tan \theta = \dfrac{\text{opposite}}{\text{adjacent}}$

$$= \frac{28}{18}.$$

Since $\tan \theta$ is known, we can now determine the value of θ by using $\tan^{-1}$. This is located on most calculators *above* the $\boxed{\tan}$ button and is accessed by pressing $\boxed{\text{shift}}$ $\boxed{\tan}$ or $\boxed{\text{inv}}$ $\boxed{\tan}$.

Therefore, to find θ we have

$$28 \boxed{\div} 18 \boxed{=} \boxed{\tan^{-1}} \boxed{57.264774}$$

Thus, $\theta = 57°$ (to the nearest whole degree).

To find the *exact* size of θ (in degrees, minutes and seconds) we would press

$$28 \boxed{\div} 18 \boxed{=} \boxed{\tan^{-1}} \boxed{\text{shift}} \boxed{° ' "} \boxed{57° 15° 53.19}$$

Therefore, more exactly, $\theta = 57° 15' 53.19"$.

EXAMPLE 10.9: Word problems

(a) The angle of elevation of the sun is 20° and the length of the shadow cast by a particular tree is 35 metres on ground level. Find the height of the tree in metres, correct to 1 decimal place.

(b) A ladder reaches 16 metres up a vertical wall and has its foot on level ground 7 metres from the base of the wall. Determine the angle the ladder makes with the wall.

Solution

(a)

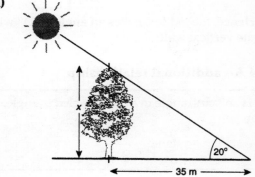

From the diagram, we require the length of x. Using the tangent ratio, we know that

$$\tan \theta = \frac{\text{opposite}}{\text{adjacent}}.$$

Hence, $\tan 20° = \frac{x}{35}$

or $x = 35 \tan 20°$ (making x the subject).

Using the calculator to evaluate $35 \tan 20°$, we have

$$35 \boxed{\times} 20 \boxed{\text{tan}} \boxed{=} \boxed{12.738958}$$

Therefore, $x = 12.738958$.

Thus, the height of the tree in question is 12.7 metres.

(b)

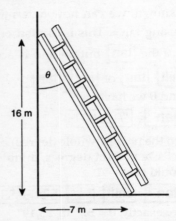

As shown in the diagram, we are required to find θ, the angle the ladder makes with the wall.

Using the tangent ratio, we have

$$\tan \theta = \frac{\text{opposite}}{\text{adjacent}}$$

$$= \frac{7}{16}.$$

Using the calculator to find θ, we have

$$7 \boxed{\div} 16 \boxed{=} \boxed{\text{tan}^{-1}} \boxed{23.629378}$$

Therefore, $\theta = 23.629378°$ or $23° \ 37' \ 46''$ using $\boxed{\text{shift}} \ \boxed{° \, ' \, ''}$.

Hence, the ladder makes an angle of 24° with the vertical wall.

10.2.4 An additional relationship

If θ is an acute angle in a right-angled triangle, then

$$\tan \theta = \frac{\sin \theta}{\cos \theta}.$$

Why?
We already know from our trigonometric ratios that

(i) $\sin \theta = \dfrac{\text{opp.}}{\text{hyp.}} = \dfrac{a}{c}$

(ii) $\cos \theta = \dfrac{\text{adj.}}{\text{hyp.}} = \dfrac{b}{c}$

(iii) $\tan \theta = \dfrac{\text{opp.}}{\text{adj.}} = \dfrac{a}{b}$

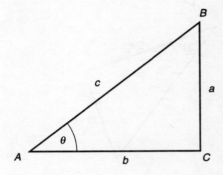

Now, $\dfrac{\sin \theta}{\cos \theta} = \dfrac{\frac{a}{c}}{\frac{b}{c}} = \dfrac{a}{c} \times \dfrac{c}{b} = \dfrac{a}{b} = \tan \theta.$

Therefore, we know that

$$\tan \theta = \frac{\sin \theta}{\cos \theta}.$$

For example: If $\theta = 50°$, then (from the calculator)

(i) $\sin 50° = 0.7660444$

(ii) $\cos 50° = 0.6427876$

(iii) $\tan 50° = 1.1917536$

Now, $\dfrac{\sin 50°}{\cos 50°} = \dfrac{0.7660444}{0.6427876} = 1.1917535 = \tan 50°$

Therefore, as expected, $\tan 50° = \dfrac{\sin 50°}{\cos 50°}$.

Final note:

Since $\tan \theta = \dfrac{\sin \theta}{\cos \theta}$, using simple algebraic manipulation we can say that

(i) $\sin \theta = \tan \theta \cdot \cos \theta$, and

(ii) $\cos \theta = \dfrac{\sin \theta}{\tan \theta}$.

In each of the following exercises you will first need to determine which of the three trigonometric ratios should be used to yield a solution.

10.1 In each of the following, determine the length of the unknown side in centimetres (correct to 1 decimal place).

(a)

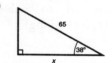

(b)

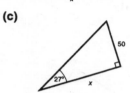

(c)

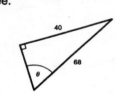

10.2 In each of the following, determine the size of the unknown angle, to the nearest degree.

(a)

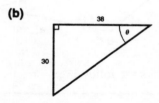

(b)

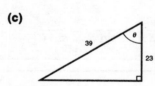

(c)

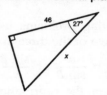

10.3 In each of the following, determine the length of the unknown side in millimetres (correct to 2 decimal places).

(a)

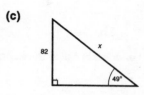

10.3 *Cont.*

(b)

(c)

10.4 In △ABC, ∠A = 42°, ∠B = 90° and BC = 80 mm. Draw a diagram marking the given information and hence find the length of CA to the nearest millimetre.

10.5 In △XYZ, XY = 46.4 cm., YZ = 54.2 cm. and ∠Y = 90°. Draw a diagram showing the given information and hence determine
  (i) the size of angle Z, to the nearest degree
  (ii) the length of side XZ, correct to the nearest centimetre.

10.6 An isosceles triangle ABC has sides AB = BC = 36 cm. and a perpendicular height BD of 25 cm. Find the size of angle ABC, to the nearest degree.

  (**Hint:** First draw a diagram and mark the information given.)

10.7 A ladder has its foot 3.5 metres from the base of a wall on level ground. If it makes an angle of 65° with the horizontal, how far up the wall does it reach ?

  (Answer to the nearest centimetre.)

10.8 A flagpole is supported by a wire stay that is attached 9 metres above the base of the pole. Find the length of the wire if it makes an angle of 58° with the horizontal.

10.9 Calculate the angle of elevation of the sun when a tree 15 metres high casts a shadow 23 metres long on the ground. (Answer to the nearest degree.)

10.10 Refer back to exercise 10.9. Determine the length of the shadow cast by the tree when the sun has sunk to *half* the angle of elevation just determined.

  (Answer correct to 2 decimal places.)

10.3 Reciprocal ratios

In *Section 2* we defined what are usually referred to as the three basic trigonometric ratios – namely, sin θ, cos θ and tan θ.

We can now define another set of very useful trigonometric ratios usually referred to as the **reciprocal ratios.**

Definitions

1. The reciprocal of $\sin\theta$ is called the **cosecant** of θ ($\operatorname{cosec}\theta$).

$$\therefore\ \operatorname{cosec}\theta = \frac{1}{\sin\theta}$$

2. The reciprocal of $\cos\theta$ is called the **secant** of θ ($\sec\theta$).

$$\therefore\ \sec\theta = \frac{1}{\cos\theta}$$

3. The reciprocal of $\tan\theta$ is called the **cotangent** of θ ($\cot\theta$).

$$\therefore\ \cot\theta = \frac{1}{\tan\theta}.$$

Note:

$\dfrac{1}{\sin\theta}$ is called the **reciprocal** of $\sin\theta$, while

$\sin^{-1}\theta$ is called the **inverse** of $\sin\theta$.

Similarly, $\dfrac{1}{\cos\theta}$ and $\dfrac{1}{\tan\theta}$ are the **reciprocals** of $\cos\theta$ and $\tan\theta$ respectively, while $\cos^{-1}\theta$ and $\tan^{-1}\theta$ are their corresponding **inverses.**

Beware:
With all trigonometric functions, reciprocals and inverses are *not* equal. In fact, they represent two totally different concepts.

Hence, (i) $\operatorname{cosec}\theta = \dfrac{1}{\sin\theta} \neq \sin^{-1}\theta$

(ii) $\sec\theta = \dfrac{1}{\cos\theta} \neq \cos^{-1}\theta$

(iii) $\cot\theta = \dfrac{1}{\tan\theta} \neq \tan^{-1}\theta$

For example: Consider the angle $\theta = 35°$.

Then, (i) $\tan 35° = 0.7002075$

(ii) $\dfrac{1}{\tan 35°} = 1.428148$

(iii) $\tan^{-1} 35 = 88.363423$.

Clearly, $\dfrac{1}{\tan 35°} \neq \tan^{-1} 35$.

Final note:

Since $\tan\theta = \dfrac{\sin\theta}{\cos\theta}$ (see *Section 2*), we have that

$$\cot\theta = \frac{1}{\tan\theta} = \frac{\cos\theta}{\sin\theta}.$$

10.4 Non-right-angled triangles

In the preceding sections all the trigonometric ratios and associated formulae that were developed and presented apply only to right-angled triangles, and hence are rather limiting in the types of practical real-life problems that can be solved.

However, we are now able to build upon these concepts in order to develop further rules and formulae that allow us to solve problems associated with any type of triangle.

10.4.1 The sine rule

In *any* triangle *ABC*,

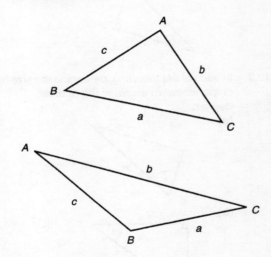

the **sine rule** states the following:

Sine rule 1:

$$\frac{a}{\sin A} = \frac{b}{\sin B} = \frac{c}{\sin C}$$

Sine rule 2:

$$\frac{\sin A}{a} = \frac{\sin B}{b} = \frac{\sin C}{c}$$

Note:
When actually solving problems using the sine rule, usually only two of the three expressions are required, e.g. $\dfrac{a}{\sin A} = \dfrac{b}{\sin B}$ or $\dfrac{\sin B}{b} = \dfrac{\sin C}{c}$ etc. The choice depends on the information given.

When is the sine rule applicable?

a. **Sine rule 1** enables us to determine a side, given a side and two internal angles.

b. **Sine rule 2** enables us to determine an angle, given two sides and an angle that is *not* the included angle.

EXAMPLE 10.10

Find the length of side b in the following triangles. (Answer correct to 2 decimal places.)

(a)

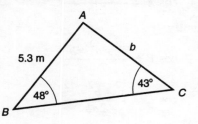

(b)

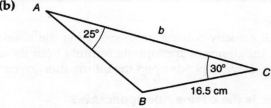

Solution

(a) Here we are required to find side b given side c and opposite angles B and C.

Hence, **sine rule 1** is appropriate, using

$$\frac{b}{\sin B} = \frac{c}{\sin C}.$$

On substitution, we have

$$\frac{b}{\sin 48°} = \frac{5.3}{\sin 43°}$$

or $\quad b = \dfrac{5.3 \sin 48°}{\sin 43°}$

Using the calculator, we have

5.3 $\boxed{\times}$ 48 $\boxed{\sin}$ $\boxed{\div}$ 43 $\boxed{\sin}$ $\boxed{=}$ $\boxed{5.7751863}$

Therefore, the length of side b in $\triangle ABC$ is 5.78 metres.

(b) Here we are again required to find side b. However, before we can apply **sine rule 1**, we must first find the size of angle B, since the rule involves sides and their *opposite* angles. Since the internal angles of a triangle must total 180°, we have that

$$\angle B = 180 - (25 + 30) = 125°.$$

Then, applying the **sine rule 1**, we have

$$\frac{b}{\sin B} = \frac{a}{\sin A},$$

giving, on substitution,

$$\frac{b}{\sin 125°} = \frac{16.5}{\sin 25°}.$$

Hence, $b = \dfrac{16.5 \sin 125°}{\sin 25°}$

$\qquad = \dfrac{16.5\,(0.819152)}{(0.4226182)}$

$\qquad = 31.9816$

Therefore, the length of side b in $\triangle ABC$ is 31.98 cm.

EXAMPLE 10.11

Find the size of angle Y in the following triangles.

(a)

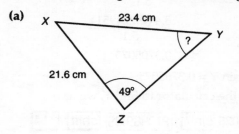

(b)

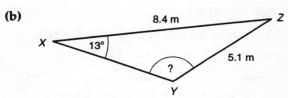

Solution

(a) Here we are required to find angle Y given the opposite side y together with another angle and its opposite side.

Hence, **sine rule 2** is appropriate, using

$$\frac{\sin Y}{y} = \frac{\sin Z}{z}.$$

On substitution, we have

$$\frac{\sin Y}{21.6} = \frac{\sin 49°}{23.4}$$

or $\quad \sin Y = \dfrac{21.6 \sin 49°}{23.4}.$

Using the calculator to find $\dfrac{21.6 \sin 49°}{23.4}$, we have

21.6 $\boxed{\times}$ 49 $\boxed{\sin}$ $\boxed{\div}$ 23.4 $\boxed{=}$ $\boxed{0.696655}$

Hence, $\sin Y = 0.696655$.

Using the calculator to find Y, we have

.696655 $\boxed{\sin^{-1}}$ $\boxed{44.159246}$

Therefore, the size of angle Y in $\triangle XYZ$ is
$\angle Y = 44.16°$ or $\angle Y = 44° \, 9' \, 33''$.

(That is, as a decimal, $\angle Y = 44.16°$, but in degrees, minutes and seconds, using

$\boxed{\text{shift}}$ $\boxed{° \, ' \, ''}$, $\angle Y = 44°9' \, 33''$.)

b) Again, the **sine rule 2** is appropriate for finding angle Y with the information given.

Here, $\quad \dfrac{\sin Y}{y} = \dfrac{\sin X}{x}$

becomes $\dfrac{\sin Y}{8.4} = \dfrac{\sin 13°}{5.1}$, on substitution.

Hence, $\quad \sin Y = \dfrac{8.4 \sin 13°}{5.1}$

$$= \dfrac{8.4 \, (0.224951)}{5.1}$$

$$= 0.3705075$$

Thus, $\sin Y = 0.3705075$.

Using the calculator to find Y, we have

0.3705075 $\boxed{\sin^{-1}}$ $\boxed{21.7469195}$ $\boxed{\text{shift}}$ $\boxed{° \, ' \, ''}$

$\boxed{21° \, 44° \, 48.91}$

Therefore, the size of angle Y in $\triangle XYZ$ is
$\angle Y = 21.75°$ or $\angle Y = 21°44' \, 49''$.

Note:

In decimal form, $\angle Y = 21.75°$, while in expanded form, $\angle Y = 21° \, 44' \, 49''$.

10.4.2 The cosine rule

In *any* triangle *ABC*,

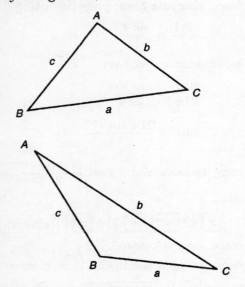

the **cosine rule** states the following:

Cosine rule 1:
$$a^2 = b^2 + c^2 - 2bc \cos A$$
or equivalently
$$b^2 = c^2 + a^2 - 2 \, ca \cos B$$
$$c^2 = a^2 + b^2 - 2 \, ab \cos C$$

Cosine rule 2:
$$\cos A = \dfrac{b^2 + c^2 - a^2}{2bc}$$
or equivalently
$$\cos B = \dfrac{c^2 + a^2 - b^2}{2ca}$$
$$\cos C = \dfrac{a^2 + b^2 - c^2}{2ab}$$

Note:
When actually solving problems using the cosine rule, the choice of appropriate formula from those given above depends upon the information given.

When is the cosine rule applicable?

a. Given the lengths of any two sides of a triangle *and* the size of the *included* angle, the **cosine rule 1** enables us to determine the length of the third side.

b. Given the lengths of all three sides of a triangle, the **cosine rule 2** enables us to determine the size of any internal angle.

EXAMPLE 10.12
Find the length of side x in the following triangles. (Answer correct to 2 decimal places.)

(a)

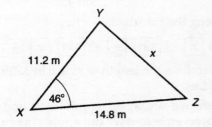

(b)

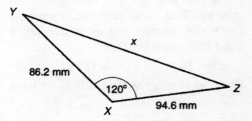

Solution

(a) Here we are required to find side x, given the lengths of sides y and z and the size of the included angle.

Hence, **cosine rule 1** is appropriate, using

$$x^2 = y^2 + z^2 - 2yz \cos x.$$

On substitution, we have

$$x^2 = 14.8^2 + 11.2^2 - 2(14.8)(11.2)(\cos 46°)$$
$$= 219.04 + 125.44 - 331.52 \cos 46°$$
$$= 344.48 - 230.2934$$
$$= 114.18686$$

Thus, $x = \sqrt{114.18686} = 10.685825$.

Therefore, the length of side x in $\triangle XYZ$ is 10.69 metres.

(b) Again, given two sides and the included angle, the **cosine rule 1** is appropriate for determining the length of side x.

From the diagram,

$$x^2 = 86.2^2 + 94.6^2 - 2(86.2)(94.6) \cos 120°$$
$$= 7430.44 + 8949.16 - 16309.04 \cos 120°$$
$$= 16379.6 - (-8154.52)$$
$$= 24534.12$$

Thus, $x = \sqrt{24534.12} = 156.63$.

Therefore, the length of side x in $\triangle XYZ$ is 156.63 mm.

EXAMPLE 10.13

Find the size of angle C in the following triangles.

(a)

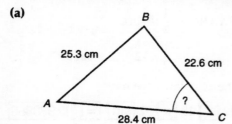

(b)

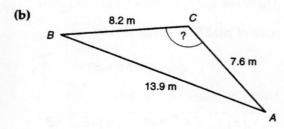

Solution

(a) Since we are required to find the size of $\angle C$ given the lengths of the three sides, the

cosine rule 2 is appropriate, with

$$\cos C = \frac{a^2 + b^2 - c^2}{2ab}.$$

On substitution, we have

$$\cos C = \frac{22.6^2 + 28.4^2 - 25.3^2}{2(22.6)(28.4)}$$
$$= \frac{677.23}{1283.68}$$
$$= 0.5275691$$

Hence, $\cos C = 0.5275691$.

Using the calculator to find C, we have

$$0.5275691 \ \boxed{\cos^{-1}} \ \boxed{58.158645} \ \boxed{\text{shift}} \ \boxed{° \ ' \ ''}$$
$$\boxed{58° \ 9° \ 31.1}$$

Therefore, the size of $\angle C$ in triangle ABC is $\angle C = 58.16°$ or $\angle C = 58° \ 9' \ 31.1''$.

(b) Again, with the information given, **cosine rule 2** is appropriate for determining the size of angle C.

Here, $\cos C = \dfrac{a^2 + b^2 - c^2}{2ab}$

$$= \frac{8.2^2 + 7.6^2 - 13.9^2}{2(8.2)(7.6)}$$
$$= \frac{67.24 + 57.56 - 193.21}{124.64}$$
$$= \frac{-68.21}{124.64}$$
$$= -0.5472561$$

Hence, $\cos C = -0.5472561$

Now, using $\boxed{\cos^{-1}}$ and $\boxed{\text{shift}}$ $\boxed{° \ ' \ ''}$, we have that

$$C = 123.1789729° \ (= 123° \ 10' \ 44.3'').$$

Therefore, the size of angle C in triangle ABC is

$$\angle C = 123.18° \text{ or } \angle C = 123°10'.$$

10.4.3 Area of a triangle

In this sub-section we examine an alternative method for calculating the area of any given triangle.

Recall:

Until now, the only method available for calculating the area of a triangle is to apply the formula $A = \dfrac{1}{2}bh$,

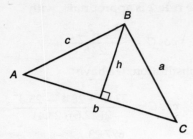

where A = area

b = length of triangle base

h = perpendicular height.

However, the trigonometric ratios defined in *Section 2*, now allow us to determine the area of any triangle without relying on information about the perpendicular height. In fact, provided we know the length of any two sides and the size of the *included* angle, then

> **Area of triangle** $= \dfrac{1}{2}$ x product of lengths of any two sides x sine of *included* angle.

Hence, given the following triangles

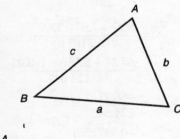

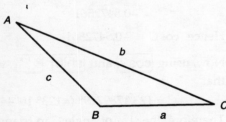

$$\text{Area of } \triangle ABC = \frac{1}{2}ab\sin C$$

$$= \frac{1}{2}ac\sin B$$

$$= \frac{1}{2}cb\sin A$$

(See Example 10.15 for a simple proof).

Note:
The formula given above involves multiplying half the product of two sides by the sine of the *included* angle. Using the correct angle is *essential*, and by *included* angle we mean the angle lying between two given *adjacent* sides.

EXAMPLE 10.14
Find the area of the following triangles.

(a)

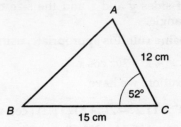

(b)

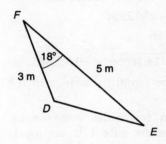

(c)

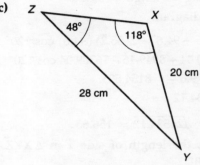

Solution

(a) $\text{Area of } \triangle ABC = \dfrac{1}{2}ab\sin C$

$$= \frac{1}{2} \times 12 \times 15 \times \sin 52°$$

Using the calculator, we have

$$.5 \boxed{\times} \; 12 \; \boxed{\times} \; 15 \; \boxed{\times} \; 52 \; \boxed{\sin} \; \boxed{=} \; \boxed{70.920968}$$

Therefore, the required area is 70.92 (cm)^2.

(b) $\text{Area of } \triangle DEF = \dfrac{1}{2}de\sin F$

$$= \frac{1}{2} \times 5 \times 3 \times \sin 18°$$

Using the calculator, we have

$$.5 \boxed{\times} \; 5 \; \boxed{\times} \; 3 \; \boxed{\times} \; 18 \; \boxed{\sin} \; \boxed{=} \; \boxed{2.3176275}$$

Therefore, the required area is 2.32 (metres)^2.

(c) From the diagram, neither $\angle X$ nor $\angle Z$ (both given) are *included* angles. We need angle Y. However, since the three internal angles of a

triangle must total 180°, we have that

$$Y = 180 - (118 + 48)$$
$$= 14°$$

Hence, area of $\triangle XYZ = \frac{1}{2}xz \sin Y$

$$= \frac{1}{2} \times 28 \times 20 \times \sin 14°$$

Using the calculator, we have

 .5 ⊠ 28 ⊠ 20 ⊠ 14 sin = 67.738131

Therefore, the required area is 67.74 (cm)².

EXAMPLE 10.15

For $\triangle ABC$ below, show that its area given by
$A = \frac{1}{2}bh$, is exactly equivalent to its area given by
$A = \frac{1}{2}ab \sin C$ or $A = \frac{1}{2}bc \sin A$.

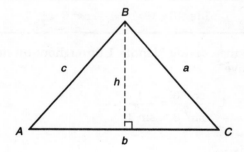

Solution

From *Section 2* we know that, in $\triangle ABC$ above,

(i) $\sin C = \dfrac{\text{opposite}}{\text{hypotenuse}} = \dfrac{h}{a}$, and

(ii) $\sin A = \dfrac{\text{opposite}}{\text{hypotenuse}} = \dfrac{h}{c}$

Hence, from (i) and (ii) above, we have

(iii) $h = a \sin C$ (by multiplying both sides by a), and

(iv) $h = c \sin A$ (by multiplying both sides by c).

Therefore, substituting for h in the formula
$A = \frac{1}{2}bh$, we have that

$$A = \frac{1}{2}b(a \sin C) = \frac{1}{2} ab \sin C, \text{ and}$$

$$A = \frac{1}{2}b(c \sin A) = \frac{1}{2} bc \sin A.$$

Hence, for any triangle ABC, labelled as above, we have that

$$\text{Area of } \triangle ABC = \begin{cases} \dfrac{1}{2}bh \\[6pt] \dfrac{1}{2}ab \sin C \\[6pt] \dfrac{1}{2}bc \sin A \\[6pt] \dfrac{1}{2}ac \sin B \end{cases}$$

YOUR TURN

10.11 In each of the following triangles find the length of the side labelled y, correct to 2 decimal places. (*Hint:* use sine rule 1).

(a)

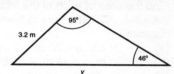

(b)

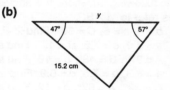

10.12 In each of the following triangles find the size of the angle labelled θ in degrees and minutes, correct to the nearest minute. (*Hint:* use sine rule 2.)

(a)

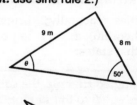

(b)

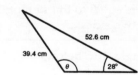

10.13 Evaluate d, correct to 1 decimal place, in each of the following triangles.

(a)

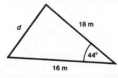

(b)

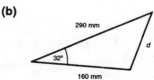

10.14 Using cosine rule 2, evaluate θ (in degrees and minutes) in the following triangles.

(a)

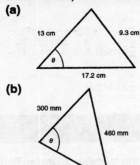

13 cm 9.3 cm

θ

17.2 cm

(b)

300 mm

460 mm

θ

325 mm

10.15 Find the area of each of the following triangles, *ABC*. (Be sure to draw a diagram.)
  (a) $a = 24$ cm, $b = 32$ cm, $C = 65°$.
  (b) $b = 28$ m, $c = 18$ m, $A = 150°$.
  (c) $c = 15$ mm, $a = 12$ mm, $B = 49°$.

10.16 In the following questions the angles and sides refer to △ *ABC*.
  (a) $b = 4$, $c = 6$, $C = 60°$. Find ∠ *B*.
  (b) $a = 3.9$, $c = 7.8$, $B = 45°$. Find side *b*.
  (c) $A = 37°$, $B = 48°$, $b = 16$. Find side *a*.
  (d) $a = 8.3$, $b = 7.5$, $c = 6.9$. Find ∠ *B*.
  (e) $C = 32°$, $A = 67°$, $c = 12$. Find side *b*.

10.5 The Pythagorean identities

These identities actually originated from **Pythagoras' theorem**, a very important mathematical result that states:

> In any right-angled triangle, the square of the hypotenuse is equal to the sum of the squares of the other two sides.

Hence, in triangle *ABC* below,

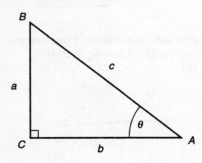

$$c^2 = a^2 + b^2,$$ by Pythagoras' theorem.

Now, if $c^2 = a^2 + b^2$, then dividing both sides by c^2 gives:

$$1 = \frac{a^2 + b^2}{c^2}$$

$$= \frac{a^2}{c^2} + \frac{b^2}{c^2}$$

$$= \left(\frac{a}{c}\right)^2 + \left(\frac{b}{c}\right)^2$$

In △ *ABC* above, let θ be the acute angle, as shown. Then, from our knowledge of trigonometric ratios (see *Section 2*), we know that

(i) $\sin\theta = \dfrac{a}{c}$

(ii) $\cos\theta = \dfrac{b}{c}$

Therefore, since $\left(\dfrac{a}{c}\right)^2 + \left(\dfrac{b}{c}\right)^2 = 1$, we have that

> **Identity 1:** $\sin^2\theta + \cos^2\theta = 1$

If we then divide Identity 1 throughout by $\sin^2\theta$, we have

$$1 + \frac{\cos^2\theta}{\sin^2\theta} = \frac{1}{\sin^2\theta}.$$

That is,

> **Identity 2:** $1 + \cot^2\theta = \operatorname{cosec}^2\theta$

On dividing Identity 1 above throughout by $\cos^2\theta$, we have

$$\frac{1}{\cos^2\theta} = \frac{\sin^2\theta}{\cos^2\theta} + 1.$$

That is,

> **Identity 3:** $\sec^2\theta = \tan^2\theta + 1$

Note:

$\sin^2\theta$, $\cos^2\theta$ ... $\sec^2\theta$ and $\cot^2\theta$, are all abbreviates of $(\sin\theta)^2$, $(\cos\theta)^2$ ... $(\sec\theta)^2$ and $(\cot\theta)^2$. Be very careful not to confuse $\sin^2\theta$ with $\sin\theta^2$, where only θ is squared.

To sum up, the Pythagorean identities may be written

  1. $\sin^2\theta + \cos^2\theta = 1$
  2. $\operatorname{cosec}^2\theta = 1 + \cot^2\theta$
  3. $\sec^2\theta = 1 + \tan^2\theta$

EXAMPLE 10.16

Using your calculator, check that

$$\sin^2 34.56789° + \cos^2 34.56789° = 1.$$

Solution

(i) From the calculator,

$$\sin 34.56789° = 0.5673823$$

$$\therefore \sin^2 34.56789° = 0.3219227$$

(ii) From the calculator,

$$\cos 34.56789° = 0.8234544$$

$$\therefore \cos^2 34.56789° = 0.6780773$$

Therefore, $\sin^2 34.56789° + \cos^2 34.56789°$

$$= 0.3219227 + 0.6780773$$

$$= 1, \text{ as expected.}$$

EXAMPLE 10.17

Find the acute angle, A, (in degrees, minutes and seconds) that satisfies

$$2 \cos^2 A = 1.864.$$

Solution

If $2 \cos^2 A = 1.864$, then

$$\cos^2 A = 0.932$$

and $\cos A = \sqrt{0.932}$

$$= 0.9654014$$

Hence, using the calculator to find A, we have

0.9654014 $\boxed{\cos^{-1}}$ $\boxed{\text{shift}}$ $\boxed{° ' ''}$ $\boxed{15° 6° 56.3}$

Therefore, if $2 \cos^2 A = 1.864$, then

$$A = 15° 6' 56.3''.$$

EXAMPLE 10.18

Simplify the following, expressing each in terms of a single trigonometric function, or as a constant:

(a) $1 - \cos^2 \theta$

(b) $(\sin A + \cos A)^2 - 2 \sin A \cos A$

Solution:

(a) From the Pythagorean identity 1, we know that

$$\sin^2 \theta + \cos^2 \theta = 1$$

Therefore, $\sin^2 \theta = 1 - \cos^2 \theta$.

(b) Expanding $(\sin A + \cos A)^2$ we have

$(\sin A + \cos A)^2$

$= \sin^2 A + 2\sin A \cos A + \cos^2 A$

Hence, $(\sin A + \cos A)^2 - 2 \sin A \cos A$

$= \sin^2 A + 2 \sin A \cos A + \cos^2 A - 2 \sin A \cos A$

$= \sin^2 A + \cos^2 A$

However, from the Pythagorean identity 1, we know that $\sin^2 A + \cos^2 A = 1$.

Therefore, $(\sin A + \cos A)^2 - 2 \sin A \cos A = 1$.

10.17 Using your calculator, check that
(a) $\sec^2 \theta = 1 + \tan^2 \theta$, where θ is any angle of your choice.
(b) $\csc^2 B = 1 + \cot^2 B$, where B is any angle of your choice.

10.18 Using your calculator, evaluate the following, correct to 3 decimal places.
(a) $\tan^2 38.7°$
(b) $\cos^2 43.8° - \tan^2 52.6°$
(c) $\sec^2 74°$

10.19 Find the acute angle that satisfies each of the following.

(Use your calculator and give your answer in degrees, minutes and seconds.)
(a) $\sin^2 \theta = 0.6425$
(b) $3.5 \tan^2 A = 6.395$
(c) $\sec^2 C = 3.752$

10.20 Simplify the following, expressing each in terms of a single trigonometric ratio, or as a constant.
(a) $\sec^2 \theta - \tan^2 \theta$
(b) $2 - 2 \sin^2 A$
(c) $\sqrt{1 - \cos^2 x}$
(d) $\tan \theta \sqrt{1 - \sin^2 \theta}$
(e) $\dfrac{1 - \cos^2 B}{\cos^2 B}$

Note:

Even though this is only the first of two chapters addressing the branch of mathematics called *trigonometry*, there are a number of additional topics that are best left to more advanced books. These topics include:

- the sum and difference of sin and cos
- double angles
- exact ratios
- angles of any magnitude
- trigonometric equations of the first degree
- trigonometric equations of the second degree, etc.

Summary

This is the first of two chapters dealing with topics in trigonometry, and presents the more basic trigonometric ratios and rules, together with some associated applications.

Under the heading **terminology and notation**, we revised the right-angled triangle, carefully defining the adjacent and opposite sides and the hypotenuse, and pointing out the conventional method of labelling all sides and angles.

The term **ratio** was also re-defined, in the trigonometric context, as a fraction $\frac{a}{b}$ rather than $a : b$ (as in Chapter 1).

In *Section 2* we presented the three basic trigonometric ratios, all associated with right-angled triangles.

1. The sine ratio: $\sin \theta = \dfrac{\text{opposite}}{\text{hypotenuse}}$

2. The cosine ratio: $\cos \theta = \dfrac{\text{adjacent}}{\text{hypotenuse}}$

3. The tangent ratio: $\tan \theta = \dfrac{\text{opposite}}{\text{adjacent}}$

Each of the above ratios allows us to determine an unknown side or angle in a right-angled triangle, with the choice of appropriate formula depending upon the information given.

Finally in this section we pointed out an additional and very useful relationship arising from the three basic ratios – namely,

$$\tan \theta = \frac{\sin \theta}{\cos \theta}.$$

With some algebraic manipulation, this relationship leads to

(i) $\sin \theta = \tan \theta \,.\, \cos \theta$, and

(ii) $\cos \theta = \dfrac{\sin \theta}{\tan \theta}$.

In *Section 3* we defined the *reciprocal* ratios, each of which is very useful in future trigonometric manipulations. These reciprocals are defined as

1. $\dfrac{1}{\sin \theta} = \operatorname{cosec} \theta$

2. $\dfrac{1}{\cos \theta} = \sec \theta$

3. $\dfrac{1}{\tan \theta} = \cot \theta$.

Here we also noted that, for all trigonometric functions, the *reciprocal* and the *inverse* represent totally different concepts.

Hence, $\dfrac{1}{\sin \theta}$, called the *reciprocal* of $\sin \theta$, is *not* the same as $\sin^{-1} \theta$, called the *inverse* of $\sin \theta$.

Similarly, with the reciprocals and inverses of $\cos \theta$ and $\tan \theta$, we have

$$\frac{1}{\cos \theta} \neq \cos^{-1} \theta$$

$$\frac{1}{\tan \theta} \neq \tan^{-1} \theta.$$

The main purpose of *Section 4* was to develop several formulae and procedures for solving problems associated with **non-right-angled triangles.**

These include:

A.   The **sine rule:** This states that

(i) $\dfrac{a}{\sin A} = \dfrac{b}{\sin B} = \dfrac{c}{\sin C}$, and

(ii) $\dfrac{\sin A}{a} = \dfrac{\sin B}{b} = \dfrac{\sin C}{c}$.

This rule allows us to determine

(a) an unknown side *given* a side and two internal angles, or

(b) an unknown angle *given* two sides and a *non-included* angle.

B.   The **cosine rule**: This states that

(i) $a^2 = b^2 + c^2 - 2bc \cos A$

$b^2 = c^2 + a^2 - 2ca \cos B$

$c^2 = a^2 + b^2 - 2ab \cos C$, and

(ii) $\cos A = \dfrac{b^2 + c^2 - a^2}{2bc}$

$\cos B = \dfrac{c^2 + a^2 - b^2}{2ca}$

$\cos C = \dfrac{a^2 + b^2 - c^2}{2ab}$

This rule allows us to determine

(a) the unknown third side *given* the other two sides and the *included* angle, or

(b) an unknown angle *given* all three sides.

C. Finding the area of a triangle

Using trigonometry to find an alternative expression for the perpendicular height, h, of a triangle we found that, instead of using

$$A = \frac{1}{2}bh,$$

we can determine the area of any triangle ABC using the formula

$$\text{Area of } \Delta\, ABC = \begin{cases} \dfrac{1}{2}ab\sin C \\[2mm] \dfrac{1}{2}ac\sin B \\[2mm] \dfrac{1}{2}bc\sin A \end{cases}$$

In words, this formula says that the area of a triangle is $\dfrac{1}{2}$ x product of the lengths of any two sides x sine of the *included* angle.

Remember: The *included* angle is the angle lying between two adjacent sides.

In the final section (*Section 5*) we presented results called **Pythagorean identities** that actually originated from an important mathematical theorem – namely, Pythagoras' theorem.

This theorem states:

> For any right-angled triangle, the square of the hypotenuse is equal to the sum of the squares of the other two sides.

The trigonometric Pythagorean identities, all very useful in manipulating and simplifying trigonometric expressions, may now be written as

1. $\sin^2\theta + \cos^2\theta = 1$
2. $\operatorname{cosec}^2\theta = 1 + \cot^2\theta$
3. $\sec^2\theta = 1 + \tan^2\theta$

Remember: $\sin^2\theta$ is an abbreviation of $(\sin\theta)^2$ and should *not* be confused with $\sin\theta^2$ (where only θ is squared).

The final chapter of this text revisits the branch of trigonometry and delves a little more deeply into some more specific areas of trigonometric concepts and associated applications.

10.1 Given that $\sin\alpha = \dfrac{3}{5}$, find

(a) $\cos\alpha$ (b) $\tan\alpha$ (c) $\sec\alpha$

(d) $\operatorname{cosec}\alpha$ (e) $\cot\alpha$

10.2 Find the length, x, of each triangle, correct to one decimal place.

(a)

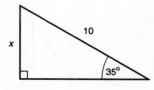

(b)

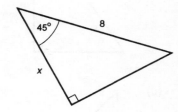

(c)

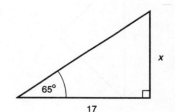

(d)

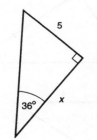

(e)

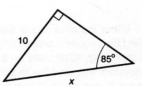

(f)

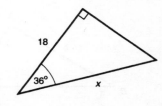

10.3 In each of the following find the unknown angle, α, correct to the nearest degree.

(a)

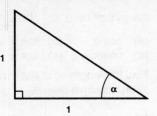

(b)

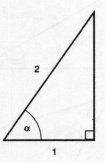

(c)

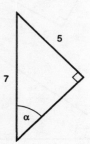

(d)

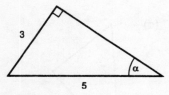

(e)

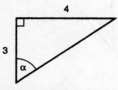

10.4 Use either the sine rule or the cosine rule to find the required side or angle in each of the following. (Calculate an angle to the nearest minute and the length of a side correct to one decimal place.)

(a) Find angle B

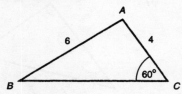

10.4 *Cont.*

(b) Find angle A

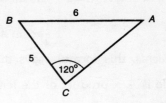

(c) Find side a

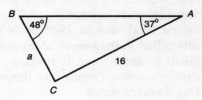

(d) Find side c

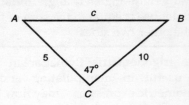

(e) Find side a

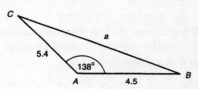

(f) Find angle C

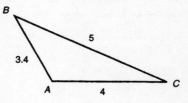

(g) Find **(i)** angle B
      (ii) side AC
      (iii) side AD

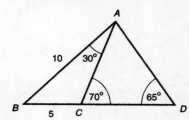

10.5 Find the area of the triangle, *ABC*, defined by the angles and sides given in each of the following:

    (a) $a = 6$ $b = 9$ $C = 62°$

    (b) $a = 13$ $c = 17$ $B = 29° \ 30'$

    (c) $c = 29.4$ $a = 18.3$ $B = 170°$

    (d) $b = 22$ $c = 15$ $A = 100°$

    (e) $a = 6$ $b = 8$ $C = 39°$

For each of questions **10.6** to **10.10** below, draw the appropriate diagram showing all the given information.

10.6 A person travels 10 km north, then 10 km east. Determine the angle from the starting point.

*10.7 A person travels 20 km due north, 10 km due west and then 5 km due south. Determine the angle from the starting point.

*10.8 A person walking on flat ground towards a flagpole observes that the top of the flagpole makes an angle of $10°$ from the horizontal. After proceeding a further 50 m, the angle is $15°$. Find the height of the flagpole.

10.9 Find the perimeter of a triangle if two sides are 32 m and 74 m, and the included angle is $120°$.

*10.10 A person walks 25 m due east of *A*. Another person walks 20 m at an angle of $31° \ 37'$ from *A*. Determine the distance between the two people.

*10.11 Simplify each of the following:

    (a) $\cos^2 \alpha - \sin^2 \alpha$

    (b) $2 \sin 4\theta \cdot \cot 4\theta$

    (c) $2 - 2 \sin^2 B$

    (d) $\dfrac{2 \tan \beta}{1 + \tan^2 \beta}$

10.12 Given that $\sec^2 B = 1$, find *B*, where $0 \le B < 90°$.

11 Extras – An introduction to matrix algebra

Introduction

In this chapter we will explore another branch of mathematics that is widely used in most of the science and mathematics disciplines as well as in the areas of economics, marketing and business management – namely, **matrix algebra.**

Matrix algebra is, in fact, a generalisation of elementary algebra. Where **elementary** algebra is concerned with operations on *single* values, **matrix** algebra describes operations on whole *arrays* of values, and is therefore an extremely powerful tool.

As you will see as you work through the chapter, matrices (plural of matrix) are a very convenient means of organising, displaying and manipulating large quantities of data and hence are widely used in any discipline involving very large amounts of data.

In *Section 1* we begin our discussion of matrices with some relevant definitions, notation and mathematical conventions. This is followed, in *Section 2*, by a study of the rules associated with the matrix operations of addition, subtraction and multiplication.

Sections 3 and *4* explore further some special matrices and associated matrix operations, while *Sections 5* and *6* discuss ways that we can use our knowledge of matrices to solve some practical problems.

Note: It is also very important to keep in mind that the main purpose of this chapter is to provide an introduction to matrices together with some of their more common uses in solving real-life problems, rather than to undertake a detailed study of matrix algebra with all its underlying mathematical theory.

11.1 Definitions, notation and conventions

A **matrix** is a rectangular array of numbers and/or letters.

By convention, the given values (numbers and/or letters) are enclosed in either *square* or *round* brackets to denote a matrix. In this text, we use square brackets in most cases.

For example:

$$A = \begin{bmatrix} 1 & 0 & -3 \\ 2 & 5 & 6 \\ -1 & 1/4 & 4 \end{bmatrix} \quad B = \begin{bmatrix} 2 & 3 \\ 0 & -5 \\ -1 & 1 \end{bmatrix} \quad C = \begin{bmatrix} -3 & 1 & 0 \end{bmatrix}$$

Each value in a matrix is called an **element** of that matrix.

The elements are arranged in horizontal **rows** and vertical **columns** and, as you will see later, the partitioning of matrices into rows and columns plays an important part in matrix operations.

Matrices are generally denoted by a **capital letter**, e.g. matrix A above, and their elements are denoted by the same **small** letter with two subscripts. These subscripts together indicate the unique position of the element within the matrix – the first subscript indicates the *row number* while the second subscript indicates the *column number*.

Thus, the element in the 2nd row and 3rd column of matrix A above, is denoted a_{23} and its value is 6.

Similarly, $b_{12} = 3$, $c_{13} = 0$, $b_{11} = 2$ and $a_{32} = 1/4$.

The **size** of any matrix is given by quoting its number of rows and columns – **always in that order.**

Therefore, matrix A above is a [3 x 3], read as 'three-by-three', matrix and, by convention, is said to be of **size** or **order** [3 x 3].

Similarly, matrix B is of size [3 x 2] while matrix C is of size [1 x 3].

Any matrix whose elements are all located in either **one row** or **one column** is called a **row** or **column vector.**

Hence, matrix C above, with only one row, is an example of a **row vector**.

Similarly, matrix $D = \begin{bmatrix} 8 \\ 4 \\ -3 \end{bmatrix}$ is an example of a

column vector.

11.2 Matrix operations

11.2.1 Equality

Two matrices, say A and B, are said to be **equal** $(A = B)$ if

(i) they are of the same size, and

(ii) their corresponding elements are equal.

For example:
Suppose

$$A = \begin{bmatrix} 4 & -1 \\ 5 & 2 \end{bmatrix} \text{ and } B = \begin{bmatrix} x & -1 \\ 5 & y \end{bmatrix},$$

then A and B are of the **same** size and $A = B$ **only if** $x = 4$ and $y = 2$.

If $C = \begin{bmatrix} 4 & -1 & 0 \\ 5 & 2 & 0 \end{bmatrix}$ then $A \neq C$, since A and C are

not of the same order (size).

11.2.2 Addition and subtraction

Matrices can be **added** or **subtracted**, provided they are of the **same size**.

To add (or subtract) two matrices, we simply add (or subtract) the elements in corresponding pairs and write the result of the addition (or subtraction) in the same position in the **result** matrix.

Note: The result matrix $(A + B)$ or $(A - B)$ will be of the **same** size as the individual matrices A and B.

EXAMPLE 11.1

If $A = \begin{bmatrix} 1 & 2 \\ 3 & 4 \end{bmatrix}$ and $B = \begin{bmatrix} 5 & 6 \\ 7 & 8 \end{bmatrix}$,

then the matrix $C = [A + B]$ is given by

$$C = \begin{bmatrix} 1 & 2 \\ 3 & 4 \end{bmatrix} + \begin{bmatrix} 5 & 6 \\ 7 & 8 \end{bmatrix}$$

$$= \begin{bmatrix} 1+5 & 2+6 \\ 3+7 & 4+8 \end{bmatrix}$$

$$= \begin{bmatrix} 6 & 8 \\ 10 & 12 \end{bmatrix}$$

As expected, C is size [2 x 2], as is A and B.

Note: If A and B are any two matrices of the same size, then $[A + B] = [B + A]$ **always**. (Prove this yourself using A and B in example 11.1 above.)

EXAMPLE 11.2

If $A = \begin{bmatrix} 3 & 2 \\ 6 & 4 \\ 1 & 5 \end{bmatrix}$ and $B = \begin{bmatrix} 1 & 5 \\ 7 & 3 \\ 6 & 8 \end{bmatrix}$,

then the matrix $C = A - B$ is given by

$$C = \begin{bmatrix} 3 & 2 \\ 6 & 4 \\ 1 & 5 \end{bmatrix} - \begin{bmatrix} 1 & 5 \\ 7 & 3 \\ 6 & 8 \end{bmatrix} = \begin{bmatrix} 3-1 & 2-5 \\ 6-7 & 4-3 \\ 1-6 & 5-8 \end{bmatrix}$$

So, $C = \begin{bmatrix} 2 & -3 \\ -1 & 1 \\ -5 & -3 \end{bmatrix}$

Again, matrix C is of size [3 x 2], as are A and B.

EXAMPLE 11.3

If $A = \begin{bmatrix} 1 & 2 \\ 3 & 4 \end{bmatrix}$ and $B = \begin{bmatrix} 1 & 5 \\ 7 & 3 \\ 6 & 8 \end{bmatrix}$,

then $C = [A + B]$ cannot be determined.

Why? Because A is of size $[2 \times 2]$ and B is of size $[3 \times 2]$, they *cannot* be added or subtracted as they are **not** of the same size, as required.

11.2.3 Scalar multiplication

Multiplication by a **scalar** (that is, a single number) is accomplished by multiplying each element in the matrix by the scalar and then writing each product in the corresponding position of the result matrix.

In more general terms,

> If A is any matrix and k is any number, then the matrix kA is obtained by multiplying each element of A by the number k.
> *Note:* The matrix kA is the **same size** as the matrix A.

EXAMPLE 11.4

If $\quad A = \begin{bmatrix} 3 & 2 \\ 6 & 4 \\ 1 & 5 \end{bmatrix}$, then

(a) $\quad 3A = 3 \times \begin{bmatrix} 3 & 2 \\ 6 & 4 \\ 1 & 5 \end{bmatrix} = \begin{bmatrix} 3 \times 3 & 2 \times 3 \\ 6 \times 3 & 4 \times 3 \\ 1 \times 3 & 5 \times 3 \end{bmatrix}$

$\qquad\qquad = \begin{bmatrix} 9 & 6 \\ 18 & 12 \\ 3 & 15 \end{bmatrix}$

(b) $\quad \frac{1}{2}A = \frac{1}{2}\begin{bmatrix} 3 & 2 \\ 6 & 4 \\ 1 & 5 \end{bmatrix} = \begin{bmatrix} 1\frac{1}{2} & 1 \\ 3 & 2 \\ \frac{1}{2} & 2\frac{1}{2} \end{bmatrix}$

(c) $\quad -5A = -5\begin{bmatrix} 3 & 2 \\ 6 & 4 \\ 1 & 5 \end{bmatrix} = \begin{bmatrix} -15 & -10 \\ -30 & -20 \\ -5 & -25 \end{bmatrix}$

Note: In each of **(a)**, **(b)** and **(c)** above, the **result** matrix is the **same** size as the original matrix A – namely, $[3 \times 2]$.

EXAMPLE 11.5
Consider a company with February sales

$$F = \begin{matrix} \text{Prod. X} & \text{Prod. Y} & \text{Prod. Z} \\ \begin{bmatrix} 110 & 70 & 90 \\ 60 & 100 & 140 \end{bmatrix} & & \begin{matrix} \text{Area 1} \\ \text{Area II} \end{matrix} \end{matrix}$$

If sales increase by 10%, then the **new** sales are given by the matrix $[F + 0.1F] = 1.1F$, where

$$1.1F = 1.1\begin{bmatrix} 110 & 70 & 90 \\ 60 & 100 & 140 \end{bmatrix}$$

$$= \begin{matrix} \text{Prod. X} & \text{Prod. Y} & \text{Prod. Z} \\ \begin{bmatrix} 121 & 77 & 99 \\ 66 & 110 & 154 \end{bmatrix} & & \begin{matrix} \text{Area 1} \\ \text{Area II} \end{matrix} \end{matrix}$$

11.2.4 Matrix multiplication

Multiplying one matrix by another matrix is only possible if **the number of columns** of the **first** matrix, A, is equal to the **number of rows** of the **second** matrix, B.

Thus, if A is of size $[3 \times 4]$, then $A \cdot B$ (where the $\cdot$ denotes 'multiplied by') will be possible **only if** matrix B has 4 rows – that is, if B is of size $[4 \times 1]$, $[4 \times 2]$, or $[4 \times$ anything$]$.

Note: One of the main differences between matrix algebra and elementary algebra is that matrix-multiplication is generally **not** commutative. Therefore, because in most cases $A \cdot B \neq B \cdot A$, the **order** of the multiplication is important. You may also find that one of the multiplications, $A \cdot B$ or $B \cdot A$, is not even possible.

How to find the product $A \cdot B$
Suppose we have

$$A = \begin{bmatrix} 2 & 3 & 2 \\ 3 & 1 & 4 \end{bmatrix} \text{ and } B = \begin{bmatrix} 4 & 2 \\ 3 & 5 \\ 1 & 6 \end{bmatrix}.$$

There are two basic rules to follow when multiplying matrices such as A and B above.

(i) If A is any matrix of size $[R_1 \times m]$ and B is any matrix of size $[m \times C_2]$, then the *product* $A \cdot B$ exists and is a matrix of size $[R_1 \times C_2]$.

The following illustration may be useful:

When an $[R_1 \times m]$ matrix is multiplied by an $[m \times C_2]$ matrix
— these must be equal —
— the product is an $[R_1 \times C_2]$ matrix —

(ii) The value of the element in the ith row and jth column of the result matrix, $A \cdot B$, is calculated as follows:

Multiply each element of the ith row of the **first matrix** by the corresponding element of the jth column of the **second matrix** (that is, first by first, second by second, ..., mth by mth) and then **add** these multiplied pairs together to give the single value of the appropriate element in the result matrix.

EXAMPLE 11.6

Suppose we now take the matrices A and B introduced above – namely,

$$A = \begin{bmatrix} 2 & 3 & 2 \\ 3 & 1 & 4 \end{bmatrix} \text{ and } B = \begin{bmatrix} 4 & 2 \\ 3 & 5 \\ 1 & 6 \end{bmatrix}.$$

Then the product, $A \cdot B$, is

$$A \cdot B = \begin{bmatrix} 2 & 3 & 2 \\ 3 & 1 & 4 \end{bmatrix} \cdot \begin{bmatrix} 4 & 2 \\ 3 & 5 \\ 1 & 6 \end{bmatrix}$$

$$= \begin{bmatrix} (2 \cdot 4) + (3 \cdot 3) + (2 \cdot 1) & (2 \cdot 2) + (3 \cdot 5) + (2 \cdot 6) \\ (3 \cdot 4) + (1 \cdot 3) + (4 \cdot 1) & (3 \cdot 2) + (1 \cdot 5) + (4 \cdot 6) \end{bmatrix}$$

$$= \begin{bmatrix} 19 & 31 \\ 19 & 35 \end{bmatrix}$$

Note: As expected, with A of size [2 x 3] and B of size [3 x 2], we have that the product, $A \cdot B$, is of size [2 x 2].

EXAMPLE 11.7

Let $A = \begin{bmatrix} 1 & 2 \\ 3 & 4 \end{bmatrix}$ and $B = \begin{bmatrix} 5 & 6 & 7 \\ 8 & 9 & 0 \end{bmatrix}$.

Now A is of size [2 x 2] and B is of size [2 x 3].

Since the number of **columns** (=2) of A is equal to the number of **rows** (=2) of B, the product matrix, $A \cdot B$, exists, and

$$A \cdot B = \begin{bmatrix} 1 & 2 \\ 3 & 4 \end{bmatrix} \cdot \begin{bmatrix} 5 & 6 & 7 \\ 8 & 9 & 0 \end{bmatrix}$$

$$= \begin{bmatrix} (1 \cdot 5) + (2 \cdot 8) & (1 \cdot 6) + (2 \cdot 9) & (1 \cdot 7) + (2 \cdot 0) \\ (3 \cdot 5) + (4 \cdot 8) & (3 \cdot 6) + (4 \cdot 9) & (3 \cdot 7) + (4 \cdot 0) \end{bmatrix}$$

$$= \begin{bmatrix} 5 + 16 & 6 + 18 & 7 + 0 \\ 15 + 32 & 18 + 36 & 21 + 0 \end{bmatrix}$$

$$= \begin{bmatrix} 21 & 24 & 7 \\ 47 & 54 & 21 \end{bmatrix}.$$

Note: Each of the two **rows** of A has been multiplied by each of the three **columns** of B and the resulting matrix, $A \cdot B$, is of size [2 x 3], as expected.

Comment:

In each of examples 11.6 and 11.7 above, A was the first matrix and B was the second. By convention, it is usually said that we **pre**-multiplied by A and **post**-multiplied by B.

It is important to differentiate between the pre- and post-multiplier because, as mentioned earlier in this section, the *commutative law* of basic algebra generally cannot be applied in matrix algebra.

Hence, in most instances, $A \cdot B \neq B \cdot A$.

In example 11.6 above, the result matrix, $B \cdot A$, would be of size [3 x 3], while in example 11.7, the product, $B \cdot A$, cannot be determined. **Why?**

Interpretation:

We can now multiply together two matrices, A and B, provided certain conditions are met, **but** how do we give the values in the result matrix, $A \cdot B$, a meaning?

As you would expect, the interpretation will depend upon what the original data in A and B actually represent in a practical situation. This is best explained by means of an example.

EXAMPLE 11.8

Suppose John goes to his local supermarket and buys 4 kg sugar at 60 cents per kilogram, 2 kg of lamb chops at $4.00 per kilogram and 3 loaves of bread at $1.20 per loaf.

Another shopper, Janice, goes to her local supermarket and buys 2 kg sugar at 75 cents per kilogram, 1 kg of lamb chops at $3.50 per kilogram and 4 loaves of bread at $1.00 per loaf.

We might decide to organise this so that all similar information is grouped together. If we put all the shopping quantities together in a matrix, we would have:

$$A = \begin{array}{c} \\ \\ \end{array} \begin{array}{ccc} \text{sugar} & \text{lamb} & \text{bread} \\ \begin{bmatrix} 4 & 2 & 3 \\ 2 & 1 & 4 \end{bmatrix} & & \end{array} \begin{array}{c} \text{John} \\ \text{Janice} \end{array}$$

where the first row represents John's quantities, the second row represents Janice's quantities and the columns represent sugar, lamb and bread respectively.

Similarly, we could summarise the prices of the items at each supermarket using another matrix, as follows:

$$A = \begin{array}{cc} \text{John's market} & \text{Janice's market} \\ \begin{bmatrix} 0.60 & 0.75 \\ 4.00 & 3.50 \\ 1.20 & 1.00 \end{bmatrix} \end{array} \begin{array}{c} \text{sugar} \\ \text{lamb} \\ \text{bread} \end{array}$$

Now the product, $A \cdot B$, can be understood since quantity (A) times price (B) should represent the total amount spent.

With A and B above, we have

$$A \cdot B = \begin{array}{c} \\ \text{John} \\ \text{Janice} \end{array} \begin{array}{ccc} \text{sugar} & \text{lamb} & \text{bread} \\ \begin{bmatrix} 4 & 2 & 3 \\ 2 & 1 & 4 \end{bmatrix} \end{array} \begin{array}{cc} \text{John's} \\ \text{market} & \begin{array}{c} \text{Janice's} \\ \text{market} \end{array} \\ \begin{bmatrix} 0.60 & 0.75 \\ 4.00 & 3.50 \\ 1.20 & 1.00 \end{bmatrix} \end{array} \begin{array}{c} \text{sugar} \\ \text{lamb} \\ \text{bread} \end{array}$$

$$= \begin{bmatrix} \begin{matrix} \text{John's market} \\ 2.40 + 8 + 3.60 \\ \\ 1.2 + 4 + 4.8 \end{matrix} & \begin{matrix} \text{Janice's market} \\ 3 + 7 + 3 \\ \\ 1.5 + 3.5 + 4 \end{matrix} \end{bmatrix} \begin{matrix} \text{John's total cost} \\ \\ \text{Janice's total cost} \end{matrix}$$

$$= \begin{bmatrix} \begin{matrix} \text{John's market} \\ 2.40 + 8 + 3.60 \\ 1.2 + 4 + 4.8 \end{matrix} & \begin{matrix} \text{Janice's market} \\ 3 + 7 + 3 \\ 1.5 + 3.5 + 4 \end{matrix} \end{bmatrix} \begin{matrix} \text{John's total cost} \\ \text{Janice's total cost} \end{matrix}$$

$$= \begin{bmatrix} 14 & 13 \\ 10 & 9 \end{bmatrix}$$

If you check the individual calculations, you will see that $14.00 is the amount John spent doing his shopping at his own supermarket, and $13.00 is the amount he would have spent on the same items at Janice's supermarket.

Similarly, if Janice shopped at her own supermarket she would spend $9.00, whereas she would spend $10.00 for those same goods at John's supermarket.

YOUR TURN

11.6 If $A = \begin{bmatrix} 5 & 1 & -4 \\ 9 & 2 & 1 \\ 0 & 6 & 0 \end{bmatrix}$ and $B = \begin{bmatrix} 6 & -1 & 0 \\ 9 & -2 & -4 \\ 0 & -3 & 6 \end{bmatrix}$

find
(a) $A + B$
(b) $B + A$ (Comment on your resulting matrices in (a) and (b).)

11.7 If $A = \begin{bmatrix} 1 & 2 \\ 3 & 4 \end{bmatrix}$ and $B = \begin{bmatrix} 5 & 6 \\ 7 & 8 \end{bmatrix}$, find

(a) $A - B$
(b) $B - A$

11.8 Suppose a company has set the following *targets* for sales in January:

	Prod. X	Prod. Y	Prod. Z	
$T =$	120	70	105	Region I
	65	100	145	Region II

and that the *actual* sales in January are:

	Prod. X	Prod. Y	Prod. Z	
$J =$	120	50	100	Region I
	80	75	150	Region II

Find the differences between the target sales and the actual sales in the 2 regions for each of the 3 products.

11.9 If $A = \begin{bmatrix} 3 & 2 & 6 & 8 \\ 0 & 1 & 4 & 2 \\ 4 & 2 & 3 & 6 \end{bmatrix}$, find

(a) $2A$
(b) $0.5A$
(c) $5A$

11.10 If $A = \begin{bmatrix} 3 & 2 \\ 6 & 4 \\ 1 & 5 \end{bmatrix}$ and $B = \begin{bmatrix} 1 & 5 \\ 7 & 3 \\ 6 & 8 \end{bmatrix}$, find

(a) $A + B$
(b) $A - B$
(c) $B + A$
(d) $B - A$
(e) $2A$
(f) $2A + B$

11.11 A stock broker sold a customer 200 shares of stock A, 300 shares of stock B, 500 shares of stock C and 250 shares of stock D. The prices per share of A, B, C and D are $100, $150, $200 and $300, respectively.
   (a) Write a **row** matrix representing the number of shares of each stock bought.
   (b) Write a **column** matrix representing the price per share of each stock.
   (c) Using matrix multiplication, find the total cost of the stocks.

11.12 If $A = \begin{bmatrix} 3 & 2 \\ 0 & 1 \\ -1 & 6 \end{bmatrix}$ $B = \begin{bmatrix} -1 & 3 & 0 \\ 2 & 6 & 4 \end{bmatrix}$

$C = \begin{bmatrix} 0 & 1 \\ 3 & -4 \end{bmatrix}$ $D = \begin{bmatrix} 9 & 3 \\ 2 & -1 \end{bmatrix}$

$E = \begin{bmatrix} 2 \\ 4 \end{bmatrix}$ $F = \begin{bmatrix} 3 \\ -1 \\ 0 \end{bmatrix}$

find (if possible)
(a) $A \cdot B$ (b) $A \cdot C$
(c) $B \cdot C$ (d) $B \cdot F$
(e) $C \cdot D$ (f) $C \cdot E$
(g) $A \cdot E$

11.13 Suppose a building contractor accepts orders for 4 houses, 3 flats and 9 duplexes. These orders may be represented by the following row matrix,

	house	flat	duplex
$Q =$	4	3	9

11.13 *Cont.*

If the quantities, in appropriate units, of raw materials used in the construction of each type of dwelling are given by the following matrix,

$$R = \begin{bmatrix} 100 & 30 & 50 & 30 & 120 \\ 40 & 80 & 20 & 40 & 100 \\ 60 & 50 & 70 & 20 & 100 \end{bmatrix} \begin{matrix} \text{house} \\ \text{flat} \\ \text{duplex} \end{matrix}$$

bricks timber glass concrete labour

then the quantity of each raw material required to fulfil the orders can be found from the product matrix, $Q \cdot R$. Find $Q \cdot R$.

11.3 Special matrices and further matrix operations

11.3.1 The square matrix

Any matrix with the same number of rows and columns is called a **square matrix**.

For example: Matrix A, where $A = \begin{bmatrix} 5 & 0 & 1 \\ 0 & 1 & 4 \\ 7 & 2 & 3 \end{bmatrix}$, with 3 rows and 3 columns, is a **square** matrix of size [3 x 3] or, more commonly, just *size 3* (or *order 3*).

In a **square** matrix, the diagonal running from top left to bottom right is called the **principal** or **leading diagonal**.

For example: In matrix A above, the elements of the principal diagonal are 5, 1 and 3.

11.3.2 The identity matrix

A **square** matrix in which the principal diagonal elements are all 1's and every other element is 0 is called an **identity matrix** and is usually given the label I.

There is an identity matrix of each size – the identity matrices of sizes two, three and four are as follows:

$$\begin{bmatrix} 1 & 0 \\ 0 & 1 \end{bmatrix} \quad \begin{bmatrix} 1 & 0 & 0 \\ 0 & 1 & 0 \\ 0 & 0 & 1 \end{bmatrix} \quad \begin{bmatrix} 1 & 0 & 0 & 0 \\ 0 & 1 & 0 & 0 \\ 0 & 0 & 1 & 0 \\ 0 & 0 & 0 & 1 \end{bmatrix}$$

Note: An important feature of the identity matrix is that it behaves, for matrices, in the same way as the number 1 for numbers – that is, the

multiplication (pre- or post-) of any *square matrix* by its corresponding identity matrix of the *same* size, gives the original matrix as the result. Hence,

If A is any matrix, then $I \cdot A = A \cdot I = A$

EXAMPLE 11.9

If $A = \begin{bmatrix} 1 & 2 & 3 \\ 4 & 5 & 6 \\ 7 & 8 & 9 \end{bmatrix}$, then

$$A \cdot I = \begin{bmatrix} 1 & 2 & 3 \\ 4 & 5 & 6 \\ 7 & 8 & 9 \end{bmatrix} \begin{bmatrix} 1 & 0 & 0 \\ 0 & 1 & 0 \\ 0 & 0 & 1 \end{bmatrix}$$

$$= \begin{bmatrix} (1 \cdot 1) + (2 \cdot 0) + (3 \cdot 0) & (1 \cdot 0) + (2 \cdot 1) + (3 \cdot 0) \\ (4 \cdot 1) + (5 \cdot 0) + (6 \cdot 0) & (4 \cdot 0) + (5 \cdot 1) + (6 \cdot 0) \\ (7 \cdot 1) + (8 \cdot 0) + (9 \cdot 0) & (7 \cdot 0) + (8 \cdot 1) + (9 \cdot 0) \end{bmatrix}$$

$$\begin{bmatrix} (1 \cdot 0) + (2 \cdot 0) + (3 \cdot 1) \\ (4 \cdot 0) + (5 \cdot 0) + (6 \cdot 1) \\ (7 \cdot 0) + (8 \cdot 0) + (9 \cdot 1) \end{bmatrix}$$

$$= \begin{bmatrix} 1 & 2 & 3 \\ 4 & 5 & 6 \\ 7 & 8 & 9 \end{bmatrix}$$

$$= A$$

Similarly, $I \cdot A = A$ (prove this for yourself).

11.3.3 The zero matrix

Any matrix whose elements are **all** zero is called the **zero matrix**.

For example:

$E = \begin{bmatrix} 0 & 0 & 0 \\ 0 & 0 & 0 \end{bmatrix}$ is a zero matrix of size [2 x 3].

11.3.4 The transpose of a matrix

The **transpose** of a matrix A, denoted A^T (read as 'A transpose') is the matrix that is obtained if the rows and columns of A are interchanged.

For example:

If $A = \begin{bmatrix} 1 & 2 & 3 \\ 6 & 7 & 8 \end{bmatrix}$, then $A^T = \begin{bmatrix} 1 & 6 \\ 2 & 7 \\ 3 & 8 \end{bmatrix}$.

Note: If A is of size [2 x 3], then A^T is of size [3 x 2].

In general, if any matrix A is of size [m x n], then its transpose, A^T, is of size [n x m].

Comments

1. Matrix A is said to be **symmetric** if the corresponding elements around the leading diagonal are the same – that is, if $a_{ij} = a_{ji}$ for all $i \neq j$.

2. If A is any **square** matrix such that it is **equal** to its **transpose** – that is, such that $A = A^T$, then A is said to be **symmetric**.

For example:

$$\text{If } A = \begin{bmatrix} 1 & -1 & 3 \\ -1 & 2 & 4 \\ 3 & 4 & 0 \end{bmatrix}, \quad \text{then } A^T = \begin{bmatrix} 1 & -1 & 3 \\ -1 & 2 & 4 \\ 3 & 4 & 0 \end{bmatrix}.$$

Hence, since $A = A^T$, A must be **symmetric**.

Furthermore, we can see that A is symmetric since

$$a_{12} = a_{21} = -1$$
$$a_{13} = a_{31} = 3$$
$$a_{23} = a_{32} = 4$$

Note: Clearly, all **identity matrices** of **any** size are **symmetric**.

11.3.5 The inverse matrix

If A and B are two **square** matrices of the **same** size, and if the *product* of A and B (namely, $A \cdot B$ or $B \cdot A$), is equal to the **identity matrix** (of the **same** size), then matrix B is called the **inverse** of matrix A. (Similarly, matrix A is also the *inverse* of matrix B.)

Now, the **inverse** of matrix A is denoted A^{-1} (read as 'A inverse'). Hence, $A \cdot A^{-1} = A^{-1} \cdot A = I$.

Note:

a. The superscript, –1, is **not** used as a power in the usual algebraic sense. Thus, A^{-1} does **not** mean $\frac{1}{A}$ – in fact, in matrix algebra there is no such thing as division. Here the superscript, –1, is used purely as a means of denoting the inverse of a matrix, if it exists.

b. If a square matrix A **has** an inverse, then matrix A is said to be **non-singular** – and it can be shown that the inverse of any non-singular matrix is **unique**.

c. Not every square matrix has an inverse – furthermore, any square matrix that does **not** have an inverse is said to be **singular**.

EXAMPLE 11.10

If $A = \begin{bmatrix} 0 & 1 & 1 \\ 1 & 2 & 2 \\ 1 & 1 & 3 \end{bmatrix}$ is a square matrix of size [3 x 3],

verify that $A^{-1} = \begin{bmatrix} -2 & 1 & 0 \\ \frac{1}{2} & \frac{1}{2} & -\frac{1}{2} \\ \frac{1}{2} & -\frac{1}{2} & \frac{1}{2} \end{bmatrix}$

is indeed the inverse of A, by carrying out the following multiplications:

(a) $A \cdot A^{-1}$, and

(b) $A^{-1} \cdot A$

If A^{-1} is indeed the true inverse of A, then the result matrix in (a) and (b) above should be the **identity matrix** of size [3 x 3].

(a) $A \cdot A^{-1} = \begin{bmatrix} 0 & 1 & 1 \\ 1 & 2 & 2 \\ 1 & 1 & 3 \end{bmatrix} \begin{bmatrix} -2 & 1 & 0 \\ \frac{1}{2} & \frac{1}{2} & -\frac{1}{2} \\ \frac{1}{2} & -\frac{1}{2} & \frac{1}{2} \end{bmatrix}$

$$= \begin{bmatrix} 0+\frac{1}{2}+\frac{1}{2} & 0+\frac{1}{2}-\frac{1}{2} & 0-\frac{1}{2}+\frac{1}{2} \\ -2+1+1 & 1+1-1 & 0-1+1 \\ -2+\frac{1}{2}+1\frac{1}{2} & 1+\frac{1}{2}-1\frac{1}{2} & 0-\frac{1}{2}+1\frac{1}{2} \end{bmatrix}$$

$$= \begin{bmatrix} 1 & 0 & 0 \\ 0 & 1 & 0 \\ 0 & 0 & 1 \end{bmatrix} = I, \text{ of size [3 x 3].}$$

Similarly,

(b) $A^{-1} \cdot A = \begin{bmatrix} -2 & 1 & 0 \\ \frac{1}{2} & \frac{1}{2} & -\frac{1}{2} \\ \frac{1}{2} & -\frac{1}{2} & \frac{1}{2} \end{bmatrix} \cdot \begin{bmatrix} 0 & 1 & 1 \\ 1 & 2 & 2 \\ 1 & 1 & 3 \end{bmatrix}$

$$= \begin{bmatrix} 0+1+0 & -2+2+0 & -2+2+0 \\ 0+\frac{1}{2}-\frac{1}{2} & \frac{1}{2}+1-\frac{1}{2} & \frac{1}{2}+1-1\frac{1}{2} \\ 0-\frac{1}{2}+\frac{1}{2} & \frac{1}{2}-1+\frac{1}{2} & \frac{1}{2}-1+1\frac{1}{2} \end{bmatrix}$$

$$= \begin{bmatrix} 1 & 0 & 0 \\ 0 & 1 & 0 \\ 0 & 0 & 1 \end{bmatrix} = I, \text{ of size [3 x 3].}$$

Therefore, since $A \cdot A^{-1} = A^{-1} \cdot A = I$, we have verified that A^{-1} is indeed the unique inverse of A.

Note: In *Section 4* following, we discuss a reasonably simple procedure for determining the inverse of a given square matrix, if it exists.

11.3.6 The determinant of a matrix

> A **determinant** is a unique number associated with (or characteristic of) a **square matrix**. The determinant of a square matrix, A, is denoted $|A|$.

Note: Although the notation $|A|$ resembles the absolute value symbol, the vertical bars around a matrix are used **solely** to denote its determinant.

The determinant of a matrix has many uses, the most important of which involve:

(i) determining whether a square matrix A is **singular** or **non-singular**, and

(ii) finding the **inverse** of a non-singular square matrix.

Unfortunately, calculation of the determinant is rather tedious by hand for all except the very small matrices and hence is usually carried out by a computer.

However, fairly simple techniques for calculating the determinant of any square matrix of sizes [2 x 2] or [3 x 3] will be presented here.

a.   The determinant of a [2 x 2] matrix

> If A is any [2 x 2] matrix,
>
> $$A = \begin{bmatrix} a_{11} & a_{12} \\ a_{21} & a_{22} \end{bmatrix}, \text{ then}$$
>
> the **determinant** of A, denoted $|A|$, is the **single number** defined by
>
> $$|A| = \begin{vmatrix} a_{11} & a_{12} \\ a_{21} & a_{22} \end{vmatrix} = a_{11} \cdot a_{22} - a_{12} \cdot a_{21}$$

Therefore, to compute the determinant of a [2 x 2] matrix, multiply together the elements of the principal diagonal and then subtract from this the product of the elements in the off-diagonal.

EXAMPLE 11.11

(a)   Let $A = \begin{bmatrix} 2 & 3 \\ 1 & 4 \end{bmatrix}$ be a square matrix of size [2 x 2].

Then, $|A| = \begin{vmatrix} 2 & 3 \\ 1 & 4 \end{vmatrix} = (2 \cdot 4) - (3 \cdot 1)$

$$= 8 - 3 = 5$$

(b)   Let $B = \begin{bmatrix} 5 & -3 \\ 4 & 2 \end{bmatrix}$ be a square matrix of size [2 x 2].

Then, $|B| = \begin{vmatrix} 5 & -3 \\ 4 & 2 \end{vmatrix} = (5 \cdot 2) - (-3 \cdot 4)$

$$= 10 + 12 = 22$$

(c)   Let $C = \begin{bmatrix} 8 & 4 \\ 4 & 2 \end{bmatrix}$ be a square matrix of size [2 x 2].

Then, $|C| = \begin{vmatrix} 8 & 4 \\ 4 & 2 \end{vmatrix} = (8 \cdot 2) - (4 \cdot 4)$

$$= 16 - 16 = 0$$

Important note: In *Section 4* following, when we come to *calculating* the inverse of a matrix, A, the first step will be to check whether the inverse of A, namely A^{-1}, actually exists. Remember that if A^{-1} exists, then A is said to be **non-singular**, otherwise it is said to be **singular**.

Also, in our discussion of determinants so far we have said that one of its main purposes is to determine whether a square matrix, A, is **non-singular** – implying that A^{-1} exists.

In fact, the following rule (to be explained further in 11.4) may be applied:

> A square matrix, A, is said to be **non-singular**, implying that A^{-1} exists, if the determinant of A, $|A|$, is **not zero**.
> So, if A is a square matrix, then A^{-1} exists **provided** $|A| \neq 0$.
> Similarly, if $|A| = 0$, then A is said to be **singular**, implying that A^{-1} does **not** exist.

Therefore, in example 11.11 above, matrix A and matrix B are both **non-singular** since $|A|$ and $|B|$ are non-zero. Hence, A^{-1} and B^{-1} both exist.

However, matrix C is a **singular** matrix since $|C| = 0$, implying that C^{-1} does **not** exist.

b.   The determinant of a [3 x 3] matrix

Since we already know how to calculate the determinant of a [2 x 2] matrix, the simplest procedure for finding the determinant of any higher-order matrix is to decompose it into a series of computations of [2 x 2] determinants.

Before we look specifically at a [3 x 3] determinant, some terminology is necessary.

Definition
Let a_{ij} be an element of a [3 x 3] matrix, A. Then the **cofactor** of a_{ij}, denoted A_{ij}, is defined as $(-1)^{i+j}$ times the **determinant** of the [2 x 2] matrix obtained by mentally deleting the ith row and the jth column of the original matrix.

For example: In the [3 x 3] matrix

$$\begin{bmatrix} a_{11} & a_{12} & a_{13} \\ a_{21} & a_{22} & a_{23} \\ a_{31} & a_{32} & a_{33} \end{bmatrix}$$

the cofactor of a_{12} is

$$A_{12} = (-1)^{1+2} \begin{vmatrix} a_{21} & a_{23} \\ a_{31} & a_{33} \end{vmatrix} = - \begin{vmatrix} a_{21} & a_{23} \\ a_{31} & a_{33} \end{vmatrix}$$

and the cofactor of a_{13} is

$$A_{13} = (-1)^{1+3} \begin{vmatrix} a_{21} & a_{22} \\ a_{31} & a_{32} \end{vmatrix} = \begin{vmatrix} a_{21} & a_{22} \\ a_{31} & a_{32} \end{vmatrix}.$$

We can now define the determinant of a [3 x 3] matrix in terms of its cofactors.

Definition
Let A be a square matrix of size [3 x 3]. Then its determinant, using the elements of row 1, is given by

$$|A| = a_{11}A_{11} + a_{12}A_{12} + a_{13}A_{13}$$

where A_{11}, A_{12} and A_{13} are the **cofactors** of a_{11}, a_{12} and a_{13} respectively.

Note: The cofactors of any row or any column may be used to calculate the determinant.

EXAMPLE 11.12

Let $A = \begin{bmatrix} 1 & 2 & -3 \\ 0 & 1 & -2 \\ 2 & -1 & 0 \end{bmatrix}$

Compute the determinant, $|A|$, by expansion along the first row.

Solution
According to the definition, since A is a square matrix of size [3 x 3], by expanding along the first row we have:

$$|A| = a_{11}A_{11} + a_{12}A_{12} + a_{13}A_{13}$$

$$= 1 \cdot (-1)^{1+1} \begin{vmatrix} 1 & -2 \\ -1 & 0 \end{vmatrix} + 2 \cdot (-1)^{1+2} \begin{vmatrix} 0 & -2 \\ 2 & 0 \end{vmatrix}$$

$$+ (-3)(-1)^{1+3} \begin{vmatrix} 0 & 1 \\ 2 & -1 \end{vmatrix}$$

$$= 1(0-2) - 2(0+4) - 3(0-2)$$
$$= -2 - 8 + 6$$
$$= -4$$

EXAMPLE 11.13
Repeat example 11.12 above, but here compute $|A|$ by expansion down the first column.

Solution
Again applying the definition, we have

$$|A| = a_{11}A_{11} + a_{21}A_{21} + a_{31}A_{31}$$

$$|A| = 1 \cdot (-1)^{1+1} \begin{vmatrix} 1 & -2 \\ -1 & 0 \end{vmatrix}$$

$$+ 0 \cdot (-1)^{2+1} \begin{vmatrix} 2 & -3 \\ -1 & 0 \end{vmatrix} + 2 \cdot (-1)^{3+1} \begin{vmatrix} 2 & -3 \\ 1 & -2 \end{vmatrix}$$

$$= 1(0-2) - 0 + 2(-4-(-3)) = -2 - 2$$
$$= -4, \text{ as expected.}$$

Note: When finding the determinant of a [3 x 3] matrix, computations will be greatly reduced if we expand along a row or column that contains a zero (where applicable) – **why?**

EXAMPLE 11.14

Let $A = \begin{bmatrix} 2 & -1 & 3 \\ -2 & 6 & 2 \\ 4 & 5 & 0 \end{bmatrix}$

Calculate the determinant, $|A|$, by expansion along the last **row** (since it contains a zero element).

Solution
By definition,

$$|A| = a_{31}A_{31} + a_{32}A_{32} + a_{33}A_{33}$$

$$= 4(-1)^{3+1} \begin{vmatrix} -1 & 3 \\ 6 & 2 \end{vmatrix} + 5(-1)^{3+2} \begin{vmatrix} 2 & 3 \\ -2 & 2 \end{vmatrix}$$

$$+ 0(-1)^{3+3} \begin{vmatrix} 2 & -1 \\ -2 & 6 \end{vmatrix}$$

$$= 4(-2-18) - 5(4-(-6)) + 0$$
$$= -80 - 50$$
$$= -130$$

Note: Clearly, the matrices in examples 11.12, 11.13 and 11.14 are all **non-singular**, since $|A| \neq 0$, and thus a unique inverse, A^{-1}, exists for each matrix, A.

YOUR TURN

Answer exercises 11.14 to 11.23 using the following matrices

$$A = \begin{bmatrix} 3 & 2 \\ 0 & 1 \\ -1 & 6 \end{bmatrix} \qquad B = \begin{bmatrix} 1 & 0 & 0 \\ 2 & 3 & -1 \\ -5 & 4 & 2 \end{bmatrix}$$

$$C = \begin{bmatrix} -1 & 3 & 0 \\ 2 & 6 & 4 \end{bmatrix} \qquad D = \begin{bmatrix} 9 & 3 \\ 2 & -1 \end{bmatrix}$$

$$I = \begin{bmatrix} 1 & 0 & 0 \\ 0 & 1 & 0 \\ 0 & 0 & 1 \end{bmatrix}$$

11.14 Write down
   (i) the elements of the principal diagonal
   (ii) the elements of the off-diagonal
      of
      (a) matrix B
      (b) matrix D.

11.15 Calculate $A \cdot C$ and state its size.

11.16 Calculate $C \cdot A$ and state its size.

11.17 Show that $B \cdot I = I \cdot B = B$.

11.18 Write down the elements of C^T.

11.19 Calculate $C^T \cdot D$ and state its size.

11.20 Write down the elements of A^T.

11.21 Calculate $A^T \cdot B$ and state its size.

11.22 Calculate $|D|$.

11.23 Calculate $|B|$ by expanding along
   (i) the first row
   (ii) the first column.

11.24 Show that

$$C = \begin{bmatrix} 1 & 1 \\ 1 & 1 \end{bmatrix} \text{ is a } \textbf{singular} \text{ matrix.}$$

11.25 Show that if

$$A = \begin{bmatrix} -4 & 1 & 2 \\ 7 & -1 & -4 \\ -\frac{1}{2} & 0 & \frac{1}{2} \end{bmatrix}, \text{ then } A^{-1} = \begin{bmatrix} 1 & 1 & 4 \\ 3 & 2 & 4 \\ 1 & 1 & 6 \end{bmatrix}$$

11.4 Calculation of the inverse matrix

Finding the inverse of a square matrix is certainly not difficult although it can be lengthy and extremely tedious and therefore would normally be determined by computer.

In this section, although we restrict our discussion to finding the inverse of a square matrix of size 2 only, the same procedure may be applied to determine the inverse of any **non-singular** square matrix.

Procedure
The most common method of calculating the inverse matrix proceeds in *three* stages. We will illustrate each step with a *general* [2 x 2] matrix, before re-inforcing the procedure with several practical examples.

Step 1:
Form the cofactor matrix, A^c, where A^c is a matrix obtained by replacing each element a_{ij} of A by its cofactor A_{ij}.

Remember:
The *cofactor* A_{ij} is defined as $(-1)^{i+j}$ times the determinant of the remaining elements of A after deleting (mentally) the ith row and the jth column. Therefore, if

$$A = \begin{bmatrix} a_{11} & a_{12} \\ a_{21} & a_{22} \end{bmatrix}, \text{ then } A^c = \begin{bmatrix} A_{11} & A_{12} \\ A_{21} & A_{22} \end{bmatrix}.$$

Here,

$$A_{11} = (-1)^{1+1}|a_{22}| = a_{22}$$

$$A_{12} = (-1)^{1+2}|a_{21}| = -a_{21}$$

$$A_{21} = (-1)^{2+1}|a_{12}| = -a_{12}$$

$$A_{22} = (-1)^{2+2}|a_{11}| = a_{11}$$

Hence, if $A = \begin{bmatrix} a_{11} & a_{12} \\ a_{21} & a_{22} \end{bmatrix}$, then $A^c = \begin{bmatrix} a_{22} & -a_{21} \\ -a_{12} & a_{11} \end{bmatrix}$

Step 2:
Transpose the cofactor matrix.
That is, find $(A^c)^T$.

Remember:
The *transpose* of any matrix is obtained by interchanging the rows and columns.
Therefore, if

$$A^c = \begin{bmatrix} a_{22} & -a_{21} \\ -a_{12} & a_{11} \end{bmatrix}, \text{ then } (A^c)^T = \begin{bmatrix} a_{22} & -a_{12} \\ -a_{21} & a_{11} \end{bmatrix}.$$

Step 3:
Divide $(A^c)^T$ by the determinant of A, $|A|$.
This gives the inverse matrix, A^{-1}.
Therefore, if

$$(A^c)^T = \begin{bmatrix} a_{22} & -a_{12} \\ -a_{21} & a_{11} \end{bmatrix}, \text{ then } A^{-1} = \frac{1}{|A|}\begin{bmatrix} a_{22} & -a_{12} \\ -a_{21} & a_{11} \end{bmatrix}.$$

Summarising steps 1 to 3 we can now define the inverse matrix symbolically as follows:

> If A is any **non-singular square** matrix, then its **inverse**, A^{-1}, may be found using
>
> $$A^{-1} = \frac{1}{|A|}(A^c)^T$$
>
> where **(i)** $|A|$ is the determinant of A
>
> **(ii)** $(A^c)^T$ is the transpose of the cofactor matrix.

Important note:
It should be clear from the above summary that the existence of an inverse depends upon the value of $|A|$.

As mentioned in the previous section, if $|A| = 0$, then **no** inverse exists (since division by zero is impossible), and A is said to be **singular**.

Whenever $|A| \neq 0$, then a *unique* inverse exists and A is said to be **non-singular**.

A point of interest
For any non-singular square matrix, A,

$$(A^c)^T \cdot A = \begin{bmatrix} |A| & & \\ & |A| & \bigcirc \\ & & \cdot \\ \bigcirc & & \\ & & |A| \end{bmatrix}$$

That is, pre-multiplying a square matrix A by its transposed cofactor matrix produces a result matrix with $|A|$ along the leading diagonal and zero elsewhere.

EXAMPLE 11.15
Let A be the [2 x 2] matrix,

$$A = \begin{bmatrix} 4 & 1 \\ 3 & 2 \end{bmatrix}$$

Find the inverse, A^{-1}, if it exists.

Solution
Firstly, we should check that A is *non-singular*, implying that there exists a unique inverse.

By definition, $|A| = (4 \cdot 2) - (1 \cdot 3)$
$$= 8 - 3$$
$$= 5$$
Therefore, since $|A| \neq 0$, we know that A^{-1} exists.

Calculation of A^{-1}
Step 1: **Form the cofactor matrix, A^c**

Now, $A^c = \begin{bmatrix} A_{11} & A_{12} \\ A_{21} & A_{22} \end{bmatrix} = \begin{bmatrix} a_{22} & -a_{21} \\ -a_{12} & a_{11} \end{bmatrix}$

So, in this example,

$$A^c = \begin{bmatrix} 2 & -3 \\ -1 & 4 \end{bmatrix}$$

Step 2: **Transpose the cofactor matrix**

$$(A^c)^T = \begin{bmatrix} 2 & -1 \\ -3 & 4 \end{bmatrix}$$

Step 3: **Divide $(A^c)^T$ by $|A|$ to find A^{-1}**

That is, $A^{-1} = \frac{1}{|A|}(A^c)^T$

$$= \frac{1}{5}\begin{bmatrix} 2 & -1 \\ -3 & 4 \end{bmatrix}, \text{ since } |A| = 5$$

$$= \begin{bmatrix} \frac{2}{5} & -\frac{1}{5} \\ -\frac{3}{5} & \frac{4}{5} \end{bmatrix}$$

Check:
Always check that you have determined A^{-1} correctly, using the fact that $A \cdot A^{-1} = I$.

Here, $A \cdot A^{-1} = \begin{bmatrix} 4 & 1 \\ 3 & 2 \end{bmatrix} \cdot \begin{bmatrix} \frac{2}{5} & -\frac{1}{5} \\ -\frac{3}{5} & \frac{4}{5} \end{bmatrix}$

$$= \begin{bmatrix} \frac{8}{5} - \frac{3}{5} & \frac{-4}{5} + \frac{4}{5} \\ \frac{6}{5} - \frac{6}{5} & \frac{-3}{5} + \frac{8}{5} \end{bmatrix}$$

$$= \begin{bmatrix} 1 & 0 \\ 0 & 1 \end{bmatrix}$$

$$= I$$

Therefore, we may conclude that if $A = \begin{bmatrix} 4 & 1 \\ 3 & 2 \end{bmatrix}$, then its unique inverse is $A^{-1} = \begin{bmatrix} \dfrac{2}{5} & -\dfrac{1}{5} \\ -\dfrac{3}{5} & \dfrac{4}{5} \end{bmatrix}$.

EXAMPLE 11.16
Let A be the [2 x 2] matrix

$$A = \begin{bmatrix} 3 & -1 \\ -5 & 4 \end{bmatrix}$$

Find the inverse, A^{-1}, if it exists.

Solution
Does A^{-1} exist? $|A| = (3 \cdot 4) - (-1 \cdot -5)$
$$= 12 - 5 = 7$$

Hence A^{-1} exists, since A is non-singular.

By definition,

$$A^{-1} = \frac{1}{|A|}(A^c)^T = \frac{1}{7}\begin{bmatrix} 4 & 5 \\ 1 & 3 \end{bmatrix}^T$$

$$= \frac{1}{7}\begin{bmatrix} 4 & 1 \\ 5 & 3 \end{bmatrix} = \begin{bmatrix} \dfrac{4}{7} & \dfrac{1}{7} \\ \dfrac{5}{7} & \dfrac{3}{7} \end{bmatrix}$$

Check
Show that $A \cdot A^{-1} = I$

Now, $A \cdot A^{-1} = \begin{bmatrix} 3 & -1 \\ -5 & 4 \end{bmatrix} \cdot \begin{bmatrix} \dfrac{4}{7} & \dfrac{1}{7} \\ \dfrac{5}{7} & \dfrac{3}{7} \end{bmatrix}$

$$= \begin{bmatrix} \dfrac{12}{7} - \dfrac{5}{7} & \dfrac{3}{7} - \dfrac{3}{7} \\ \dfrac{-20}{7} + \dfrac{20}{7} & \dfrac{-5}{7} + \dfrac{12}{7} \end{bmatrix}$$

$$= \begin{bmatrix} 1 & 0 \\ 0 & 1 \end{bmatrix}$$

$$= I$$

Therefore, we may conclude that if $A = \begin{bmatrix} 3 & -1 \\ -5 & 4 \end{bmatrix}$,

then its unique inverse is $A^{-1} = \begin{bmatrix} \dfrac{4}{7} & \dfrac{1}{7} \\ \dfrac{5}{7} & \dfrac{3}{7} \end{bmatrix}$.

11.5 Using matrices to formulate and represent a system of equations

Many situations in business and in the fields of scientific and medical research can be modelled mathematically using what is called a **system of linear equations,** where

(i) a **system** of equations refers to a situation where a given problem may be expressed using **two or more consistent** equations, and

(ii) a **linear** equation is one in which all the unknown variables are of **order 1.** *For example*: $y = 3 + 2x$ is a linear equation whereas $y = 3 + 2x^2$ is *not*.

Note: Two or more equations are said to be **consistent** provided:

(i) the number of equations is the same as the number of unknown variables, and

(ii) the equations are all mathematically different (implying that one is not just a constant multiple of another).

Furthermore, it can be shown that any system of **consistent** linear equations has a **unique** solution and in *Section 6* following, one method of obtaining this solution is presented.

However, for now, let us explore further the notion of a system of linear equations.

For example:

(i) A system of two consistent linear equations in two unknown variables, x and y, may be written:
$$3x - y = 8$$
$$x + 2y = 5$$

(ii) A system of three consistent linear equations in three unknowns – x, y and z – may be written:

$$y + z = 10$$
$$x + 2y + 2z = 18$$
$$x + y + 3z = 12$$

Now, as you will see as you read on, we can also represent any system of linear equations, regardless of size, far more compactly in **matrix form.**

For example: Consider again the system of two linear equations in two unknowns, x and y – namely,

$$3x - y = 8$$
$$x + 2y = 5$$

Firstly, we can replace these two algebraic equations by *one* **matrix** equation,

$$\begin{bmatrix} 3x - y \\ x + 2y \end{bmatrix} = \begin{bmatrix} 8 \\ 5 \end{bmatrix}$$

remembering that any two matrices of the same order (here [2 x 1]) are equal if their corresponding elements are equal.

Then, using the rules of matrix multiplication, we can re-write the first [2 x 1] matrix as the product of two *separate* matrices, one containing the coefficients of the unknowns and the other containing the unknowns themselves.

Thus, we have $\begin{bmatrix} 3 & -1 \\ 1 & 2 \end{bmatrix} \cdot \begin{bmatrix} x \\ y \end{bmatrix} = \begin{bmatrix} 8 \\ 5 \end{bmatrix}$.

(Carry out the multiplication of the two matrices on the left hand side to prove to yourself that the result will be the left hand side of the original set of equations.)

Now, if we define the matrices A, X and B as

$$A = \begin{bmatrix} 3 & -1 \\ 1 & 2 \end{bmatrix}, X = \begin{bmatrix} x \\ y \end{bmatrix} \text{ and } B = \begin{bmatrix} 8 \\ 5 \end{bmatrix},$$

then our matrix equation may be written in the form

$$A \cdot X = B.$$

Generalising the above explanation, we can now state the following result.

> Any system of linear equations may be expressed in **matrix form** as
>
> $$A \cdot X = B, \text{ where}$$
>
> $A =$ the **coefficient** matrix whose elements are the coefficients of the unknown variables
>
> $X =$ the column **variable** matrix whose elements are the unknown variables
>
> $B =$ the column matrix of **constants** – often called **constraints.**

EXAMPLE 11.17

Express the following system of linear equations in matrix form.

$$y + z = 10$$
$$x + 2y + 2z = 18$$
$$x + y + 3z = 12$$

Solution

Since the three equations above form a system of linear equations in three unknowns, we should be able to represent them in matrix form as

$$A \cdot X = B, \text{ where}$$

A is the square matrix of coefficients

X is the column matrix of variables

B is the column matrix of right-hand side constants (constraints).

So, from the given equations we have

$$A = \begin{bmatrix} 0 & 1 & 1 \\ 1 & 2 & 2 \\ 1 & 1 & 3 \end{bmatrix}, X = \begin{bmatrix} x \\ y \\ z \end{bmatrix} \text{ and}$$

$$B = \begin{bmatrix} 10 \\ 18 \\ 12 \end{bmatrix}.$$

Putting these together in matrix form, we may now represent the original system of equations as

$$\begin{bmatrix} 0 & 1 & 1 \\ 1 & 2 & 2 \\ 1 & 1 & 3 \end{bmatrix} \cdot \begin{bmatrix} x \\ y \\ z \end{bmatrix} = \begin{bmatrix} 10 \\ 18 \\ 12 \end{bmatrix}$$

(As an exercise, you should perform the multiplication $A \cdot X$ to check that you obtain the left hand side of each equation in the original set.)

Comment:

Although we have now seen how to represent a given system of linear equations in matrix form, there is one problem we have **not** addressed – namely, *how to set up the system of equations in the first place.* That is, given a problem in words, how do we formulate a system of equations to represent that problem?

This is generally the most difficult part of solving any given problem. However, if you keep in mind the following procedure, the task may become a little simpler.

Step 1:

Identify the **unknowns** by *name* and **count** them.

Step 2:

Identify the **conditions** or **constraints** by name and check that the number of conditions is **equal** to the number of unknowns.

Step 3:

Set up one equation for **each** condition.

Now let us apply these procedural steps to a particular problem.

EXAMPLE 11.18

A hardware firm trades as

- **(i)** a wholesaler,
- **(ii)** a retailer, and
- **(iii)** a supplier to tradespeople.

It produces three types of a left-handed knob, one for each of the above markets.

Each type of knob is made on the same production line, passing through

- **(a)** the casting stage,
- **(b)** the assembly stage, and
- **(c)** the polishing stage.

The type for wholesalers takes 1 hour, 3 hours and 2 hours to pass through each stage, respectively.

The type for retailers takes 2 hours, 3 hours and 3 hours at each stage, respectively.

The type for tradespeople takes 2 hours, 2 hours and 3 hours, respectively.

In any one week there are 180 hours of casting, 280 hours of assembly and 290 hours of polishing available for producing these knobs, and all available hours must be used up.

Formulate a set of equations that will allow us to determine how many of each type of knob the firm can make.

Solution

Step 1:

Identify the unknowns.

Here there are three unknowns, namely the quantity of each of the three types of knob.

So, let

x = no. of knobs for wholesalers.

y = no. of knobs for retailers.

z = no. of knobs for tradespeople.

Step 2:

There are three 'conditions', one for each stage in the production of a knob – namely, casting, assembly and polishing.

(*Note:* The number of conditions is equal to the number of unknowns, as required).

Step 3:

Take each condition (constraint) in turn and set up an equation using the information given. Here we are told how long each type of knob takes to pass through a given stage with the overall constraint being the total weekly hours available for that stage.

This information allows us to set up the following system of equations.

(a) **Casting:** $x + 2y + 2z = 180$

Why? – the type for wholesalers, x, takes 1 hour for casting; the type for retailers, y, takes 2 hours; and the type for tradespeople, z, takes 2 hours. Finally, there are 180 hours available for casting per week.

Similarly,

(b) **Assembly:** $3x + 3y + 2z = 280$, and

(c) **Polishing:** $2x + 3y + 3z = 290$.

So, we can represent the solution to the problem with the following set of linear equations:

$$x + 2y + 2z = 180$$
$$3x + 3y + 2z = 280$$
$$2x + 3y + 3z = 290$$

These in turn may be re-written in **matrix form** as
$A \cdot X = B$, **or**

$$\begin{bmatrix} 1 & 2 & 2 \\ 3 & 3 & 2 \\ 2 & 3 & 3 \end{bmatrix} \cdot \begin{bmatrix} x \\ y \\ z \end{bmatrix} = \begin{bmatrix} 180 \\ 280 \\ 290 \end{bmatrix}$$

YOUR TURN

11.28 Write each of the following systems of equations in **matrix form** – that is, in the form $A \cdot X = B$.

(a) $3x + 3y + 6z = 3$
$2y + 7z = 4$
$2x + 2y + 4z = 4$

(b) $x - 2y + 3z = 5$
$2x + 3y - z = 6$
$2z = 4$

11.29 A mine ships its ore through 3 ports, because 3 widely separated countries of destination are involved and the customers insist on the use of particular shipping routes. A total of 1000 tonnes of ore leaves the mine each day by rail, some going to each port, at a total daily rail freight cost of $3 100. The cost of rail transport from the mine to Afriport is $2 per tonne of ore, to Bulgaport is $3 per tonne and to Calciport is $5 per tonne.

By rail, Afriport is 50 kilometres from the mine, Bulgaport is 200 kilometres and Calciport is 1000 kilometres. Daily freight capacities correspond to 340 000 tonne-kilometres.

(a) Formulate a system of equations that will allow you to find the number of tonnes of ore shipped through each port.

(b) Represent your system of equations in matrix form $A \cdot X = B$.

11.6 Using matrices to solve a system of equations

In the previous section we learned how a system of equations may be formulated in order to solve a particular problem, and how matrices may be used to represent those equations in a more compact form – namely,

$$A \cdot X = B$$

where
- A = the coefficient matrix
- X = the variable matrix
- B = the constraint matrix.

Our next step is to use our knowledge of matrix operations to *solve* the system of equations – that is, to find values for the unknown variables in the variable matrix. We will, in fact, present *two* methods for solving a system of linear equations.

Method 1: Using the inverse matrix.

Provided the coefficient matrix, A, has an inverse, A^{-1}, (and it will if $|A| \neq 0$), then if we *pre-multiply* both sides of the equation $A \cdot X = B$ by A^{-1}, we obtain $A^{-1} \cdot A \cdot X = A^{-1} \cdot B$.

However, since $A^{-1} \cdot A = I$, the left-hand side becomes $I \cdot X$ which equals X. Therefore, the equation can be expressed as $X = A^{-1} \cdot B$.

Since both A^{-1} and B are composed only of *known* elements, and A^{-1} (provided it exists) is square with the same number of columns as B has rows, then $A^{-1} \cdot B$ may be found by ordinary matrix multiplication. Furthermore, $A^{-1} \cdot B$ will be the same size as B, namely a column vector, with an element for each equation – that is, one element corresponding to each unknown element in the column vector, X.

Note: You will find the above explanation much easier to understand if you work carefully through the following example.

EXAMPLE 11.19

Solve the system of linear equations

$$3x + 2y = 13$$
$$x + y = 6$$

Solution

In matrix form, the above system of equations may be written

$$\begin{bmatrix} 3 & 2 \\ 1 & 1 \end{bmatrix} \cdot \begin{bmatrix} x \\ y \end{bmatrix} = \begin{bmatrix} 13 \\ 6 \end{bmatrix},$$

so that $\begin{bmatrix} x \\ y \end{bmatrix} = \begin{bmatrix} 3 & 2 \\ 1 & 1 \end{bmatrix}^{-1} \cdot \begin{bmatrix} 13 \\ 6 \end{bmatrix}$.

Our first step is to find the elements of the inverse matrix $A^{-1} = \begin{bmatrix} 3 & 2 \\ 1 & 1 \end{bmatrix}^{-1}$.

From *Section 4* we know that

$$A^{-1} = \frac{1}{|A|} \left(A^c \right)^T$$

$$= \frac{1}{|A|} \begin{bmatrix} a_{22} & -a_{12} \\ -a_{21} & a_{11} \end{bmatrix}$$

So, with $|A| = 3 - 2 = 1$, we have that

$$A^{-1} = \begin{bmatrix} 1 & -2 \\ -1 & 3 \end{bmatrix}$$

Hence, we may now write $X = A^{-1} \cdot B$, or

$$\begin{bmatrix} x \\ y \end{bmatrix} = \begin{bmatrix} 1 & -2 \\ -1 & 3 \end{bmatrix} \cdot \begin{bmatrix} 13 \\ 6 \end{bmatrix}$$

$$= \begin{bmatrix} 13 - 12 \\ -13 + 18 \end{bmatrix}.$$

Thus, $\begin{bmatrix} x \\ y \end{bmatrix} = \begin{bmatrix} 1 \\ 5 \end{bmatrix}$.

Since these two matrices of the same size are equal, their corresponding elements must also be equal. Therefore, the solution to the original system of equations is that $x = 1$ and $y = 5$.

Check:

It is **always** advisable to check your solutions by substituting back into the original equations. So, with $x = 1$ and $y = 5$, we have

$$3x + 2y = 3(1) + 2(5) = 13$$
$$x + y = 1 + 5 = 6,$$

as required.

EXAMPLE 11.20

Solve the following system of linear equations:

$$y + z = 6$$
$$x + 2y + 2z = 10$$
$$x + y + 3z = 2$$

Solution

In *matrix form* the above system of equations may be written $A \cdot X = B$, or

$$\begin{bmatrix} 0 & 1 & 1 \\ 1 & 2 & 2 \\ 1 & 1 & 3 \end{bmatrix} \cdot \begin{bmatrix} x \\ y \\ z \end{bmatrix} = \begin{bmatrix} 6 \\ 10 \\ 2 \end{bmatrix}$$

so that $\begin{bmatrix} x \\ y \\ z \end{bmatrix} = A^{-1} \cdot \begin{bmatrix} 6 \\ 10 \\ 2 \end{bmatrix}$.

Now, if you carried out the rather tedious calculations you would find that,

if $A = \begin{bmatrix} 0 & 1 & 1 \\ 1 & 2 & 2 \\ 1 & 1 & 3 \end{bmatrix}$, then $A^{-1} = \begin{bmatrix} -2 & 1 & 0 \\ \frac{1}{2} & \frac{1}{2} & -\frac{1}{2} \\ \frac{1}{2} & -\frac{1}{2} & \frac{1}{2} \end{bmatrix}$.

Thus, we may now write $X = A^{-1} \cdot B$, or

$$\begin{bmatrix} x \\ y \\ z \end{bmatrix} = \begin{bmatrix} -2 & 1 & 0 \\ \frac{1}{2} & \frac{1}{2} & -\frac{1}{2} \\ \frac{1}{2} & -\frac{1}{2} & \frac{1}{2} \end{bmatrix} \cdot \begin{bmatrix} 6 \\ 10 \\ 2 \end{bmatrix}$$

$$= \begin{bmatrix} -12 + 10 + 0 \\ 3 + 5 - 1 \\ 3 - 5 + 1 \end{bmatrix}.$$

So, $\begin{bmatrix} x \\ y \\ z \end{bmatrix} = \begin{bmatrix} -2 \\ 7 \\ -1 \end{bmatrix}$.

Hence, the solution to the original system of equations is $x = -2$, $y = 7$ and $z = -1$.

Check:
By substitution we have:

(i) $y + z = 7 - 1 = 6$ (**true**)

(ii) $x + 2y + 2z = -2 + 14 - 2 = 10$ (**true**)

(iii) $x + y + 3z = -2 + 7 - 3 = 2$ (**true**)

Method 2: Using Cramer's rule

Cramer's rule provides an alternative method for solving a set of linear equations using **determinants**.
We will discuss this rule only as it applies to solving two equations in two unkowns, although it can easily be extended to solve larger systems, provided the determinants can be found.
Suppose we have two equations:

$$a_{11}x + a_{12}y = c_1$$

$$a_{21}x + a_{22}y = c_2$$

where x and y are the unknowns; a_{11}, a_{12}, a_{21} and a_{22} of the coefficients and c_1 and c_2 are the known constraints.
In **matrix form** these equations may be written as

$$A \cdot X = B$$

where $A = \begin{bmatrix} a_{11} & a_{12} \\ a_{21} & a_{22} \end{bmatrix}$ is the coefficient matrix

$X = \begin{bmatrix} x \\ y \end{bmatrix}$ is the variable matrix

$B = \begin{bmatrix} c_1 \\ c_2 \end{bmatrix}$ is the constraint matrix.

Then,

> **Cramer's rule** states:
> Each unknown is the *quotient* of two **determinants,** where
> (i) the *denominator* is always $|A|$
> (ii) the numerator is $|A|$ with *one* column replaced by B – namely, the first column for the first unknown, the second column for the second unknown, and so on.

Thus, for a system of two equations in two unknowns,

with $A = \begin{bmatrix} a_{11} & a_{12} \\ a_{21} & a_{22} \end{bmatrix}$ and $B = \begin{bmatrix} c_1 \\ c_2 \end{bmatrix}$, we have (by

Cramer's rule) $x = \dfrac{\begin{vmatrix} c_1 & a_{12} \\ c_2 & a_{22} \end{vmatrix}}{|A|}$ and $y = \dfrac{\begin{vmatrix} a_{11} & c_1 \\ a_{21} & c_2 \end{vmatrix}}{|A|}$,

where $|A| = a_{11}a_{22} - a_{12}a_{21}$.

Note: Since **Cramer's rule** involves division by $|A|$, it is clearly *not* applicable if A is a **singular** matrix (i.e. if $|A| = 0$).

Once again we will clarify the above explanation with an example.

EXAMPLE 11.21
Solve the system of linear equations given in example 11.19 using **Cramer's rule**. Check that your solutions agree.

$$3x + 2y = 13$$

$$x + y = 6$$

Solution
In matrix form, the above system of equations may be written

$$\begin{bmatrix} 3 & 2 \\ 1 & 1 \end{bmatrix} \cdot \begin{bmatrix} x \\ y \end{bmatrix} = \begin{bmatrix} 13 \\ 6 \end{bmatrix}.$$

Using **Cramer's rule** we know that

$$x = \frac{\begin{vmatrix} c_1 & a_{12} \\ c_2 & a_{22} \end{vmatrix}}{|A|} \text{ and } y = \frac{\begin{vmatrix} a_{11} & c_1 \\ a_{21} & c_2 \end{vmatrix}}{|A|}$$

Now, $\quad |A| = 3 - 2 = 1,$

$$\begin{vmatrix} c_1 & a_{12} \\ c_2 & a_{22} \end{vmatrix} = \begin{vmatrix} 13 & 2 \\ 6 & 1 \end{vmatrix} = 13 - 12 = 1,$$

$$\begin{vmatrix} a_{11} & c_1 \\ a_{21} & c_2 \end{vmatrix} = \begin{vmatrix} 3 & 13 \\ 1 & 6 \end{vmatrix} = 18 - 13 = 5,$$

Therefore, $x = \dfrac{1}{1} = 1$ and $y = \dfrac{5}{1} = 5.$

As expected, this agrees with the values for x and y obtained in example 11.19 and hence further checking is not required.

EXAMPLE 11.22
Solve the following system of equations using

    (i) the inverse matrix

    (ii) Cramer's rule.

$$x + 3y = 5$$
$$2x - y = -4$$

Be sure to check your answer using substitution.

Solution
(i) Using the inverse matrix
In matrix form, the given system of equations may be written

$$\begin{bmatrix} 1 & 3 \\ 2 & -1 \end{bmatrix} \cdot \begin{bmatrix} x \\ y \end{bmatrix} = \begin{bmatrix} 5 \\ -4 \end{bmatrix},$$

so that $\begin{bmatrix} x \\ y \end{bmatrix} = \begin{bmatrix} 1 & 3 \\ 2 & -1 \end{bmatrix}^{-1} \cdot \begin{bmatrix} 5 \\ -4 \end{bmatrix}$

Since $\quad A^{-1} = \dfrac{1}{|A|} \left(A^c \right)^T$

$$= \frac{1}{|A|} \begin{bmatrix} a_{22} & -a_{12} \\ -a_{21} & a_{11} \end{bmatrix},$$

we have $\begin{bmatrix} 1 & 3 \\ 2 & -1 \end{bmatrix}^{-1} = \dfrac{1}{-7} \begin{bmatrix} -1 & -3 \\ -2 & 1 \end{bmatrix} = \begin{bmatrix} \frac{1}{7} & \frac{3}{7} \\ \frac{2}{7} & -\frac{1}{7} \end{bmatrix}$

Hence, $\begin{bmatrix} x \\ y \end{bmatrix} = \begin{bmatrix} \frac{1}{7} & \frac{3}{7} \\ \frac{2}{7} & -\frac{1}{7} \end{bmatrix} \cdot \begin{bmatrix} 5 \\ -4 \end{bmatrix}$

$$= \begin{bmatrix} \frac{5}{7} - \frac{12}{7} \\ \frac{10}{7} + \frac{4}{7} \end{bmatrix}$$

$$= \begin{bmatrix} -1 \\ 2 \end{bmatrix}$$

Therefore, the solution to the original system of equations is $x = -1$ and $y = 2.$

Check:

$$x + 3y = -1 + 6 = 5 \text{ (true)}$$
$$2x - y = -2 - 2 = -4 \text{ (true)}$$

(ii) Using Cramer's rule
We have already seen that the given system, in matrix form, may be written

$$\begin{bmatrix} 1 & 3 \\ 2 & -1 \end{bmatrix} \cdot \begin{bmatrix} x \\ y \end{bmatrix} = \begin{bmatrix} 5 \\ -4 \end{bmatrix}.$$

By **Cramer's rule** we know that

$$x = \frac{\begin{vmatrix} c_1 & a_{12} \\ c_2 & a_{22} \end{vmatrix}}{|A|} = \frac{\begin{vmatrix} 5 & 3 \\ -4 & -1 \end{vmatrix}}{-1 - 6} = \frac{-5 + 12}{-7} = -1$$

$$y = \frac{\begin{vmatrix} a_{11} & c_1 \\ a_{21} & c_2 \end{vmatrix}}{|A|} = \frac{\begin{vmatrix} 1 & 5 \\ 2 & -4 \end{vmatrix}}{-7} = \frac{-4 - 10}{-7} = 2$$

Therefore, using **Cramer's rule**, the solution to the original system of equations is $x = -1$ and $y = 2$, as expected.

YOUR TURN

In each of the questions 11.30 and 11.31, represent the given system of equations in matrix form, $A \cdot X = B$, and then use

    i) the inverse matrix

    ii) Cramer's rule

to obtain a solution.

Finally, be sure to check your answer by substituting back into the original equations.

11.30 $2x + 5y = 17$
$\quad\quad\quad 3x - y = 17$

11.31 $\begin{aligned} 6x \quad\quad &= 6 \\ 3x + 2y &= 3 \end{aligned}$

In exercises 11.32 and 11.33, represent the given system of equation in matrix form, $A \cdot X = B$, and then use the inverse matrix provided to obtain a solution.

Again be sure to check your answer by substituting back into the original equations.

11.32 $\begin{aligned} x + 2y + 5z &= 1 \\ y + 4z &= 1 \\ 2x + y \quad\quad &= 1 \end{aligned}$

Here, take $A^{-1} = \begin{bmatrix} -2 & \frac{5}{2} & \frac{3}{2} \\ 4 & -5 & -2 \\ -1 & \frac{3}{2} & \frac{1}{2} \end{bmatrix}$.

11.33 $\begin{aligned} 3x + y + 3z &= 2 \\ 3x + 3y + z &= 2 \\ 2x + \quad 3z &= 1 \end{aligned}$

Here, take $A^{-1} = \begin{bmatrix} \frac{9}{2} & -\frac{3}{2} & -4 \\ -\frac{7}{2} & \frac{3}{2} & 3 \\ -3 & 1 & 3 \end{bmatrix}$.

Summary

The main purpose of this chapter was to introduce a very convenient mathematical procedure, called **matrix algebra,** often used for organising and manipulating large amounts of data and for solving fairly complex problems that arise in a variety of disciplinary fields.

We have spent much of the chapter discussing definitions, notation and rules associated with matrix operations, and it is only in the last couple of sections that we have actually used this information to formulate and solve some rather simple problems.

In practice, most 'real-life' problems involve quite large amounts of data and, although they could be solved using the procedures discussed here, it would be a somewhat tedious and inefficient process. So, unless only small amounts of data are involved, it is far more common and efficient to use computers to arrive at the required solution.

The following brief outline covers the most important aspects of this chapter.

A **matrix**, A, is a rectangular array of **elements**, a_{ij} (numbers), and is said to be of **size** $[m \times n]$, where m = number of **rows** and n = number of **columns.**

Any $[1 \times n]$ or $[m \times 1]$ matrix is called a **row** or **column vector.**

Suppose A and B are two matrices, then the following points are true:

(a) $A = B$ if

   (i) size of A = size of B.

   (ii) $a_{ij} = b_{ij}$ for all i and j.

(b) To find $(A + B)$ or $(A - B)$ just add (or subtract) the corresponding elements.

   Then, *size of* $(A + B)$ = *size of* A = *size of* B.

(c) If k is any **scalar** (number), then kA is found by multiplying each element, a_{ij}, by k.

   Then, *size of* kA = *size of* A.

(d) If A is of size $[R \times m]$ and B is of size $[m \times C]$ then $A \cdot B$ (that is, A times B) exists and is of size $[R \times C]$.

   To multiply A and B, the number of **columns** of A *must* **equal** the number of **rows** of B.

   To find a_{ij} (the i-j element in $A \cdot B$), multiply each element in the ith **row** of A by the corresponding element in the jth **column** of B and *add* these products.

   Note: In general, $A \cdot B \neq B \cdot A$.

(e) A matrix A is **square** if the no. of rows is equal to the no. of columns.

   Also, the diagonal running from **top left** to **bottom right** is called the **principal diagonal.**

(f) The **identity matrix,** I, is a **square** matrix whose principal diagonal consists of 1's only, with all other elements, zero.

$$I \cdot A = A \cdot I = A$$

(g) The **zero matrix** has all elements = 0.

(h) If A is of size $[m \times n]$, then its **transpose** A^T is of size $[n \times m]$ and is found by inter-changing rows and columns.

   Note: If $A = A^T$, A is said to be **symmetric.**

(i) The **inverse** of A – namely, A^{-1}, if it exists – is the matrix giving the following result:

$$A^{-1} \cdot A = A \cdot A^{-1} = I.$$

   If A^{-1} exists, A is said to be **non-singular.**

   If A^{-1} does **not** exist, A is said to be **singular.**

(j) If A is a *square* matrix, then its **determinant,** $|A|$, is

   (i) $a_{11} \cdot a_{22} - a_{12} \cdot a_{21}$, if A is of size $[2 \times 2]$

(ii) $a_{11}A_{11} + a_{12}A_{12} + a_{13}A_{13}$, if A is of size [3 x 3],

where A_{ij}, called the **cofactor** of a_{ij}, is $(-1)^{i+j}$ times the **determinant** of the [2 x 2] matrix that remains after the ith row and j^{th} column of the original matrix A have been (mentally) deleted.

(k) Finding the inverse, A^{-1}

If A is any **square** matrix, then its **inverse**, A^{-1}, is given by

$$A^{-1} = \frac{1}{|A|}\left(A^c\right)^T$$

where

(i) $|A|$ = the determinant of A

(ii) A^c = the cofactor matrix whose elements, A_{ij}, are the cofactors of the corresponding elements a_{ij} in the original matrix, A

(iii) $\left(A^c\right)^T$ = the transpose of A^c.

Note: Clearly, A^{-1} exists **only if** $|A| \neq 0$. Therefore, if $|A| = 0$, A is said to be **singular** and if $|A| \neq 0$, A is said to be **non-singular**.

(l) Any system of linear equations may be expressed in **matrix form** as $A \cdot X = B$, where

A = the **coefficient** matrix whose elements are the variable coefficients.

X = the **variable** matrix (column vector) whose elements are the unknown variables.

B = the matrix of **constants** (or constraints) – also a column vector.

(m) The solution of a system of equations represented in matrix form, $A \cdot X = B$, may be obtained using

(i) the **inverse matrix method**, where

$X = A^{-1} \cdot B$ provides the solution.

(ii) Cramer's rule, where each unknown is the *quotient* of two determinants with

1. the denominator always $|A|$

2. the numerator being $|A|$ with the 1st column replaced by the 1st unknown, *or* the 2nd column replaced by the 2nd unknown, and so on.

Note: Both solution methods require that A is **non-singular** – that is, that $|A| \neq 0$.

11.1 Consider the following matrices:

$$A = \begin{bmatrix} 4 & 3 & 2 \\ 7 & 0 & 5 \\ 2 & 9 & 8 \\ 3 & 3 & 4 \end{bmatrix} \qquad B = \begin{bmatrix} 4 & 7 & 2 & 3 \\ 3 & 0 & 9 & 3 \\ 1 & 5 & 8 & 4 \end{bmatrix}$$

What value is given to the following elements in A and B above?
a_{33} a_{21} b_{23} a_{32} b_{22}

11.2 If $A = \begin{bmatrix} 1.7 & 2.7 \\ 2.6 & 4.9 \\ -0.8 & 1.5 \\ 3.5 & -1.5 \end{bmatrix}$

then A is of size _____.

11.3 If $B = \begin{bmatrix} 2.0 & 4.1 & 3.0 \\ 1.9 & 3.2 & 0 \\ 4.5 & -4.3 & 1.8 \end{bmatrix}$

then B is of size _____.

11.4 If $C = \begin{bmatrix} 5.0 \\ 9.0 \\ 7.0 \\ -2.3 \end{bmatrix}$

then C is of size _____.

11.5 If $D = [17\ 24\ 51\ 74\ 38]$
then D is of size _____.

11.6 $A = \begin{bmatrix} 2 & 3 & 1 \\ 4 & 2 & 5 \end{bmatrix}$ $B = \begin{bmatrix} 3 & -2 & 4 \\ 1 & 0 & 6 \end{bmatrix}$

$C = \begin{bmatrix} 4 & 2 & 1 \\ 3 & 1 & 2 \end{bmatrix}$ $D = \begin{bmatrix} 1 & 3 \\ 2 & 4 \\ 2 & 1 \end{bmatrix}$

Evaluate, where possible,
(a) $A + B$ **(b)** $A - B$
(c) $C + D$ **(d)** $C^T + D$
(e) $B - D^T$

11.7 Refer back to Example 11.8.
In that example, matrix A represented the matrix of quantities, while B was the matrix of prices.
Therefore, the product $A \cdot B$ represented both the actual and potential shopping bills of John and Janice.
Now multiply $B \cdot A$ (which is possible) and interpret the meaning of the 9 elements.
(***Note:*** Here $A \cdot B \neq B \cdot A$)

11.8 Given the following matrices:

$$A = \begin{bmatrix} 1 & 3 & 6 \\ 2 & 7 & 4 \end{bmatrix} \qquad B = \begin{bmatrix} 5 & 7 & 2 \\ 0 & 6 & 8 \end{bmatrix}$$

$$C = \begin{bmatrix} 18 & 5 & 9 \\ 11 & 3 & 2 \end{bmatrix} \qquad D = \begin{bmatrix} 4 & 1 & 6 \\ 2 & 5 & 3 \end{bmatrix}$$

$$E = \begin{bmatrix} 1 & 3 & 8 \\ 7 & 4 & 5 \\ 2 & 6 & 9 \end{bmatrix} \qquad F = \begin{bmatrix} 12 & 2 & 0 \\ 3 & 7 & 5 \\ 9 & 11 & 4 \end{bmatrix}$$

$$G = \begin{bmatrix} 6 & 2 & 8 \\ 5 & 19 & 7 \\ 13 & 15 & 9 \end{bmatrix}$$

find:
- **(a)** $A + B$
- **(b)** $A - B$
- **(c)** $E - G$
- **(d)** $G + F - E$
- **(e)** $A + B + C + D$
- **(f)** $7 \times E$
- **(g)** $\frac{1}{6} \times D$
- **(h)** $5C + B - 2D$
- **(i)** $\frac{1}{4}(F - 3E)$

11.9 Multiply together the pairs of matrices below, in whichever order is possible.

- **(a)** $A = \begin{bmatrix} 1 & 2 & 8 \\ 5 & 1 & 2 \\ 0 & 9 & 6 \\ 1 & 1 & 3 \end{bmatrix} \quad B = \begin{bmatrix} 7 & 8 \\ 6 & 5 \\ 4 & 3 \end{bmatrix}$

- **(b)** $C = \begin{bmatrix} 1 \\ 2 \\ 3 \end{bmatrix} \quad D = \begin{bmatrix} 4 & 5 & 6 \end{bmatrix}$

- **(c)** $E = \begin{bmatrix} 4 & 0 \\ 1 & -1 \end{bmatrix} \quad F = \begin{bmatrix} 0 & 3 \\ -1 & -2 \end{bmatrix}$

- **(d)** $G = \begin{bmatrix} 1 & 0 & 2 & 0 & 3 \\ 0 & 2 & 1 & 1 & 0 \end{bmatrix} \quad H = \begin{bmatrix} 1 & 2 \\ 3 & 4 \end{bmatrix}$

- **(e)** $J = \begin{bmatrix} 1 & 2 \\ 3 & 4 \\ 5 & 6 \end{bmatrix} \quad K = \begin{bmatrix} -1 & 0 \\ 4 & 1 \\ 3 & 2 \end{bmatrix}$

11.10 A paint store operator has 2 main colour sellers: white and green. He stocks economy and super quality for each colour. There are 300 tins of economy green, 400 tins of economy white and 100 of each colour in the super quality.

Normal retail prices are $8 and $6 for the super and economy respectively, but they are going on special sale at $1 off per tin.

11.10 *Cont.*

What would be the gross amount realised if
- **(a)** all tins are sold at the normal price?
- **(b)** all are sold at the sale price?

Answer these questions using matrices.

11.11 Matrix C below shows the cost for labour, materials and overheads (in that order) per unit of products W (col 1), X (col 2), Y (col 3) and Z (col 4).

$$C = \begin{bmatrix} 0.70 & 3.00 & 1.55 & 0.90 \\ 2.00 & 4.20 & 0.20 & 3.10 \\ 2.10 & 8.00 & 2.00 & 1.30 \end{bmatrix}$$

- **(a)** If all costs rise by 50%, find the new cost matrix by scalar multiplication.
- **(b)** Production of W, X, Y and Z in a certain period is given by the vector, P, where

$$P = \begin{bmatrix} 1000 \\ 500 \\ 800 \\ 50000 \end{bmatrix}$$

Find $C \cdot P$ and interpret the result of this multiplication.

11.12 Given the matrices

$$A = \begin{bmatrix} 3 & 4 \\ 2 & 3 \end{bmatrix} \quad B = \begin{bmatrix} 3 & 4 \\ 2 & -3 \end{bmatrix} \quad C = \begin{bmatrix} 3 & -4 \\ -2 & 3 \end{bmatrix}$$

Use matrix multiplication to determine which 2 matrices are inverses of each other.

11.13 For each of the following matrices find
- (i) the determinant,
- (ii) the inverse, if it exists.

- **(a)** $A = \begin{bmatrix} 2 & 5 \\ 3 & -1 \end{bmatrix}$
- **(b)** $B = \begin{bmatrix} 3 & 4 \\ 2 & -3 \end{bmatrix}$
- **(c)** $C = \begin{bmatrix} 1 & 2 & -1 \\ 0 & -3 & 2 \\ 5 & 1 & 0 \end{bmatrix}$
- **(d)** $D = \begin{bmatrix} 1 & 2 & 5 \\ 0 & 1 & 4 \\ 2 & 1 & 0 \end{bmatrix}$

11.14 Use the fact that

$$\begin{bmatrix} -2 & 3\frac{1}{2} \\ 1 & -1\frac{1}{2} \end{bmatrix} \text{ is the inverse of } \begin{bmatrix} 3 & 7 \\ 2 & 4 \end{bmatrix}$$

to solve the following systems of equations.
- **(a)** $3x + 7y = 5$
  $2x + 4y = 4$
- **(b)** $2x + 4y = 16$
  $3x + 7y = 27$
- **(c)** $3x + 7y = 15$
  $2y + x = 4$

11.15 Solve each of the following systems of equations by:

    (i) representing them in matrix form $A \cdot X = B$

    (ii) finding the appropriate inverse and carrying out the multiplication, $A^{-1} \cdot B$.

(a)
$$2x + 3y + z = 5$$
$$2x + 2y + 2z = 12$$
$$2y - z = 8$$

(b)
$$2x + 3y + 4z = 5$$
$$4x + 3y + z = 5$$
$$x + 2y + 4z = 5$$

11.16 At the end of each working week, an accounting log of all computer runs is submitted by different departments. The two matrices below summarize this information for March and April. Each entry is measured in dollars.

	Payroll	Engineering	Marketing
Week 1	38	48	95
Week 2	110	52	70
Week 3	45	75	78
Week 4	135	80	45

$M = $ (above)

	Payroll	Engineering	Marketing
Week 1	42	50	85
Week 2	115	90	55
Week 3	40	40	25
Week 4	100	25	35

$A = $ (above)

Use matrix addition to find the total computer expenditures for the two departments during corresponding weeks of March and April.

*11.17 Suppose that a law firm has a hierarchy of lawyers. Matrix A records the cost per hour for each of four levels or grades of legal assistance (I, II, III, IV).

Cost per hour

$$A = \begin{bmatrix} \$30 \\ \$50 \\ \$75 \\ \$100 \end{bmatrix} \begin{matrix} I \\ II \\ III \\ IV \end{matrix}$$

Matrix B records the total number of hours reported by each level during each week of one particular month.

	I	II	III	IV
Week 1	70	65	73	60
Week 2	64	55	30	60
Week 3	40	20	40	0
Week 4	20	15	25	20

$B = $ (above)

(a) Use matrix multiplication to find the amount of money that was generated during each of the 4 weeks of that month.

(b) Find the total amount of money generated by the law firm in that month.

*11.18 A certain company manufactures nuts and bolts. Machine A requires 15 minutes to produce 100 nuts and 20 minutes to produce 100 bolts. Machine B grinds each item to remove rough edges, taking 10 minutes to grind the 100 nuts and 15 minutes to grind the 100 bolts. If each machine operates 8 hours per day, how many nuts and bolts can be produced in a day?

*11.19 Three shoppers, Larry, Todd and Anne, go into a shop to buy fruit. Larry purchases 12 oranges, 5 grapefruit, 20 apples, 6 bananas and 3 lemons. Todd purchases 20 oranges, 3 grapefruit, 10 apples and 4 bananas. Anne purchases 10 oranges, 10 grapefruit and 12 bananas. Suppose that oranges cost 10 cents each, grapefruit 20 cents each, apples 8 cents each, bananas 6 cents each and lemons 5 cents each.

(a) Represent the purchases of *each* shopper by means of a row matrix.

(b) Represent the prices by means of a column matrix.

(c) Use matrix multiplication to find **each** shopper's bill.

(d) Use matrix addition to find the total quantities of each kind of fruit purchased by the shoppers.
Use this result and matrix multiplication to find the total amount spent by the shoppers.

*11.20 A chemist has to make a compound using 3 basic elements, A, B and C, so that it has 10 litres of A, 12 litres of B and 20 litres of C. He makes this compound by mixing 3 compounds, I, II and III.

Each unit of compound I has 4 litres of A, 3 of B and **no** C. The ingredients of a unit of compound II are 1 litre of A, 2 litres of B and 4 litres of C. Finally, a unit of compound III has **no** A, 1 litre of B and 5 litres of C. How many units of each compound are required to meet the minimal requirements of the basic elements?
(You must formulate and then solve the appropriate system of linear equations.)

12 Extras – Calculus

Introduction

Calculus, the topic for discussion in this chapter, is a particular branch of mathematics concerned basically with the mathematical analysis of change – that is, with describing and predicting changing variables – for example, rates of growth and decay.

Actually, **calculus** was developed independently in the seventeenth century by two prominent mathematicians – Sir Isaac Newton (English, 1642 – 1727) and Gottfried Leibnitz (German, 1646 – 1716). Newton developed calculus for his studies in physics and astronomy, while Leibnitz developed calculus for his studies in geometry.

Today, calculus provides us with a means of describing many occurrences in the physical world. *For example:* In economics, calculus is a powerful means of investigating marginal costs and revenues while in physics, calculus may be used to predict, say, the rate at which radio-active elements decay.

Using the theory and rules of calculus, various quantities may be obtained to solve a particular problem, but it is the **interpretation** of these quantities that is dependent upon **who** is trying to solve the problem.

For example:

1. Suppose that a postage stamp costs 45c. Then the total cost of x stamps could be expressed as

   $$y = 45x$$

   where y = total cost

   x = no. of stamps purchased.

   So, if I buy 6 stamps (i.e., $x = 6$),

   total cost = $y = (45 \times 6) = \$2.70$.

   Now,

   a. an *economics* student may say that the **marginal cost** of a stamp is 45c. Having bought any number of stamps, the cost of *one extra* is 45c.

   b. an *algebra* student may say that the **gradient** (slope) of the line, $y = 45x$, is 45.

   c. a **calculus** student may say that the **differential** of $y = 45x$ is 45 – that is, that the function is changing at a rate of 45 cents per extra stamp purchased.

2. Suppose that a repairman charges \$12 per call plus \$8 for each 15-minute period.

   If q = no. of 15-minute periods, and
   p = total cost,

   then the relationship can be expressed as

   $$p = 12 + 8q.$$

   So, for a 1 hour call, $q = 4$, and thus

   $$p = 12 + (8 \times 4) = \$44.$$

   Now,

   a. an *economics* student may say that the **marginal cost** of each 15-minute period is \$8.

   b. an *algebra* student may say that the **gradient** (slope) of the line $p = 12 + 8q$ is 8.

   c. a *calculus* student may say that the **differential** of $p = 12 + 8q$ is 8 – that is, that the function is changing at a rate of \$8 per unit of time.

From the above two illustrations it should be clear that, although the actual *values* are the same, and are, in fact, obtained in the same mathematical way, in a real life situation the interpretation depends upon whether the problem is being viewed by an economist, a physicist, a mathematician, etc.

As you will see as you read on, calculus can be divided into two main areas – namely,

1. **differential calculus**, using the technique of differentiation to find derivatives, and

2. **integral calculus**, using the technique of integration to find integrals.

Basically speaking, **differentiation** is used to compute the rate at which one variable changes in relation to another at any particular instant. **Integration**, on the other hand, is the reverse process, and involves using information about change to describe the variables that are changing.

In the *first* section of this chapter we discuss the mathematical rules of differentiation associated with finding the first derivative, as well as the second and higher order derivatives.

In the *second* section we present some practical examples where the rules of differential calculus, studied in *section 1*, may be applied.

For example:

a. Finding maximum and minimum values.

b. Determining marginal cost, marginal revenue and marginal profit, etc.

Section 3 begins our discussion of integration and examines firstly the notion of definite integrals, followed by a brief look at indefinite integrals.

The *last* section presents some practical problems whose solutions require us to apply our knowledge of integral calculus.

As notation is very important in all areas of mathematics, we will present several alternative forms of notation in this chapter, all of which are important and commonly used in the study of calculus.

Note:

Our rather limited examination of calculus is included here as an **extra** chapter because, at the tertiary level of study, it is not an area of mathematics that is required by all disciplines.

In fact, our main aim here is to assist students who wish to pursue studies in areas such as economics, mathematics and science, by providing some necessary but somewhat specialised mathematical tools rather than to delve deeply into the pure mathematical theory of calculus and related topics.

12.1 Differentiation – mathematical rules

12.1.1 The first derivative

In calculus, given $y = $ some function of x, or $y = f(x)$, then the expression $\frac{dy}{dx}$ (pronounced dee-y; dee-x), is called the **differential coefficient** of y with respect to x, and means the **rate** at which y is changing with respect to x.

You will also find that the expression $\frac{dy}{dx}$ is commonly called the **first derivative** or, more simply, the **derivative** of y with respect to x.

If we can find the derivative of a function, $f(x)$, then the function is said to be **differentiable**. The process of finding the derivative is called **differentiation**, and the instruction '**differentiate**'

is used when a derivative is to be found.

It should be realised from the outset that $\frac{dy}{dx}$ is **not** a fraction, but consists of two parts, $\frac{d}{dx}$ and y, where $\frac{d}{dx}$ may be thought of as an abbreviation of the words: "the differential coefficient of..., with respect to x".

Similarly, if p is some function of q – that is, $p = f(q)$ – then the expression $\frac{dp}{dq}$ (pronounced dee-p; dee-q) is the first derivative of p with respect to q, and means the **rate** at which p is changing with respect to q.

Now the derivative of a function may, in fact, be determined from 'first principles' – that is, 'from scratch'. However, this can become rather cumbersome and time-consuming.

Fortunately for the non – mathematicians among us, there are a number of simple **rules of differentiation** that can be used to find the derivative of various types of functions.

Although these rules have themselves been obtained from first principles, in this chapter they will be stated only, with no proof or explanation, and then applied in a series of examples.

> **Rule 1**
> If $y = a$, where 'a' is a constant, then
> $$\frac{dy}{dx} = 0.$$
> That is, the derivative of a constant is *zero*.

This is easily seen *geometrically*, since $y = $ 'a' is a horizontal line cutting the y axis at 'a'. Because any horizontal line has a gradient of zero, it follows that the derivative must also be zero.

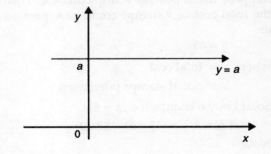

EXAMPLE 12.1

(a) If $p = 2\,000\,000$, then $\frac{dp}{dq} = 0$.

(b) If $y = 30$, then $\frac{dy}{dx} = 0$.

(c) If $y = 0$, then $\dfrac{dy}{dx} = 0$.

(d) If $t = -17$, then $\dfrac{dt}{dw} = 0$.

(Here $\dfrac{dt}{dw}$ is the first derivative of t with respect to w.)

Rule 2

If $y = ax + b$, where a and b are constant, then
$$\dfrac{dy}{dx} = a.$$

Remember:
The derivative of $ax = a$, while the derivative of 'b' is zero. So, the derivative of $ax + b$ is just 'a', and is the amount by which y will change if x increases by **one** unit.

EXAMPLE 12.2

(a) If $y = 13x + 100$, then $\dfrac{dy}{dx} = 13$.

(b) If $y = -\dfrac{1}{4x} + 1$, then $\dfrac{dy}{dx} = \dfrac{-1}{4}$.

(c) If $p = 36q$, then $\dfrac{dp}{dq} = 36$.

(d) If $m = 5.3n - 12$, then $\dfrac{dm}{dn} = 5.3$.

(e) If $y = 7 - 4x$, then $\dfrac{dy}{dx} = -4$.

Rule 3

If $y = x^n$, where n is any real number, then
$$\dfrac{dy}{dx} = nx^{n-1}, n \neq 0.$$

That is, the derivative of the nth power of variable is just n times the $(n-1)^{\text{th}}$ power of the variable.

EXAMPLE 12.3

(a) If $y = x^4$, then $\dfrac{dy}{dx} = 4x^3$.

(b) If $y = x^{14}$, then $\dfrac{dy}{dx} = 14x^{13}$.

(c) If $p = q^{100}$, then $\dfrac{dp}{dq} = 100q^{99}$.

(d) If $p = q^2$, then $\dfrac{dp}{dq} = 2q$.

(e) If $y = x^{-5}$, then $\dfrac{dy}{dx} = -5x^{-6}$.

Rule 4

If $y = ax^n$, where 'a' is a constant, then
$$\dfrac{dy}{dx} = a \cdot nx^{n-1}$$
where $a \cdot n$ means a times n.

In words: **Rule 4** says that the first derivative of a constant times a function is just the constant times the first derivative of the function.

EXAMPLE 12.4

(a) If $y = 2x^{11}$, then
$$\dfrac{dy}{dx} = 2\,(11x^{10}) = 22x^{10}.$$

(b) If $y = -5x^3$, then
$$\dfrac{dy}{dx} = -5\,(3x^2) = -15x^2.$$

(c) If $y = 10x^{10}$, then
$$\dfrac{dy}{dx} = 10\,(10x^9) = 100x^9.$$

(d) If $y = \dfrac{1}{3}x^{24}$, then
$$\dfrac{dy}{dx} = \dfrac{1}{3}\,(24x^{23}) = 8x^{23}.$$

(e) If $y = 14x^{1/2}$, then
$$\dfrac{dy}{dx} = 14\left(\dfrac{1}{2}x^{\frac{1}{2}-1}\right) = 7x^{\frac{-1}{2}}.$$

Rule 5
The sum of (or difference between) functions may be differentiated part by part, using rules 1 to 4.

a. If $y = ax^n + bx + c$, where a, b and c are constants, then
$$\dfrac{dy}{dx} = a \cdot nx^{n-1} + b.$$

b. If $y = ax^m - bx^n + cx - d$, where a, b, c and d are constants, then
$$\dfrac{dy}{dx} = a \cdot mx^{m-1} - b \cdot nx^{n-1} + c.$$

Note:
In **(a)** of **rule 5**, the first derivative of ax^n is $a \cdot nx^{n-1}$, while the first derivative of bx is b, and since the

derivative of the constant c is zero, we see that the derivative of the sum is just the sum of the derivatives.

Similarly in **(b)**, it is clear that the first derivative of a sequence of terms is just the sum of or the difference between the derivatives of each term in the sequence.

EXAMPLE 12.5

(a) If $y = 4x^7 - 3x + 5$, then

$$\frac{dy}{dx} = 28x^6 - 3.$$

(b) If $y = x^3 + x^2 + x + 1$, then

$$\frac{dy}{dx} = 3x^2 + 2x + 1.$$

(c) If $p = 0.003q^3 - 0.05q^2 + 7q + 10$, then

$$\frac{dp}{dq} = 0.009q^2 - 0.10q + 7.$$

(d) If $p = -15 - 203q^2 + \frac{1}{2}q^4$, then

$$\frac{dp}{dq} = -406q + 2q^3.$$

(e) If $y = 15x^7 + x^6 + x$, then

$$\frac{dy}{dx} = 105x^6 + 6x^5 + 1.$$

(f) If $y = 6x^3 - 2x + 4x^{-1} - 8x^{-1/2}$, then

$$\frac{dy}{dx} = 18x^2 - 2 + 4(-1x^{-1-1}) - 8\left(\frac{-1}{2}x^{\frac{-1}{2}-1}\right)$$

$$= 18x^2 - 2 - 4x^{-2} + 4x^{-3/2}.$$

Note: Be careful with the signs, especially when *negative powers* are involved.

YOUR TURN

12.1 The lease of a house calls for a fixed bond of $200 plus $80 per week.
Let w = number of weeks, and
t = total cost.
Which of the following equations allows you to calculate the total cost for a given number of weeks?
(a) $t = 200 + 80w$
(b) $w = 200 + 80t$

YOUR TURN

12.1 *Cont.*
(c) $t = 200w + 8$

12.2 For each of the following equations, find the first derivative, $\frac{dy}{dx}$.
(a) $y = 40 + 5x$
(b) $y = -50x + 23$
(c) $y = \frac{3}{4}x + \frac{1}{2}$
(d) $y = 6.2$
(e) $y = 6x^4$
(f) $y = x^{-8}$

12.3 Differentiate each of the following functions:
(a) $m = 100n^3 + n^2 + 7n + 45$
(b) $m = -13n + 5n^3$
(c) $y = 7x^4 - 10x - \frac{3}{4}$
(d) $y = x^{-2} - 3x^2 + 9$
(e) $y = x^{1/2} - x^2 + 6$

12.4 If $y = x^3 - 5x^2$, find $\frac{dy}{dx}$ when $x = 3$.

12.5 If $s = 3t + 12t^2$, find $\frac{ds}{dt}$ when $t = 5$.

12.1.2 Other notations (apart from dy/dx) used for the differential coefficient

A. Function notation

As this topic has been discussed previously in some detail, we will give a brief reminder only – namely, if the expression $3x^2 - 2x - 1$ is being examined, it may be said:
(i) let $y = 3x^2 - 2x - 1$, or
(ii) let $f(x) = 3x^2 - 2x - 1$.

Note:

1. It may be found that many other letters, apart from x may be used when applying the rules and procedures of differential calculus.

   For example: Since a person's weight depends to some extent on their height, a mathematician would say that weight is a function of height or, symbolically,

$W = f(H)$

where W = weight

H = height.

Similarly, $t = f(I)$ could be used to say that tax is a function of income.

2. It will also be found, as you read on, that other symbols, apart from f, may be used as function names.

    For example: $C(Q) = 100 + 30Q$ shows how cost, C, is related to quantity produced, Q.

    Similarly, $g(x)$ and $h(x)$ are often used to denote functions of x.

The use of the function notation, $f(x)$, $g(x)$ etc. has the added advantage that $f(1)$ means the value of $f(x)$ where $x = 1$, and $f(-3)$ means the value of $f(x)$ when $x = -3$, and so on.

So, **in general**, for any constant a, $f(a)$ would denote the value of the function, $f(x)$, evaluated when a is substituted for all x.

EXAMPLE 12.6

(a) If $f(x) = 100 - 2x$, then

> **(i)** $f(8) = 100 - (2 \times 8)$
> $= 100 - 16$
> $= 84$

> **(ii)** $f(r) = 100 - 2r$

(b) If $g(y) = y^2 + 3y - 1$, then

> **(i)** $g(6) = 6^2 + (3 \times 6) - 1$
> $= 36 + 18 - 1$
> $= 53$

> **(ii)** $g(-2) = (-2)^2 + (3 \times -2) - 1$
> $= 4 - 6 - 1$
> $= -3$

B. Alternative expressions for the first derivative

If $y = 5x^2 - 2x - 8$, then the first derivative is $\frac{dy}{dx} = 10x - 2$, where $\frac{dy}{dx}$ is the expression we have used so far to denote the differential coefficient.

However, the notation $\frac{dy}{dx}$ is rather cumbersome, especially when using a typewriter, and hence is often abbreviated to y' (*pronounced y-dash or y-prime*).

Therefore, if $y = 12x - 56$, then the first derivative may be given by $y' = 12$.

Similarly, if $p = 4q^2 - 6q + 2$, then the first derivative may be given by $p' = 8q - 6$.

Now, if we use the **function** notation and hence write $f(x) = 5x^2 - 2x - 8$, then the first derivative is either $\frac{d}{dx}f(x)$, again rather cumbersome, or, as is more commonly used, $f'(x)$ – pronounced *f-dash-x* or *f-prime-x*.

Therefore, if $f(x) = 5x^2 - 2x - 8$, then the first derivative of $f(x)$ is $f'(x) = 10x - 2$.

As we have already seen, one of the main advantages of using function notation is the ease with which we may indicate that we wish to evaluate the function at some given value.

The same advantage applies should we wish to evaluate, at some particular value, the first derivative of a function expressed in function notation.

For example: If $f(x) = 2x^2 - 5x + 10$, then

> **(i)** $f'(x) = 4x - 5$

> **(ii)** $f'(3) = (4 \times 3) - 5$
> $= 7$

> **(iii)** $f'(0) = (4 \times 0) - 5$
> $= -5$

EXAMPLE 12.7

(a) If $p = 6 - 5q - 3q^2$, then the first derivative is $p' = -5 - 6q$.

(b) If $f(x) = 3x^{-2}$, then

> $f'(x) = 3(-2x^{-2-1})$
> $= -6x^{-3}$

(c) If $g(y) = y^3 + 10y$, then

> **(i)** $g(1) = 1^3 + (10 \times 1)$
> $= 1 + 10$
> $= 11$

> **(ii)** $g'(y) = 3y^2 + 10$

> **(iii)** $g'(2) = (3 \times 2^2) + 10$
> $= 12 + 10$
> $= 22$

12.6 Given that $g(Q) = \frac{1}{2}Q^4 + 12Q$, find

(a) $g(1)$

(b) $g(10)$

12.7 Given that $f(x) = 3x^2 - 5x + 2$, find

(a) $f(-3)$

(b) $f(0)$

(c) $f(3) - f(2)$

12.8 Find $f'(x)$, the 1st derivative, of each of the following functions of x:

(a) $f(x) = 11 + x$

(b) $f(x) = 5x^2 - 3x$

(c) $f(x) = \frac{20 + x}{12}$

12.9 Given that $g(Q) = \frac{1}{2}Q^4 + 12Q$, find:

(a) $g'(Q)$

(b) $g'(3)$

12.10 If $h(t) = t^4 - 4t^2 + 2t - 5$, find $h'(2)$ – that is, the first derivative evaluated at $t = 2$.

12.1.3 Second and higher order derivatives

We have already seen earlier in this section that the first derivative, $\frac{dy}{dx} = f(x)$, of a function, $y = f(x)$, is itself a function of x (or a constant).

For example: If $y = f(x) = 3x^6 + 5x^2 + 8$, then $\frac{dy}{dx} = f'(x) = 18x^5 + 10x$ is also a function of x.

Hence, $\frac{dy}{dx} = f'(x)$ can itself be differentiated with respect to x.

The derivative of $\frac{dy}{dx} = f'(x)$ is called the **second derivative** of the original function, $y = f(x)$, and is denoted $\frac{d}{dx}\left[\frac{dy}{dx}\right] = \frac{d^2y}{dx^2}$,

(read as *dee-two-y over dee-x-squared*).

Notes:

1. Since the first derivative, $f'(x)$, is defined as

the rate of change of $y = f(x)$ with respect to x, then the second derivative may be defined as the rate of change of the **first** derivative with respect to x.

2. If the first derivative of an expression is denoted by y', then the second derivative is usually denoted y'', (pronounced *y-double-dash*).

3. If the first derivative of a function, $f(x)$, is denoted $f'(x)$, then the second derivative is usually denoted $f''(x)$ – (pronounced *f-double-dash-x*).

4. Since each of $\frac{dy}{dx}$ or y' or $f'(x)$ is defined to be the *first* derivative of some function $y = f(x)$, it follows that each of $\frac{d^2y}{dx^2}$ or y'' or $f''(x)$ may be defined as either

a. the *second* derivative of some function $y = f(x)$, or

b. the derivative of the *first* derivative $\frac{dy}{dx}$, y' or $f'(x)$.

5. Just as we can find the **second** derivative of a function by differentiating the *first* derivative, so we can extend this to find further *higher order* derivatives.

For example: $\frac{d^3y}{dx^3}$ or y''' or $f'''(x)$ is the **third** derivative of a function, $y = f(x)$, and is obtained by differentiating the *second* derivative.

Similarly, $\frac{d^4y}{dx^4}$ or y'''' or $f''''(x)$ is the **fourth** derivative of a function, $y = f(x)$, and is obtained by differentiating the *third* derivative.

EXAMPLE 12.8

(a) If $y = 3x^6 + 5x^2$, then

(i) $\frac{dy}{dx} = 18x^5 + 10x$ is the first derivative, and

(ii) $\frac{d^2y}{dx^2} = 90x^4 + 10$ is the second derivative.

(b) If $p = q^2 + 10q + 1000$, then

(i) $p' = 2q + 10$ is the first derivative

(ii) $p'' = 2$ is the second derivative

(iii) $p''' = 0$ is the third derivative (since p'' is constant).

(c) If $f(x) = x^{-3} + x^3$, then

(i) $f'(x) = -3x^{-4} + 3x^2$ is the first derivative

(ii) $f''(x) = -3(-4x^{-4-1}) + 6x$

$= 12x^{-5} + 6x$ is the second derivative

(iii) $f'''(x) = -60x^{-6} + 6$ is the third derivative.

(d) If $f(x) = x^3 + 2x^2$, then

(i) $f'(x) = 3x^2 + 4x$ is the first derivative

(ii) $f''(x) = 6x + 4$ is the second derivative

(iii) $f'(2) = f'(x)$ evaluated at $x = 2$

$= (3 \times 2^2) + (4 \times 2)$

$= 12 + 8$

$= 20$

(iv) $f''(3) = f''(x)$ evaluated at $x = 3$

$= (6 \times 3) + 4$

$= 22$

YOUR TURN

You should attempt the following questions before proceeding to the next section.

12.11 Find the second derivative, $\dfrac{d^2y}{dx^2}$, of the following expressions:

(a) $y = 4x^4 + x^3$

(b) $y = 100 - 2x^6 + x$

(c) $y = 23$

(d) $y = x^{-4} + x^4$

12.12 If $f(t) = 10t - 16t^2$, find

(a) $f''(t)$

(b) $f'(0)$

12.12 If $f(t) = 10t - 16t^2$, find

(a) $f''(t)$

(b) $f'(0)$

12.13 Given that $g(y) = y^4 - 3y^3 + 2y - 5$, find

(a) $g'(0)$

(b) $g''(y)$

(c) $g''(3)$

(d) $g'''(4)$

12.2 Differentiation – applications

12.2.1 Maximum and minimum values

Many important applications of differential calculus in the business arena involve **optimization**, and thus are concerned with finding the *maximum* or *minimum* values of a function.

For example: A manufacturer should be concerned with finding the price that *maximises* his profits. This optimum price is found by maximising the profit as a function of price.

A manufacturer should also be concerned with *minimising* the total cost of producing a given level of output.

Finding maximum and minimum values

The local maximum or minimum of a function (if it exists) may be represented graphically as follows:

(i)

(ii)

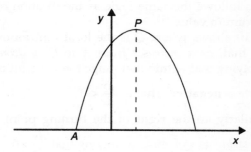

Now, P and Q are called the **turning points** (or **stationary points** or **critical points**) of the curve, and at P and Q the values of y are said to be **stationary**.

At P, the value of y is said to be a **maximum**.

At Q, the value of y is said to be a **minimum**.

1. Locating the *maximum* value

In **(i)** above, where P is the local *maximum*, we see that, as x *increases* from A to P, y also *increases* over the range. This means that the gradient (slope) of the curve to the **left** of the turning point is **positive**. That is, $\frac{dy}{dx} > 0$.

Similarly, to the **right** of the turning point, y is *decreasing* as x *increases*, implying that the gradient of the curve is **negative**.

That is, $\frac{dy}{dx} < 0$.

At the actual turning point of the curve, i.e. at P itself, the gradient of the curve is **zero** and the rate of change of y with respect to x is **zero**.

That is, $\frac{dy}{dx} = 0$.

Now, as x increases in value along the curve, $\frac{dy}{dx}$ **decreases** from positive to zero to negative. This implies that the derivative of $\frac{dy}{dx}$ is **negative**.

That is, $\frac{d^2y}{dx^2} < 0$.

2. Locating the *minimum* value

This follows the same logic as the location of the maximum value.

In **(ii)** above, where Q is the local **minimum**, we see that, as x *increases* from A to Q, y *decreases*, implying that within that range the gradient of the curve is **negative**. That is, $\frac{dy}{dx} < 0$.

Similarly, to the **right** of the turning point, y is *increasing* as x *increases*, implying that $\frac{dy}{dx} > 0$.

At Q, the gradient is **zero**. Hence, at Q, $\frac{dy}{dx} = 0$.

Now, as x increases along the curve, $\frac{dy}{dx}$ **increases** from negative to zero to positive, implying that the derivative of $\frac{dy}{dx}$ is **positive**. That is, $\frac{d^2y}{dx^2} > 0$.

We now give the following rule to define maximum and minimum values.

Rule 6

a. When a function, $y = f(x)$, is at its **maximum** value, then

  (i) $\frac{dy}{dx} = f'(x) = 0$, and

  (ii) $\frac{d^2y}{dx^2} = f''(x)$ is **negative**.

b. When a function, $y = f(x)$, is at its **minimum** value, then

  (i) $\frac{dy}{dx} = f'(x) = 0$, and

  (ii) $\frac{d^2y}{dx^2} = f''(x)$ is **positive**.

Finally, we can summarise the preceding discussion with the following steps for locating *maximum* or *minimum* values according to **rule 6** above:

Step 1: Differentiate the function $y = f(x)$.

That is, find $\frac{dy}{dx}$.

Step 2: Set $\frac{dy}{dx}$ equal to zero and then solve for x the resulting equation $\frac{dy}{dx} = 0$.

Step 3: Differentiate again. That is, find $\frac{d^2y}{dx^2}$.

Step 4: Find the value(s) of $\frac{d^2y}{dx^2}$ at the turning point value(s) found in Step 2. Then, if $\frac{d^2y}{dx^2} < 0$, $f(x)$ is a **maximum** at that value of x, while if $\frac{d^2y}{dx^2} > 0$, $f(x)$ is a **minimum** at that value of x.

EXAMPLE 12.9

Find the turning or stationary points of the function $y = x^3 - 3x^2$, and state whether the function has a maximum or minimum at each point.

Solution

Given: $y = x^3 - 3x^2$

Step 1: Find $\frac{dy}{dx}$.

$$\frac{dy}{dx} = 3x^2 - 6x$$

Step 2: Set $\frac{dy}{dx} = 0$ and solve for x.

$$\frac{dy}{dx} = 3x^2 - 6x = 0$$

So, $3x(x - 2) = 0$.

That is, $x = 0$ or $x = 2$.

Hence, the function has two turning points, one at $x = 0$ and one at $x = 2$.

Step 3: Find $\frac{d^2y}{dx^2}$.

$$\frac{d^2y}{dx^2} = 6x - 6$$

Step 4: Evaluate $\frac{d^2y}{dx^2}$ at the turning points.

When $x = 0$, $\frac{d^2y}{dx^2} = (6 \times 0) - 6 = -6$.

When $x = 2$, $\frac{d^2y}{dx^2} = (6 \times 2) - 6 = 6$.

Conclusion

(i) Since $\frac{d^2y}{dx^2}$ is *negative* when $x = 0$, we can say that $y = f(x)$ is at a **maximum** when $x = 0$.

The maximum value is $y = 0^3 - 3(0^2) = 0$.

(ii) Since $\frac{d^2y}{dx^2}$ is *positive* when $x = 2$, we can say that $y = f(x)$ is at a **minimum** when $x = 2$.

The minimum value is $y = 2^3 - 3(2^2) = -4$.

Therefore, $y = x^3 - 3x^2$ has a *maximum* at $(0,0)$ and a *minimum* at $(2,-4)$.

EXAMPLE 12.10
Divide 20 into two parts such that their **product** is a maximum.

Solution

Let $\quad x$ = one part

$\quad (20 - x)$ = the other part

$\quad y$ = the product of the two parts.

Then, $\quad y = x(20 - x)$

$\quad\quad\quad = 20x - x^2$

We want to find the value of x so that y is maximised.

Step 1: Find $\frac{dy}{dx}$.

$$\frac{dy}{dx} = 20 - 2x$$

Step 2: From **rule 6**, y will be maximised when

$$\frac{dy}{dx} = 0 \text{ and } \frac{d^2y}{dx^2} \text{ is \textbf{negative}.}$$

Now, $\frac{dy}{dx} = 0$ if $20 - 2x = 0$.

That is, if $x = 10$.

Hence, y is maximised when the two parts are 10 and 10 (giving $y = 100$).

Step 3: Check that $x = 10$ does result in y being maximised (and not minimised) by looking at the sign of $\frac{d^2y}{dx^2}$.

Here, $\frac{d^2y}{dx^2} = -2$.

Therefore, since $\frac{d^2y}{dx^2}$ is **negative**, the two parts 10 and 10 do **maximise** the product.

EXAMPLE 12.11
A certain company can sell its entire production at a price of \$10 per item. The cost, in dollars, of producing q items is given by

$$TC = q^2 + 9.$$

Find the value of q for which the profits will be a maximum.

Solution
Firstly, the total revenue from selling q items at \$10 each is given by $TR = 10q$.

We already know that the total cost of producing q items is $TC = q^2 + 9$.

Therefore, the profit is just $TR - TC$ and is given by

$$Pr = TR - TC$$
$$= 10q - (q^2 + 9)$$
$$= 10q - q^2 - 9$$

To maximise the profit we must first differentiate Pr, giving

$$\frac{dPr}{dq} = 10 - 2q .$$

Now, set $\frac{dPr}{dq} = 0$ and solve for q.

So, $\quad 10 - 2q = 0$
$$2q = 10$$
Thus, $q = 5$.

To test that $q = 5$ *maximises* profits we must look at the sign of the second derivative.

Since $\frac{dPr}{dq} = 10 - 2q$, we have that

$$\frac{d^2 Pr}{dq^2} = -2 .$$

Therefore, since $\frac{dPr}{dq} = 0$ and $\frac{d^2 Pr}{dq^2} < 0$ when $q = 5$, the profit will be *maximised* when $q = 5$.

Furthermore, the profit at its maximum is

$$Pr_{(max)} = (10 \times 5) - (5^2) - 9$$
$$= 16$$

YOUR TURN

12.14 Find any maximum or minimum values of the function $y = x^3 - 48x + 10$.

12.15 Each of the following functions, $y = f(x)$, has either a single maximum or a single minimum value.

Determine which it is, and the value of x when it occurs.

(a) $y = x^2 - 6x + 7$

(b) $y = 4x^2 + 40x$

(c) $y = 4 - 3x - x^2$

12.16 A farmer has 200 metres of fencing to build a rectangular pen. To maximise the area of the pen, he proceeds as follows:
Let x = length of the pen, and
$(100 - x)$ = breadth of the pen.

12.16 *Cont.*

Then, the area A = length × breadth
$$= x(100 - x)$$
$$= 100x - x^2$$

If $A = 100x - x^2$, then $\frac{dA}{dx} = 100 - 2x$.

Complete this problem for the farmer.

12.17 Find the level of output, q, that will maximise profits if we know that
(i) total cost of producing q items is

$$TC = q^2 + 4q$$

(ii) total revenue from selling q items is

$$TR = 24q - 4q^2 .$$

Hint: Find the value of q that maximises
$$Pr = TR - TC .$$

12.2.2 Velocity and acceleration

Let us suppose that an object (e.g. a car) is moving along a straight path covering s metres in t seconds, in such a way that $s = f(t)$, (that is, distance is a function of time).

Then it can be shown, from first principles, that $\frac{ds}{dt}$ is the **velocity** at a particular point in time.
Also, if v is the velocity at a given point in time,

then $\frac{dv}{dt} = \frac{d}{dt}\left(\frac{ds}{dt}\right) = \frac{d^2 s}{dt^2}$ is the **acceleration** at that point in time.

Therefore, if a body moves in a straight line a distance, s, in time, t, so that $s = f(t)$, then

(a) s' = the velocity

(b) s'' = the acceleration

(c) $s'(a)$ = velocity when $t = a$

(d) $s''(b)$ = acceleration when $t = b$.

EXAMPLE 12.12
A body is allowed to fall from the top of a tower. We know that the distance, s metres, through which it falls in t seconds is given by $s = 5t^2$.

(a) Determine its velocity after t seconds and also after 1, 2 and 3 seconds.

(b) Show that its acceleration is constant and find its value.

(c) After what time is its speed 25 metres per second?

Solution

(a) Given: $s = 5t^2$.

Then, $\dfrac{ds}{dt} = 10t$.

Thus, the velocity at the end of t seconds is $10t$ metres per second.

The velocity at the end of 1 second is $10 \times 1 = 10$ metres per second.

The velocity at the end of 2 seconds is $10 \times 2 = 20$ metres per second.

The velocity at the end of 3 seconds is $10 \times 3 = 30$ metres per second.

(b) Since $\dfrac{ds}{dt} = 10t$ from **(a)** above,

then $\dfrac{d^2s}{dt^2} = 10$.

Hence, the acceleration is **constant** (that is, independent of time) and its value is 10 metres per second per second.

(c) Find t when $s = 25$ metres per second.

From **(a)**, we know that velocity at the end of t seconds is $10t$ metres per second.

Thus, if $10t = 25$, we have that

$$t = 25/10 = 2\tfrac{1}{2} \text{ seconds.}$$

EXAMPLE 12.13

A car moves in a straight line covering s metres in t seconds, in such a way that $s = t^2 - 6t + 5$.

(a) Find s'. This gives the *velocity* of the car.

(b) Find s''. This gives the *acceleration* of the car.

(c) Find $s'(5)$, to give the velocity after 5 seconds.

(d) Find $s''(10)$, to give the acceleration after 10 seconds.

Solution

(a) $s' = 2t - 6$ metres per second.

(b) $s'' = 2$ metres per second per second.

(c) $s'(5) = (2 \times 5) - 6 = 4$ metres per second after 5 seconds.

(d) $s''(10) = 2$, since acceleration is constant over time.

12.18 A body moves s metres in t seconds where

$$s = 10 + 5t + 12t^2 - t^3$$

Find

(a) its speed at the end of 2 seconds

(b) its acceleration at the end of 3 seconds

(c) when its acceleration is zero (i.e. find t when $\dfrac{d^2s}{dt^2} = 0$).

12.19 A body moves s metres in t seconds in accordance with the equation

$$s = 10 + 6t + 13t^2 - t^3$$

Find

(a) its speed at the end of 3 seconds

(b) its acceleration at the end of 2 seconds

(c) when its acceleration is zero.

12.20 If $s = 2t^3 + 3t^2 + 4t + 5$, where s metres is the space a particle has moved in t seconds, find the velocity and acceleration at the end of 2 seconds and 3 seconds respectively.

12.2.3 Marginal analysis in business

The word **marginal** is used in the context of some continuously changing variables, such as cost, revenue or profit, each of which depends on the amount of activity in the firm or company.

In general, the **marginal value** of a variable is the amount of change in that variable for one unit of change in the activity.

For example: The marginal cost of producing an item is the cost incurred in producing *one more* item, whatever the current level of output.

Now, suppose we have some function, $y = f(x)$, then the *marginal* function measures the amount by which y will change with respect to a *tiny* change in x – that is, it measures the *rate of change* of y with respect to x.

As we know from our previous discussion, the rate of change of y with respect to x is just the *first derivative* of y with respect to x – namely,

$$\dfrac{dy}{dx} = f'(x).$$

It is this marginal function, $\frac{dy}{dx} = f'(x)$, that is so often used in helping to make business decisions where it is important to study the effect on y of increasing x.

The three marginal values that are of most importance in the areas of business and economics are *marginal cost*, *marginal revenue* and *marginal profit*.

a.   Marginal cost

Suppose that the cost of producing q items of output is given by the total cost function $TC = f(q)$, where $f(q)$ will include both fixed and variable costs.

Then, the **marginal cost**, MC, is

$$MC = \frac{d}{dq}(TC)$$

$$\text{or } MC = f'(q)$$

Remember that the marginal cost represents the rate of change in total cost with respect to output, and gives the additional cost required to produce one additional unit of output.

EXAMPLE 12.14

A production engineer has determined that his total costs are related to the output of items by the formula

$$TC = 0.003q^3 - 0.05q^2 + 7q + 10000$$

where TC = total costs

q = no. of items produced.

Find:

(a)   the total cost of 100 items

(b)   the marginal cost

(c)   the marginal cost when 100 items are produced

(d)   the marginal cost when 20 items are produced.

Solution

(a)   If $TC = 0.003q^3 - 0.05q^2 + 7 + 10000$,
then for 100 items,

$TC = 0.003(100^3) - 0.05(100^2) + 7(100) + 10000$

$= 3000 - 500 + 700 + 10000$

$= \$13\,200$

(b)   If $TC = 0.003q^3 - 0.05q^2 + 7q + 10000$,
then the *marginal cost* is

$MC = \frac{d(TC)}{dq} = 0.009q^2 - 0.10q + 7$

This represents the cost of producing an extra item.

(c)   If 100 items are already being produced, then

$\text{marginal cost} = 0.009(100^2) - 0.10(100) + 7$

$= 90 - 10 + 7$

$= \$87$

Hence, the approximate cost of producing one more item is $87.

(d)   If 20 items are already being produced, then

$\text{marginal cost} = 0.009(20^2) - 0.10(20) + 7$

$= \$8.60$

So, the approximate cost of producing one more item is $8.60.

b.   Marginal revenue

Suppose that the revenue from selling q items of output is given by the total revenue function, $TR = g(q)$, implying that total revenue is a function of sales, q.

Then, the **marginal revenue**, MR, is defined as the derivative of TR with respect to q.

That is,

$$MR = \frac{d}{dq}(TR)$$

$$\text{or } MR = g'(q)$$

Remember that the marginal revenue measures the rate of change in revenue with respect to output. That is, it measures the additional revenue from the sale of one additional unit of output.

EXAMPLE 12.15

Given the total revenue function

$$TR = -0.2q^2 + 50q$$

where TR = total revenue in dollars

q = no. of items sold,

find:

(a)   the total revenue for the sale of 100 items

(b)   the marginal revenue

(c)   the marginal revenue when 100 items are sold.

Solution

(a)   For 100 items, $q = 100$.

So, for the sale of 100 items, the total revenue is

$TR = -0.2(100^2) + 50(100)$

$= -2000 + 5000$

$= \$3\,000$

(b) If $TR = -0.2q^2 + 50q$, then the *marginal revenue* is $MR = \dfrac{dTR}{dq} = -0.4q + 50$.

This represents the additional revenue for selling one more item.

(c) If 100 items have already been sold, then

$$\text{marginal revenue} = -0.4(100) + 50$$
$$= -40 + 50$$
$$= \$10$$

That is, the revenue increases by \$10 from the sale of the 101st item.

c.   Marginal profit

Now, if TR is the total *revenue* function (where TR is a function of sales, q) and TC is the total *cost* function (where TC is a function of production, q) then, clearly, the **profit** function is given by

profit = total revenue − total costs.

That is, $Pr = TR - TC$,

where TR and TC are both functions of q.

Now, if we differentiate the profit function with respect to q, we have

$$\frac{dPr}{dq} = \frac{d}{dq}(TR - TC)$$
$$= \frac{dTR}{dq} - \frac{dTC}{dq}$$
$$= MR - MC$$

The first derivative of Pr, $\dfrac{dPr}{dq}$, represents the rate of change of the *profit* with respect to q, and is called the **marginal profit**.

Note: We have already learned earlier in this chapter that a given function, say $y = f(x)$, will be maximised when

(i) $\dfrac{dy}{dx} = f'(x) = 0$, and

(ii) $\dfrac{d^2y}{dx^2} = f''(x) < 0$

Therefore, since the aim of most businesses is to maximise profits, we can see that, with $\dfrac{dPr}{dq} = MR - MC$ (shown above), it follows that profits will be maximised when $\dfrac{dPr}{dq} = 0$. That is, when the marginal revenue *equals* the marginal costs.

EXAMPLE 12.16

Given that the total cost equation is $TC = 10q + 6000$ and the total revenue is $TR = -0.05q^2 + 50q$, find

(a) the marginal profit when $q = 100$ and when $q = 500$

(b) the value of q that will maximise profits. Then show that for this value of q, $MR = MC$.

Solution

(a) The profit function is given by

$$Pr = TR - TC$$
$$= (-0.05q^2 + 50q) - (10q + 6000)$$

So, $Pr = -0.05q^2 + 40q - 6000$.

Hence, the *marginal profit* is

$$\frac{dPr}{dq} = -0.10q + 40.$$

Now, when $q = 100$, the marginal profit is

$$\frac{dPr}{dq} = (-0.10 \times 100) + 40$$
$$= \$30,$$

and when $q = 500$, the marginal profit is

$$\frac{dPr}{dq} = (-0.10 \times 500) + 40$$
$$= -\$10$$

Therefore, when 100 items are sold, the extra profit from the sale of the 101st item is \$30, but when 500 items are sold, the sale of the 501st item results in a *loss* of \$10.

(b) Profits will be maximised when

$$\frac{dPr}{dq} = \text{marginal profit} = 0.$$

Now, $\dfrac{dPr}{dq} = -0.10q + 40$.

Hence, at $\dfrac{dPr}{dq} = 0$, we find that

$$-0.10q + 40 = 0$$
$$\text{or } q = \frac{40}{0.1} = 400.$$

Therefore, the profits will be maximised when 400 items are sold.

Finally, we must show that profits are maximised when $MC = MR$.

Now total costs, TC, are given by

$$TC = 10q + 6000.$$

Hence, $\dfrac{dTC}{dq} = MC = 10$.

Similarly, total revenue, TR, is given by

$$TR = -0.05q^2 + 50q.$$

Hence, $\dfrac{dTR}{dq} = MR = -0.10q + 50$.

Now, setting marginal cost equal to marginal revenue we have

$$MC = MR, \text{ or}$$
$$10 = -0.10q + 50$$

So, $q = \dfrac{40}{0.1} = 400$.

Therefore, marginal revenue = marginal cost at $q = 400$. However, we have also just seen that profit is maximised when $q = 400$.

Hence, we may conclude that the value of q for which $MR = MC$ is also the value of q that maximises the profit function, Pr – that is, profits will be maximised when $MR = MC$, as expected.

12.21 If total revenue = $TR = 60q$, where q = number of items produced, find the marginal revenue.

12.22 The total cost function of a certain company is given by

$$TC = 3000 + 0.0002q^3 - 0.03q^2 + 10q$$

where q = no. of items produced.

Find the marginal cost at
   (a) $q = 1000$
   (b) $q = 500$

12.23 Find the marginal cost, MC, if the total cost function is given by

   (a) $TC = 3q^2 + 21q + 100$

   (b) $TC = 2q^3 - 8q + 100$

   (c) $TC = 0.5q^3 + 25q + 1000$

12.24 Find the marginal revenue, MR, and evaluate MR for the indicated value of q if the total revenue function is given by

   (a) $TR = 200q - 2q^3$, $q = 5$

   (b) $TR = 300q - 75q^3$, $q = \sqrt{3}$

12.25 A firm that produces and sells a single product has the following total revenue and total cost equations

$$TR = -10q^2 + 100q$$
$$TC = 5q^2 - 200q + 1125$$

   (a) Find the marginal revenue.
   (b) Find the marginal cost.

12.25 *Cont.*
   (c) Find the firm's profit, $Pr = TR - TC$.
   (d) Find the marginal profit.
   (e) Find the value of q that maximises the firm's profits by setting
      (i) $\dfrac{dPr}{dq} = 0$, and
      (ii) $MC = MR$.

12.3 Integration – concepts and rules

This section and the next deal with problems in which the rate of change of some function is already known, and what we wish to find is the function itself.

To accomplish this, we must **reverse the process of differentiation**.

Consider the following examples.

1.   The leaking water pipe

A leak has been discovered in a water supply pipe and may take several days to repair. The water board wishes to estimate the resulting water wastage, V. They are able to find a function, $f(t)$, that they think reasonably estimates the rate of leakage by measuring the amount of water leaking per minute at different times. This rate of leakage over time may now be written $\dfrac{dV}{dt} = f(t)$.

The problem is to find V, the total quantity of water wasted as a function of time, t.

2.   Drug concentrations

Doctors treating a drug overdose patient often wish to estimate the rate at which the drug is being eliminated from the body. Since this rate can change with time, we may write $\dfrac{dC}{dt} = g(t)$,

where C is the drug concentration at time t.

Solving the above equation for C (as a function of t) may help doctors predict when the drug concentration will drop to a safe level.

To solve problems similar to those given above, we need to 'reverse' the process of differentiation. This is often called **anti-differentiation**, and involves reconstructing the '*question*' given the '*answer*' to a differentiation problem.

One method is to 'guess and check'. That is,
(i)   guess the function that, when differentiated, gives the results, and
(ii) differentiate the guess. If the result is not the given function, then alter your guess to correct it.

EXAMPLE 12.17

Lorraine was given the equation $\frac{dy}{dx} = x^2$ and asked to determine y as a function of x.

Solution

Since differentiating a power of x involves reducing the exponent by 1, Lorraine decided that, to obtain x^2 as the derivative, she must begin with x^3. This was her first guess.

To check the guess, she differentiated $y = x^3$ and obtained $\frac{dy}{dx} = 3x^2$. Since this is three times too big, she thought that if she divided her first guess by 3, she would be correct.

Thus, her second guess was $y = \frac{1}{3}x^3$, and on checking she found that $\frac{dy}{dx} = x^2$.

Hence, her second guess was correct.

Terminology

Because, over the years, many people have discussed the concept of anti-differentiation in different ways, there are a number of alternative words used to express the same notion. So, to avoid possible confusion in the future, it is important to be familiar with all terms.

So far, we have called a solution to the equation $\frac{dy}{dx} = f(x)$, an **anti-derivative** of $f(x)$, because it is obtained from $f(x)$ by reversing the process of differentiation.

However, a more widely-used term for anti-differentiation is **integration**, with an anti-derivative being called an **integral**.

Notation

There is special notation for an anti-derivative or integral of $f(x)$. We write

$$\int f(x)\, dx\ .$$

This is read as

(i) "the integral of f of $x\, dx$," or

(ii) "the integral of f of x with respect to x"

Therefore, instead of referring to "an anti-derivative of $\cos 5x$", we could simply write $\int \cos 5x\, dx$.

12.3.1 Indefinite integrals

Consider the following:

Since the derivative of x^3 is $3x^2$, then x^3 is an anti-derivative or integral of $3x^2$.

*However, it is not the **only** integral of $3x^2$.*

Since $\frac{d}{dx}(x^3 + 4) = 3x^2$ and

$$\frac{d}{dx}(x^3 - 8) = 3x^2,$$

both $(x^3 + 4)$ and $(x^3 - 8)$ are also integrals of $3x^2$. In fact, because the derivative of a constant is zero, $x^3 + C$ is also an integral of $3x^2$, for *any* constant, C.

From this we may conclude that a given function does *not* have a *single*, unique integral. Instead, an integral may be thought of as any member of a **family** of functions, any two of which differ only by a constant.

Since $x^3 + C$ describes *all* anti-derivatives (or integrals) of $3x^2$, it is often referred to as the **general** or **indefinite** integral of $3x^2$, denoted $\int 3x^2\, dx$ and read as "the indefinite integral of $3x^2$ with respect to x."

Thus, we may write

$$\int 3x^2\, dx = x^3 + C$$

where $\int$ is the integral sign

$3x^2$ is the integrand

C is the constant of integration.

The dx is part of the integral notation and indicates the variable involved. Here x is the **variable of integration**.

More generally,

The **indefinite integral** of any function f with respect to x is written $\int f(x)\, dx$, and denotes the indefinite integral of f. Since all integrals of f differ only by a constant, if F is any anti-derivative of f, then

$$\int f(x)\, dx = F(x) + C,$$

where C is constant of integration.

Using the rules of differentiation presented in *section 1*, the following list of rules of basic integration has been developed.

Basic integration rules
(C = constant of integration)

1. $\int k \, dx = kx + C$, k is a constant

2. $\int x^n \, dx = \dfrac{x^{n+1}}{n+1} + C$, for $n \neq -1$

3. $\int k f(x) \, dx = k \int f(x) \, dx$

4. $\int [f(x) \pm g(x)] \, dx = \int f(x) \, dx \pm \int g(x) \, dx$

5. $\int (ax+b)^n \, dx = \dfrac{(ax+b)^{n+1}}{a(n+1)} + C$

EXAMPLE 12.18
Using the rules of integration, find

(a) $\int 3 \, dx$

(b) $\int x^4 \, dx$

(c) $\int 10x \, dx$

(d) $\int \dfrac{1}{5x^2} \, dx$

(e) $\int (x^3 + 3x) \, dx$

(f) $\int (4x+3)^3 \, dx$

Solution

(a) Applying **rule 1**, with $k = 3$,
$$\int 3 \, dx = 3x + C \text{ (where } C = \text{constant of integration)}$$

(b) Applying **rule 2**, with $n = 4$,
$$\int x^4 \, dx = \frac{x^{4+1}}{4+1} + C$$
$$= \frac{x^5}{5} + C$$
$$= \frac{1}{5} x^5 + C$$

(c) Applying **rule 3**, with $k = 10$ and $f(x) = x$,
$$\int 10x \, dx = 10 \int x \, dx$$
Since x is just x^1, applying **rule 2** with $n = 1$,

we have $\int x^1 \, dx = \dfrac{x^{1+1}}{1+1} + C_1$
$$= \frac{1}{2} x^2 + C_1$$
where C_1 = constant of integration.

Therefore, $\int 10x \, dx = 10 \int x \, dx$
$$= 10 \left[\frac{1}{2} x^2 + C_1 \right]$$
$$= 5x^2 + 10 C_1$$

Since $10 C_1$ is just an arbitrary constant, for simplicity we will replace it with C.

Thus, $\int 10x \, dx = 5x^2 + C$.

(d) Applying **rule 3**, with $k = \dfrac{1}{5}$ and
$$f(x) = \frac{1}{x^2} = x^{-2}, \text{ we have}$$
$$\int \frac{1}{5x^2} \, dx = \frac{1}{5} \int \frac{1}{x^2} \, dx = \frac{1}{5} \int x^{-2} \, dx$$

Now, applying **rule 2** with $n = -2$, we have
$$\frac{1}{5} \int x^{-2} \, dx = \frac{1}{5} \left[\frac{x^{-2+1}}{-2+1} + C \right]$$
$$= \frac{1}{5} \left[\frac{x^{-1}}{-1} + C \right]$$
$$= -\frac{x^{-1}}{5} + C$$
$$= -\frac{1}{5x} + C$$

Therefore, $\int \dfrac{1}{5x^2} \, dx = -\dfrac{1}{5x} + C$

(e) Applying **rule 4**, with $f(x) = x^3$ and $g(x) = 3x$,
$$\int (x^3 + 3x) \, dx = \int x^3 \, dx + \int 3x \, dx$$

Now, applying **rule 2**,
$$\int x^3 \, dx = \frac{x^{3+1}}{3+1} + C_1 = \frac{x^4}{4} + C_1, \text{ and}$$
$$\int 3x \, dx = 3 \int x \, dx = 3 \left(\frac{x^2}{2} \right) + C_2$$
$$= \frac{3x^2}{2} + C_2$$

Therefore,

$$\int (x^3 + 3x)\, dx = \frac{x^4}{4} + \frac{3x^2}{2} + (C_1 + C_2)$$

$$= \frac{x^2}{2}\left(\frac{x^2}{2} + 3\right) + C$$

Note: In examples similar to this, it is conventional to eliminate the intermediate steps and simply integrate term by term.

Thus, $\int (x^3 + 3x)\, dx = \dfrac{x^4}{4} + \dfrac{3x^2}{2} + C$

$$= \frac{x^2}{2}\left(\frac{x^2}{2} + 3\right) + C$$

(f) Applying **rule 5**, with $n = 3$, $a = 4$ and $b = 3$,

$$\int (4x + 3)^3\, dx = \frac{(4x + 3)^{3+1}}{4\,(3 + 1)} + C$$

$$= \left(\frac{(4x + 3)^4}{12 + 4} + C\right)$$

$$= \frac{1}{16}(4x + 3)^4 + C$$

12.27 Using the integration rules 1 to 5, find the indefinite integral in each of the following:

(a) $\displaystyle\int 6\, dx$

(b) $\displaystyle\int 3x^{23}\, dx$

(c) $\displaystyle\int \frac{8}{x^{-5}}\, dx$

(d) $\displaystyle\int \left(2x^4 + \frac{1}{2}x^3\right) dx$

(e) $\displaystyle\int \left(4x^3 - \frac{3}{x^2}\right) dx$

(f) $\displaystyle\int \left(\frac{x^4 - 6x^3 + 5x}{x}\right) dx$

(g) $\displaystyle\int (5x + 7)^3\, dx$

12.3.2 Integration with initial conditions

If the rate of change, f', of some function, f, is known, then the function, f, is itself an anti-derivative of f'. However, as we have already seen, there are many anti-derivatives of f', with the most general one being denoted by the **indefinite integral**.

For example:

If $f'(x) = 2x$, then

$$f(x) = \int f'(x)\, dx = \int 2x\, dx = x^2 + C.$$

Hence, any function of the form $f(x) = x^2 + C$ has $2x$ as its derivative. It is because of the constant of integration, C, that we do not know the function, $f(x)$, specifically.

However, if f must take on a particular function value for a given value of x, then the value of C can easily be evaluated and hence $f(x)$ determined exactly.

For example:

If, in the preceding example, where $f(x) = x^2 + C$, we are told that $f(2) = 7$, then

$$f(2) = (2)^2 + C$$

$$= 4 + C.$$

Therefore, since $f(2) = 7$, we have

$$4 + C = 7, \text{ or}$$

$$C = 3.$$

Thus, $f(x) = x^2 + 3$ may be thought of as the particular function for which $f'(x) = 2x$ and $f(2) = 7$.

Note:

The condition $f(2) = 7$, giving a function value of f for a given x, is called an **initial condition** or **boundary value**.

EXAMPLE 12.19

If y is some function of x, such that $y' = 6x - 5$ and $y(3) = 6$, find y.

Note: $y(3) = 6$ means that $y = 6$ when $x = 3$.

Solution

Here $y(3) = 6$ is the initial condition and, since $y' = 6x - 5$, we know that y is an anti-derivative of $6x - 5$.

Hence, $y = \displaystyle\int (6x - 5)\, dx$

$$= \frac{6x^2}{2} - 5x + C$$

$$= 3x^2 - 5x + C$$

Now, by using the initial condition, we can determine the value of C.

Since $y = 6$ when $x = 3$, we have

$$6 = 3(3^2) - 5(3) + C$$

or $C = 6 - 27 + 15$

$\qquad = -6$

Therefore, if $y' = 6x - 5$ and $y(3) = 6$, then

$y = 3x^2 - 5x - 6$.

Y O U R   T U R N

12.28 Find y, subject to the given conditions.

(a) $\dfrac{dy}{dx} = 4x - 3$; $y(-1) = 8$

(b) $\dfrac{dy}{dx} = 2x^3 - \dfrac{1}{2}x$; $y(2) = 5$

(c) $y' = -\dfrac{1}{\sqrt{x}} + 2x$; $y(4) = 10$

12.3.3 Definite integrals

Let us take a few moments here to think about the 'big picture' of calculus.

As you will soon discover, there are two quite different geometrical problems that are in fact, very closely related – namely,

(i) the determination of the **gradient** (slope) of a curve, and

(ii) the determination of the **area** under a curve.

The procedure used to find the area under a curve (**integration**) is the *reverse* of the procedure used to find the gradient of a curve (**differentiation**).

Historically, it was the genius of Newton and Leibnitz, who both recognised the link between these two previously unconnected mathematical concepts, that led to what is currently termed the '**fundamental theorem of calculus**'.

Now, why are we so interested in these problems?

The **gradient** of a graph is important because it is one method of representing the rate of change of a function. Similarly, we often wish to find the **area** under a graph because it represents the total change of a function given the *rate* of change.

For example: If you know the distance travelled as a function of time, you can find the speed by differentiating. On the other hand, if you are given a graph of speed as a function of time, then the total distance travelled is just the area under the graph, and may be determined by integrating the function.

In fact, it was Leibnitz who introduced the notation for integration that we use today. Because the area under a curve was originally determined by summing many small rectangles, he used the

symbol, $\int$, an elongated S, to indicate 'sum'. Later, he introduced the symbol dx for the width of each thin rectangle of height y, giving an area of $y\,dx$. He then represented the area of the figure by $\int fy\,dx$. Today, we generally let $y = f(x)$ and hence write $\int f(x)\,dx$.

From the discussion above, we know that the area under a graph can be determined using integration.

So, if we wish to find the area under the curve $y = f(x)$ between $x = a$ and $x = b$, we first find the indefinite integral of $f(x)$, denoted $F(x)$ say, and then the required area is simply $F(b) - F(a)$, (i.e. the indefinite integral evaluated at 'b' minus the indefinite integral evaluated at 'a').

This expression, $F(b) - F(a)$, is called a **definite integral**. It is usually written as

$$\int_a^b f(x)\,dx = F(b) - F(a),$$ abbreviated to

$[\,F(x)\,]_a^b$ and read as: "the definite integral of f of $x\,dx$ from a to b". The numbers 'a' and 'b' are called the **limits of integration**.

To sum up

The fundamental theorem of integral calculus

If $F(x)$ is any indefinite integral of $f(x)$ over the interval from $x = a$ to $x = b$, then the definite integral, $F(b) - F(a)$, is written as

$$\int_a^b f(x)\,dx = [\,F(x)\,]_a^b = F(b) - F(a).$$

Note: $F(b) - F(a)$ is a constant.

EXAMPLE 12.20

Determine the following definite integrals:

(a) $\displaystyle\int_3^5 x^2\,dx$, the area under the graph of

$\qquad f(x) = x^2$, from $x = 3$ to $x = 5$

(b) $\displaystyle\int_0^2 (4 - x^2)\,dx$, the area under the graph of

$\qquad f(x) = (4 - x^2)$, from $x = 0$ to $x = 2$

(c) $\displaystyle\int_{-1}^3 (3x^2 - x + 6)\,dx$.

Solution

(a) Here $f(x) = x^2$, $a = 3$ and $b = 5$.

Hence, $\int_3^5 x^2 dx = \left[\frac{1}{3}x^3\right]_3^5$

$= \frac{125}{3} - \frac{27}{3}$

$= \frac{98}{3}$ or $32\frac{2}{3}$

(b) Here $f(x) = (4 - x^2)$, $a = 0$ and $b = 2$.

Hence, $\int_0^2 (4 - x^2)\, dx = \left[4x - \frac{1}{3}x^3\right]_0^2$

$= \left(8 - \frac{8}{3}\right) - 0$

$= \frac{16}{3}$ or $5\frac{1}{3}$

(c) Here $f(x) = 3x^2 - x + 6$, $a = -1$ and $b = 3$.
Hence,

$\int_{-1}^3 (3x^2 - x + 6)\, dx = \left[\frac{3x^3}{3} - \frac{x^2}{2} + 6x\right]_{-1}^3$

$= \left[x^3 - \frac{1}{2}x^2 + 6x\right]_{-1}^3$

$= \left(27 - \frac{9}{2} + 18\right) - \left(-1 - \frac{1}{2} - 6\right)$

$= 45 - \frac{9}{2} + \frac{15}{2}$

$= 45 + \frac{6}{2}$

$= 48$

Note:
Be very careful not to confuse indefinite and definite integrals.

(i) An indefinite integral is any one of a *family* of functions.
That is, $\int f(x)\, dx = F(x) + C$.

(ii) A definite integral is a *constant* and represents the area under a graph between two points.

That is, $\int_a^b f(x)\, dx = [F(x)]_a^b = F(b) - F(a)$
$= $ constant.

12.4 Integration – applications

12.4.1 Using indefinite integrals to solve problems, given initial conditions

EXAMPLE 12.21

The leaking water pipe
Refer back to the 'leaking water pipe' example given at the beginning of *section 3*.

Suppose that it is leaking at a rate given by the function $4 + 0.2t$. That is,

$$\frac{dV}{dt} = 4 + 0.2t$$

where V = volume of leakage (litres) after t hours.

If it will be 5 days before the pipe can be repaired, how much water will be wasted?

Solution

If $\frac{dV}{dt} = 4 + 0.2t$, then

$V = \int (4 + 0.2t)\, dt$

$= 4t + \frac{0.2t^2}{2} + C$

$= 4t + 0.1t^2 + C$

To fully answer the question asked we must

determine the value of C. If we assume that V is measured from the time the leak was discovered, then clearly $V = 0$ when $t = 0$.

Hence, $4(0) + 0.1(0)^2 + C = 0$, or $C = 0$.

Therefore, the volume of leakage after t hours is

$$V = 4t + 0.1t^2 \text{ (since } C = 0).$$

With t measured in hours, we can now calculate, by substitution, the volume of leakage after 5 days (or 5 x 24 = 120 hours).

$$V = 4(120) + 0.1(120^2)$$
$$= 1920$$

So, the quantity of water wasted from the time the leak was discovered until its repair 5 days later, is 1920 litres.

EXAMPLE 12.22

Income and education

In a mid-sized country town, a study was being conducted to determine the current average annual income, y (in dollars), that a person can expect to receive, with x years of education before looking for full-time employment.

It was estimated that the rate at which income changes with respect to education is given by

$$\frac{dy}{dx} = 15x^{3/2}, \text{ for } 4 \le x \le 17.$$

If it is known that $y = 10000$ when $x = 9$, find y.

Solution

Since $\frac{dy}{dx} = 15x^{3/2}$, we know that

$$y = \int 15x^{3/2} dx$$
$$= 15 \int x^{3/2} dx$$
$$= \left(15 \left[\frac{x^{5/2}}{5/2} \right] + C \right)$$
$$= 6x^{5/2} + C$$

Given $y = 10000$ when $x = 9$, we can determine C by substitution, as follows.

If $y = 6x^{5/2} + C$, then

$$10000 = 6(9)^{5/2} + C$$
$$\text{So, } C = 10000 - 6(243)$$
$$= 10000 - 1458$$
$$= 8542$$

Therefore, the average annual income, y, for x years of education, may be determined from

$$y = 6x^{5/2} + 8542 \text{ dollars.}$$

EXAMPLE 12.23

Determining the demand function from marginal revenue

If the marginal revenue function for a certain manufacturer's product is

$$MR = \frac{dTR}{dq} = 2000 - 24q - 6q^2$$

where TR = total revenue if q items are sold, find the demand function, $p = f(q)$, where p is the price per item.

Solution

Since $\frac{dTR}{dq}$ is the derivative of total revenue, TR,

$$TR = \int (2000 - 24q - 6q^2) \, dq$$
$$= 2000q - \frac{24q^2}{2} - \frac{6q^3}{3} + C$$
$$= 2000q - 12q^2 - 2q^3 + C$$

If we now assume that total revenue is zero when *no* items are sold – that is, $TR = 0$ when $q = 0$, we can substitute these into the equation to calculate the value of C.

Hence, $0 = 2000(0) - 12(0)^2 - 2(0)^3 + C$, giving $C = 0$.

Thus, total revenue as a function of the number of items sold, is given by

$$TR = 2000q - 12q^2 - 2q^3.$$

Now, to find the demand function we can use

(i)   the TR function obtained above, and
(ii) the relationship $TR = pq$
        where p = price per item
                q = number of items sold.

Since $TR = pq$, clearly $p = \frac{TR}{q}$ and, with

$TR = 2000q - 12q^2 - 2q^3$, we can write

$$p = \frac{TR}{q} = \frac{2000q - 12q^2 - 2q^3}{q}.$$

Therefore, the required demand function is given by

$$p = 2000 - 12q - 2q^2$$
$$= 2(1000 - 6q - q^2).$$

12.30 A motion problem.

In *section 2* you found that, if you know the position of some moving object as a function of time, you can determine its velocity by differentiating the position function, and you can determine its acceleration by differentiating the velocity function.

However, it often happens that you know the acceleration of a moving object and wish to determine its velocity and position at any time. This is best accomplished by **integration**.

A car travelling at 20 metres/second brakes suddenly, causing an acceleration of −6 metres/second?

(a) What is the significance of the negative sign in the acceleration?

(b) If V metres/second is the velocity of the car t seconds after the brakes were applied, find $\dfrac{dV}{dt}$.

(c) Integrate your equation in (b) to find V as a function of t.

(d) Use the fact that $V = 20$ when $t = 0$ to find the constant of integration, C, in your equation in (c).

(e) How long does it take the car to stop? That is, find t when $V = 0$.

(f) If the car has travelled a distance of s metres from the time the brakes were applied, find $\dfrac{ds}{dt}$.

(g) Find an expression for s as a function of t by integrating $\dfrac{ds}{dt}$ in (f).

(h) By definition, $s = 0$ when $t = 0$. Use this to calculate the value of C in your equation in (g) and hence determine the distance travelled from the time the brakes were applied until the time the car stopped.

12.31 Average cost

A manufacturer has determined that the marginal cost function for a particular item is given by

$$\frac{dTC}{dq} = 0.003q^2 - 0.4q + 40$$

where q = number of items produced

TC = total cost of producing q items.

If the marginal cost is $27.50 when $q = 50$, and if the fixed costs of production are $5000, determine the *average* cost per item of producing 100 items.

12.32 Diet for mice

A biologist studied the nutritional effects on mice that were fed on 15% protein consisting of yeast and cornflour. Over a period of time she found that a reasonable estimate of the rate of change in the average weight gain, G (in grams), of a mouse, with respect to the percentage, P, of yeast in the protein mix, is given by

$$\frac{dG}{dP} = -\frac{P}{30} + 2.5, \text{ for } 0 \le P \le 100.$$

If the biologist knows that $G = 40$ when $P = 15$, find G, the average weight gain, as a function of P, the percentage of yeast in the protein mix.

12.33 Cost function

The marginal cost function for a certain manufacturer's product is given by

$$\frac{dTC}{dq} = 10 - \frac{100}{\sqrt{q + 10}}$$

where TC = total cost (in dollars) when q items are produced.

If it is known that, when 90 items are produced, the average cost, p, is $100 per item, determine (to the nearest $1) the manufacturer's fixed cost.

12.4.2 Using definite integrals to solve problems

EXAMPLE 12.24

Area of athletics stadium

A new athletics stadium with a curved concrete roof is being designed. Each of the end walls is 25 metres long. Its height, y metres, is given by

$$y = 5\sqrt{x} - x + 4$$

where x = horizontal distance (metres) from the beginning of the wall.

Use a definite integral to find the area of these walls.

Solution

The area, A (in metres^2), of each wall is given by

$$A = \int_0^{25} (5\sqrt{x} - x + 4)\, dx$$

$$= \left[\frac{5\,(x^{3/2})}{3/2} - \frac{x^2}{2} + 4x \right]_0^{25}$$

$$= \left[\frac{10}{3}x^{3/2} - \frac{x^2}{2} + 4x\right]_0^{25}$$

$$= \left[\frac{10}{3} \times 125 - \frac{625}{2} + 100\right] - 0$$

$$= 416\frac{2}{3} - 312\frac{1}{2} + 100$$

$$= 204\frac{1}{6}$$

Therefore, the area of each wall is 204.17 metres^2.

EXAMPLE 12.25

Marginal costs

A manufacturer's marginal cost function for a particular item is

$$\frac{dTC}{dq} = 0.6q + 2$$

where TC = total cost of producing q items.

If production is currently set at $q = 80$ items per week, how much more would it cost to *increase* production to 100 units per week?

Solution

The total cost function is $TC = f(q)$ and we want to find $f(100) - f(80)$.

Since the rate of change of TC is $\frac{dTC}{dq} = 0.6q + 2$,

we have

$$f(100) - f(80) = \int_{80}^{100} (0.6q + 2)\, dq$$

$$= \left[\frac{0.6q^2}{2} + 2q\right]_{80}^{100}$$

$$= [0.3(100)^2 + 2(100)]$$
$$- [0.3(80)^2 + 2(80)]$$

$$= 3200 - 2080$$

$$= 1120$$

Therefore, the cost of increasing production from 80 items to 100 items is $1 120.

EXAMPLE 12.26

Demography

For a particular population, suppose f is a function such that $f(x)$ is the number of people who reach age x in any given year. This function is called a *life table function*.

Then the integral, $\int_n^{(x+n)} f(t)\, dt$, gives the expected number of people in the population between the ages of x and $x + n$, inclusive.

If $f(x) = 10000\sqrt{100 - x}$, determine the number of people between the ages of 36 and 64, inclusive (to the nearest integer).

Solution

The number of people between the ages of 36 and 64 is given by

$$\int_{36}^{64} 10000\sqrt{100 - x}\, dx$$

Now

$$\int_{36}^{64} 10000\sqrt{100 - x}\, dx = 10000\left[\frac{(100 - x)^{3/2}}{(-1)(3/2)}\right]_{36}^{64}$$

$$= 10000\left[-\frac{2}{3}36^{\frac{3}{2}} - \left(-\frac{2}{3}(64)^{\frac{3}{2}}\right)\right]$$

$$= 10000(-144 + 341.33333)$$

$$= 10000(197.3333)$$

$$= 1973333$$

Therefore, in the population being studied, there are 1 973 333 people between the ages of 36 and 64, inclusive.

12.34 Marginal Cost

A manufacturer's marginal cost function for a particular item is

$$\frac{dTC}{dq} = 0.2q + 3$$

where TC = the total cost of producing q items.

If TC is in dollars, determine the cost involved to increase production from 60 to 70 items.

12.35 Crime rate

In a study of crime rates in a certain city, Janet estimates that t months after the start of next year, the total number of crimes will increase at the rate of $8t + 10$ crimes per month.

(i) Determine the total number of crimes that can be expected to be committed next year.

(ii) How many crimes can be expected to be committed in the *last* 6 months of next year?

12.36 Marginal revenue

A manufacturer's marginal revenue function for a particular item is

$$\frac{dTR}{dq} = \frac{1000}{\sqrt{100q}}$$

12.36 *Cont.*

where TR = the total revenue from the sale of q items.

If TR is in dollars, find the change in the manufacturer's total revenue if production is increased from 400 to 900 items.

12.37 Areas

Find the area of the region bounded by the curve $f(x) = x^2 + 2x + 2$, the x-axis and the lines $x = -2$ and $x = 1$.

Summary

In this chapter we have examined a branch of mathematics called **calculus** that deals with the mathematical analysis of change.

The study of calculus is usually divided into two main sections:

1. **differential calculus** – this measures the rate of change in one variable given a particular value of a second related variable.

2. **integral calculus** – this is concerned with the *reverse* process, where we are given the derivative of a function and must find the original function.

1. Differential calculus

The first two sections discuss differential calculus, with *section 1* presenting relevant notation together with some theory and basic rules and *section 2* presenting some practical application.

Notation

Let $y = f(x)$ be some function of x and let the **first derivative** of the function $y = f(x)$ measure the rate at which y is changing with respect to x.

Then $\dfrac{dy}{dx}$, y' or $f'(x)$ may all be used to denote the 1st derivative and are often called the **differential coefficient**.

Similarly, $\dfrac{d^2y}{dx^2}$, y'' or $f''(x)$ each denote the 2nd derivative and may be found by differentiating $y = f(x)$ twice.

Also, $\dfrac{d^3y}{dx^3}$, y''' or $f'''(x)$ each denote the 3rd derivative and may be found by differentiating $y = f(x)$ three times.

Rules of differentiation

Rule 1

If $y = a$ (a is constant), then $\dfrac{dy}{dx} = 0$.

Rule 2

If $y = ax + b$ (a and b are constant), then $\dfrac{dy}{dx} = a$.

Rule 3

If $y = x^n$ (n is a number), then $\dfrac{dy}{dx} = nx^{n-1}$.

Rule 4

If $y = ax^n$ (a is constant), then $\dfrac{dy}{dx} = a \cdot nx^{n-1}$

Rule 5

a. If $y = ax^n + bx + c$ (a, b and c are constant), then $\dfrac{dy}{dx} = a \cdot nx^{n-1} + b$ (differentiate each term separately and then add).

b. If $y = ax^m - bx^n + cx - d$, then

$$\frac{dy}{dx} = a \cdot mx^{m-1} - b \cdot nx^{n-1} + c$$

Rule 6

a. $y = f(x)$ is at its **maximum** when

(i) $\dfrac{dy}{dx} = f'(x) = 0$, and

(ii) $\dfrac{d^2y}{dx^2} = f''(x) < 0$ **(negative).**

b. $y = f(x)$ is at its **minimum** when

(i) $\dfrac{dy}{dx} = f'(x) = 0$, and

(ii) $\dfrac{d^2y}{dx^2} = f''(x) > 0$ **(positive).**

Some practical applications

A. Finding **maximum** and **minimum** values – **apply rule 6.**

B. If $s = g(t)$ is a function relating the time, t, it takes an object to fall s metres, then

(i) $\dfrac{ds}{dt}$ measures the *velocity* of the object, and

(ii) $\dfrac{d^2s}{dt^2}$ measures the *acceleration*.

C. Finding **marginal** values.

(i) If $TC = f(q)$ is a *total cost* function given

q items produced, then $\dfrac{dTC}{dq}$ is the **marginal cost** – that is, the cost of producing one extra item.

(ii) If $TR = g(q)$ is a *total revenue* function given q items sold, then $\dfrac{dTR}{dq}$ is the **marginal revenue** – that is, the additional revenue if one extra item is sold.

(iii) If $Pr = TR - TC$ is the *profit* function given q items sold, then

$$\dfrac{dPr}{dq} = \dfrac{dTR}{dq} - \dfrac{dTC}{dq} = MR - MC \text{ is the}$$

marginal profit – that is, the additional profit if one extra item is sold.

Note: Profit, Pr, will be **maximised** when $\dfrac{dPr}{dq} = 0$; that is, when $MR = MC$.

2. Integral calculus

The last two sections of this chapter discuss integral calculus with *section 3* presenting relevant notation together with the basic mathematical theory and rules and *section 4* presenting some practical applications.

A. Indefinite integrals

An anti-derivative of a function, f, is a function, F, such that $F'(x) = f(x)$. Any two anti-derivatives of f differ at most by a constant. The most general anti-derivative of f is called the **indefinite integral** of f and is denoted $\int f(x)\,dx$.

Thus, $\int f(x)\,dx = F(x) + C$,

where C is called the constant of the integration.

Basic integration rules

(C = constant of integration)

Rule 1

$\int k\,dx = kx + C$, k is a constant.

Rule 2

$\int x^n dx = \dfrac{x^{n+1}}{n+1} + C$, for $n \neq 1$.

Rule 3

$\int kf(x)\,dx = k\int f(x)\,dx$

Rule 4

$\int [f(x) \pm g(x)]\,dx = \int f(x)\,dx \pm \int g(x)\,dx$

Rule 5

$\int (ax+b)^n dx = \dfrac{(ax+b)^{n+1}}{a(n+1)} + C$

Note:
If the indefinite integral F must take on a particular function value for a given value of x, then the value of C can easily be evaluated and hence $F(x)$ determined specifically. This additional information that allows us to evaluate C is called an **initial condition** or **boundary value**.

B. Definite integrals

Applying the *fundamental theorem of calculus* we have the following result:

If $F(x)$ is any indefinite integral of $f(x)$ over the interval from $x = a$ to $x = b$, then the **definite integral**, $F(b) - F(a)$, is written as

$$\int_a^b f(x)\,dx = \left[F(x) \right]_a^b = F(b) - F(a) .$$

Note: $F(b) - F(a)$ is a constant.

12.1 Differentiate each of the following functions:
- **(a)** $y = 3x + 1$
- **(b)** $y = -6x - 5$
- **(c)** $y = 3x^2 + 14$
- **(d)** $y = 5x^3 - 3x$
- **(e)** $y = x^3 - 2x + 3$
- **(f)** $y = x^4 - 2x^3 + 7x^2 - 3x + 1$

12.2 Evaluate $f'(6)$, where
- **(a)** $f(x) = 10x^2$
- **(b)** $f(t) = t^3 + 7t - 4$
- **(c)** $f(y) = 100y + 12$

12.3 If $g(x) = \frac{1}{2}x + x^2 + x^3$, find
- **(a)** $g(10)$
- **(b)** $g(0)$
- **(c)** $g'(x)$

12.4 Find the second and third derivatives of the following functions:
- **(a)** $y = 3x^6 + 5x^2$
- **(b)** $p = q^2 + 10q + 1000$
- **(c)** $f(x) = x^{-3} + x^3$
- **(d)** $p = 8q^2$

12.5 Find and identify any maximum or minimum values of each of the following functions:
- **(a)** $y = 64x - 2x^4 + 40$
- **(b)** $p = 2q^3 - 150q$

12.6 In a certain type of machine, the rate of expansion, r, and the number of kilograms of steam, N, used per hour are related by

$$N = 0.52r^2 - 5.2r + 32.$$

Find the value of r that gives the minimum value of N (that is, the most economic rate of expansion). Find also the corresponding value of N.

12.7 A company estimates that the expenditure of x million dollars in advertising will produce $9x - 3x^3$ million dollars profit.
- **(a)** How much should the firm spend on advertising to make the profit as large as possible?
- **(b)** How much would the maximum profit be?

12.8 A car moves in a straight line covering s metres in t seconds, in such a way that

$$s = t^2 + 6t + 5.$$
- **(a)** Find s'. This gives the velocity of the car.
- **(b)** Find s''. This gives the acceleration of the car.

12.8 *Cont.*
- **(c)** Find $s'(5)$ to give the velocity after 5 seconds.
- **(d)** Find $s'(10)$ to give the acceleration after 10 seconds.

12.9 The formula $s = 4.9t^2$, where s = distance or height in metres and t = time, may be used to determine how far an object will fall in t seconds. So, if you drop a rock down a well and hear it hit the bottom after 2 seconds, then the depth of the well is

$$4.9 \times 2^2 = 19.6 \text{ metres.}$$

Differentiate $s = 4.9t^2$ in order to determine the speed at which the rock travels.

12.10 Assume that the price of a commodity changes with time in such a way that

$$p = 0.01t + 0.01t^2 - 0.03.$$
- **(a)** Find $\dfrac{dp}{dt}$. This gives the inflation rate.
- **(b)** Find the inflation rate when $t = 4$.
- **(c)** $\dfrac{d^2p}{dt^2}$ gives the rate at which inflation is changing.
  Find this rate at $t = 4$.

12.11 A firm that produces and sells a single product has the following total revenue and total cost equations:

$$TR = 500q - 5q^2 - 50$$
$$TC = q^3 - 10q^2 + 40q$$
- **(a)** Find the marginal revenue.
- **(b)** Find the marginal cost.
- **(c)** Find the firm's profit.
- **(d)** Find the marginal profit.
- **(e)** Evaluate each of **(a)** to **(d)** above when $q = 10$ items are produced.

12.12 Given the total cost function

$$TC = q^2 + 4q$$

and the total revenue function

$$TR = 24q - 4q^2$$
- **(a)** Find the profit function, $Pr = TR - TC$.
- **(b)** Find the level of output that will **maximise** profits by setting
  - **(i)** the first derivative $\dfrac{dPr}{dq} = 0$, and
  - **(ii)** the marginal revenue, MR = marginal cost, MC (i.e. $MR = MC$).

12.13 The number of tickets sold for a certain play depends partly on the price of a ticket. Assume that the relationship between revenue (measured in thousands of dollars) and ticket price (measured in dollars) is given by

12.13 *Cont.*

$$R(x) = -x^2 + 8x - 5.$$

How much should be charged to maximise revenue?

12.14 Suppose the concentration, y, of a drug in the bloodstream (measured in milligrams per cubic centimetre), x hours after an injection is given by

$$y = 80x - 8x^2, \ 0 \le x \le 6.$$

How many hours after the injection is the concentration a maximum?

12.15 Find

(a) $\int \sqrt{x + 3} \ dx$

(b) $\int x\sqrt{x^2 - 1} \ dx$

(c) $\int x^3\sqrt{1 - x^4} \ dx$

12.16 In each of the following, find a function that satisfies all conditions.

(a) $\dfrac{dy}{dx} = x^2\sqrt{x^3 + 1}$; $y = 8$ when $x = 2$.

(b) $\dfrac{ds}{dt} = t^2 + t^{-2}$; $s = 9$ when $t = 3$.

12.17 Evaluate the following definite integrals:

(a) $\int_2^4 (x^2 - 3x + 5) \ dx$

(b) $\int_1^{16} x\sqrt{x} \ dx$

(c) $\int_0^1 (2p^3 + 5p^4) \ dp$

12.18 Oil is being pumped from a well at a rate of

$$Q'(t) = 100 + 10t - t^2$$

thousand barrels per year, where $0 \le t \le 15$ years. Assuming that $Q(0) = 0$, find the total production of this well

(a) in the first 10 years

(b) in the next 5 years.

12.19 Maintenance on equipment increases with the length of service of the equipment at the expected marginal cost

$$M'(t) = 4 + 6t + t^2$$

where t represents the number of years the equipment is in service ($0 \le t \le 10$). Find the total amount of service costs over the life of the equipment.

12.20 A shoe factory specialises in producing only one line of footwear. The management of the factory has determined that the marginal cost of the operation is

$$\frac{dC}{dx} = f'(x) = 10 + \frac{2}{\sqrt{x}}$$

where C is the cost of producing x pairs of shoes. The cost of producing 900 pairs is known to be $11 000, including fixed costs.

(a) What is the cost function that satisfies these conditions?

(b) Using your answer in (a), find the cost of producing 400 pairs of shoes.

13 Extras – Financial mathematics

Introduction

Business and financial mathematics is really just the application of some fairly basic mathematics to either a business or financial experience.

Almost everyone in today's society, at some point in their lives, will be involved in a financial transaction and/or a business venture of some description. Whether a child deposits money in a first saving's account, a young married couple borrows money to purchase a first home, or a company manager assesses the viability of a future business or investment project, some knowledge and understanding of the workings of interest rates in a wide variety of financial transactions is extremely important.

This chapter has been included primarily for those of you who are intending to pursue further study in the commerce or business area, with the aim of providing a set of quantitative techniques that may be applied to solve many of the day-to-day problems that arise in the world of business and finance.

One further intention of the chapter is to introduce the reader to several mathematical concepts associated with the time value of money. In fact, there are many different aspects to this notion of the time value of money, some of which include the following:

(i) The underlying knowledge that borrowing money and paying interest for its use results in higher total repayments than the actual value of the original loan.

(ii) The in-built relationship between the current value of an asset (such as an investment property) and the income this asset may produce over time.

(iii) The awareness that inflation implies that the same actual money value (say $500) is able to purchase fewer goods as time progresses.

(iv) The fact that most assets such as a car, a piece of equipment, a professional library etc. depreciate over time implying that a particular asset will be worth differing amounts at various points in time.

The successful operation of the financial sector depends upon the underlying ideas listed above, particularly in the area of property, valuation and development and land utilisation.

As with most of the preceding chapters, especially those covering the more specialised areas, this chapter contains many terms and symbols commonly used when discussing financial problems. These are very important and are presented in *section 1*.

Section 2 looks at the first and more basic of two types of interest – namely, **simple interest**, and discusses first the actual mechanics of calculations followed by some practical examples where this type of interest may be applied.

Section 3 presents the second and most widely-used of all forms of interest in the business and financial area – namely, **compound interest**. Again we discuss not only the concepts behind the notion of compound interest, but also examine carefully the mechanics of calculations involving compound interest by presenting some practical examples.

In *section 4* we introduce the concept of a **simple annuity** and then discuss the various quantities that may need to be calculated in an annuity problem, together with the appropriate formulae.

13.1 Terminology and notation

The following table lists each term, its corresponding symbol(s) and a definition of the meaning of the term.

Note:
The terms **simple** interest, **compound** interest, **nominal** interest rate and **effective** interest rate are defined and discussed in detail in the appropriate sections following.

Term	Symbol	Definition of term
interest (*section 1*)	*I*	The interest is either the money paid to the lender for the use of money borrowed, or it is the money earned when a given amount of capital is invested.
principal (*section 1*)	*P*	The principal is the total amount of money that is either borrowed or invested.
rate of interest (*section 1*)	*r*	The rate of interest is the ratio of interest for a stated period of time to the principal, expressed as a percentage. Hence, it is just a percentage of the principal. The period of time, unless otherwise stated, is taken to be one year.
present value (*section 1*)	*P*	The present value is just a special case of the term principal, often used when an interest-bearing transaction is concerned. It is defined to be the value or amount of the interest-bearing transaction (debt or investment) *on the day the money is borrowed or invested.*
future value (*section 1*)	*S*	The future value is the value or accumulated amount of an interest-bearing transaction (debt or investment) *on the day of maturity.* Hence, the future value of a transaction is just the principal (present value) plus the total interest involved.
compounding or interest period (*section 2*)	–	This is the length of time between 2 successive interest payments.
-	*i* (*section 2*)	This denotes the interest rate *per compounding or interest period* (as a decimal).
-	*n* (*section 2*)	This denotes the number of interest periods in the *whole transaction*.

Note: From the above terms and symbols relevant to *section 2*, we have the following results:

(i)   If r = interest *per annum* (as a decimal), and
           N = number of periods *per annum,*

         then $i = \dfrac{r}{N}$ (as a decimal).

(ii)   If N = number of periods in *one year*, and
           t = number of years in the transaction,
        then $n = Nt$ (*N* times *t*).

Term	Symbol	Definition of term
annuity (*section 3*)	–	An *annuity* is a series of equal payments made at equal time intervals.
payment interval (*section 3*)	–	The *payment interval* of an annuity is the time between successive payments.
term (*section 3*)	*t*	The *term, t*, of an annuity is the time from the beginning of the first payment interval to the end of the last payment interval.
-	*A* (*section 3*)	This refers to the *present value* of an annuity.
-	*S* (*section 3*)	This refers to the *future value* of an annuity.
-	*r* (*section 3*)	This denotes the per annum nominal rate of compound interest (in decimal form).
-	*N* (*section 3*)	This denotes the number of payment (or interest) periods in *one* year.
-	$n = Nt$ (*section 3*)	This denotes the number of payment periods in the whole term of the annuity.
-	$i = \dfrac{r}{N}$ (*section 3*)	This denotes the interest rate **per interest period** (as a decimal).
-	$s_{\overline{n}|i}$ (*section 3*)	This symbol is read '*s* angle *n* at *i*' and is an abbreviation for a term (given in *section 3*) that represents the *future value* of an annuity of \$1 per period for *n* periods at interest rate *i*. Most values of $s_{\overline{n}|i}$ may be obtained from Table I in Appendix A.
-	$a_{\overline{n}|i}$ (*section 3*)	This symbol is read '*a* angle n at *i*' and is an abbreviation for a term (given in *section 3*) that represents the *present value* of an annuity of \$1 per period for *n* periods at interest rate *i*. Most values of $a_{\overline{n}|i}$ may be obtained from Table II in Appendix A.
-	*R* (*section 3*)	This refers to the amount of the periodic payment of an annuity.

13.2 Simple interest

When only the original principal earns interest for the entire length of the transaction, the interest due at the *end* of the term is called **simple interest.**

Basic formula

The simple interest, I, earned on a principal, P, for t years at a per annum rate, r, is given by the formula:

$$I = Prt \text{ (P times r times t)}$$

where

r = interest rate per year, *expressed as a decimal*

t = length of the transaction (*in years*, since r is a per annum rate).

You will notice from the formula above, that should we wish to calculate other quantities such as the length of the term, t, or the per annum interest rate, r, this is easily achieved by simply rearranging the simple interest formula.

Hence, given I, P and r, then

$$t = \frac{I}{Pr} \quad [I \text{ divided by } (P \text{ times } r)].$$

Similarly, given I, P and t, then

$$r = \frac{I}{Pt} \quad [I \text{ divided by } (P \text{ times } t)].$$

EXAMPLE 13.1

How much interest is payable if $5 000 is borrowed at 7%p.a. simple interest for an 18 month term?

Solution

Here we are given $P = 5000$, $r = 0.07$, $t = 1.5$.
So, using the simple interest formula,

$$
\begin{aligned}
I &= Prt \\
&= 5000 \times 0.07 \times 1.5 \\
&= 525.00
\end{aligned}
$$

Therefore, $525 simple interest is earned on the given transaction.

Note: 0.07 is the per annum interest rate expressed as a decimal. 1.5 is the length of the transaction, in years (to agree with the per annum interest rate).

EXAMPLE 13.2

How much simple interest is earned if $8 000 is invested for 9 months at 12.5%p.a.?

Solution

Here we are given $P = 8000$, $r = 0.125$, $t = 0.75$ (since 9 months is 3/4 of a year).
So, using the appropriate formula,

$$
\begin{aligned}
I &= Prt \\
&= 8000 \times 0.125 \times 0.75 \\
&= 750
\end{aligned}
$$

Therefore, $750 simple interest is earned on the given transaction.

EXAMPLE 13.3

How long will it take $2 000 to earn $125 interest at a simple interest rate of 5%p.a.?

Solution

Here we are given $P = 2000$, $I = 125$, $r = 0.05$.
So, using the appropriate formula,

$$
\begin{aligned}
t &= \frac{I}{Pr} \\
&= \frac{125}{2000 \times 0.05} \\
&= 1.25
\end{aligned}
$$

Therefore, the required length of the above transaction is $1\frac{1}{4}$ years or 1 year and 3 months or 15 months.

EXAMPLE 13.4

At what rate of simple interest will a principal of $846 earn interest of $194.58 over a 2-year period?

Solution

Here we are given $P = 846$, $I = 194.58$, $t = 2$.
So, using the appropriate formula,

$$
\begin{aligned}
r &= \frac{I}{Pt} \\
&= \frac{194.58}{846 \times 2} \\
&= 0.115
\end{aligned}
$$

Therefore, the annual simple interest rate required for the above transaction is 11.5%p.a.

EXAMPLE 13.5

A student lends his friend $20 for 1 month. At the end of the month he asks for repayment of the $20 plus purchase of a chocolate bar worth 50 cents. What rate of simple interest is implied in this transaction?

Solution

Here we are given $P = 20$, $I = 0.5$, $t = \frac{1}{12}$ (in years).

So, using the appropriate formula,

$$
\begin{aligned}
r &= \frac{I}{Pt} \\
&= \frac{0.5}{20 \times \frac{1}{12}} \\
&= 0.3
\end{aligned}
$$

Therefore, the annual rate of simple interest implied in this transaction is 30%p.a.

13.2.1 Future value at simple interest

If money is borrowed or invested at simple interest, then at the end of the term, the accumulated amount, comprising the original principal plus the total interest earned, is called the **future** or **accumulated value** of the transaction, and is denoted S.

Appropriate formula

The **future value**, S, of a transaction earning simple interest, I, for t years at an annual rate, r, is given by the formula

$$S = P + I$$
$$= P + Prt \quad \text{(since } I = Prt)$$
$$= P(1 + rt)$$

where P = principal or present value

t = length of transaction (years)

r = per annum rate, expressed as a decimal.

EXAMPLE 13.6

John borrowed $2 000 from a credit union for a 2 year period at 12%p.a. simple interest. At the end of the 2 years, how much must John pay back to the credit union?

Solution

Here we are given $P = 2000$, $t = 2$, $r = 0.12$.
So, using the appropriate formula, since we are

asked for the full maturity value of the debt,

$$S = P(1 + rt)$$
$$= 2000(1 + 0.12 \times 2)$$
$$= 2000(1 + 0.24)$$

since multiplication is done before addition inside the brackets.

So, $\quad S = 2000 \times 1.24$
$\qquad = 2480$

Therefore, John must repay a total future amount of $2 480 to the credit union.

EXAMPLE 13.7

You wish to open a special savings account earning 6.25%p.a. simple interest with an initial deposit of $500.

(a) How much will be in the account at the end of 8 months?

(b) How much of the amount in **(a)** is actually interest earned?

Solution

Here we are given $P = 500$, $r = 0.0625$, $t = \frac{8}{12}$ (yrs).

(a) So, using the appropriate formula,

$$S = P(1 + rt)$$
$$= 500\left(1 + 0.0625 \times \frac{8}{12}\right)$$
$$= 500(1 + 0.0416667)$$
$$= 500 \times 1.0416667$$
$$= 520.83$$

Therefore, after 8 months you should have $520.83 in your account.

(b) If you opened the account with $P = $500 and if there was an amount $S = $520.83 in the account after 8 months, then the interest earned must be

$$I = S - P$$
$$= (520.83 - 500)$$
$$= $20.83$$

Note: The above formula is obtained by rearranging the first line of formula $S = P + I$.

EXAMPLE 13.8

At what annual rate of simple interest will a principal of $960 accumulate to a future value of $1 000 in 10 months' time?

Solution

Here we are given $P = 960$, $S = 1000$, $t = \frac{10}{12}$.

So, from the appropriate formula, we know that

$$I = S - P$$
$$= (1000 - 960)$$
$$= \$40$$

and

$$r = \frac{I}{Pt}$$

$$= \frac{40}{960 \times \dfrac{10}{12}}$$

$$= \frac{40}{800}$$

$$= 0.05$$

Therefore, the required annual rate of simple interest is 5%p.a.

YOUR TURN

13.4 To what amount will a principal of $10 000 grow when invested at a simple interest rate of
 (a) 8%p.a. for 2 years?
 (b) 11.5%p.a. for 18 months?
 (c) 6.75%p.a. for 7 months?

13.5 Find the annual rate of simple interest if an original principal of $5 000 accumulates to a future value of
 (a) $6 500 in 5 years
 (b) $6 062.50 in 30 months
 (c) $5 312.50 in 26 weeks.

 Note: In (c), assume that there are 52 weeks in one year.

13.6 Mary borrows $1 800 to buy furniture, and agrees to repay the loan in 3 years time at 9.5%p.a. simple interest. What is the maturity value of the loan? That is, how much must she pay in 3 years?

13.7 Paul deposits $800 in a special account earning 11.25%p.a. simple interest. He is saving to buy some stereo equipment costing $980. How long must he leave his money in the account in order to achieve his goal of $980?

13.2.2 Present value at simple interest

> If money is borrowed or invested at simple interest, then at the beginning of the term of the transaction, the value of this money on the day it is borrowed or invested is called the **principal** or **present value** of the transaction, and is denoted P.

Appropriate formulae

Either of *two* formulae is appropriate, depending upon the information available.

a. The **present value**, P, of a transaction earning simple interest, I, for t years at a per annum rate, r, may be obtained by rearranging the formula $I = Prt$, as follows:

$$P = \frac{I}{rt} \ [I \text{ divided by } (r \times t)]$$

where I = simple interest earned
 t = time (years)
 r = per annum rate, as a decimal.

b. The **present value**, P, of a transaction with future value, S, accumulated after t years at a per annum simple interest rate, r, may be obtained by rearranging the formula $S = P(1+rt)$, in order to make P the subject.

Thus, we have

$$P = \frac{S}{1 + rt} \ [S \text{ divided by } (1+rt)]$$

where S = the future or accumulated value
 t = time (years)
 r = per annum rate, as a decimal.

Note: When finding P, the formula in **a.** above would be easier to use if the values given were I, r and t, whereas the formula in **b.** would be preferable when S, r and t are the given values.

EXAMPLE 13.9

Joy deposited a sum of money in an account earning simple interest at a rate of 7.5%p.a. At the end of three years she had earned $225 interest. How much was her initial deposit?

Solution

Here we are given $I = 225$, $t = 3$, $r = 0.075$.
Therefore, using an appropriate formula (noting that I and not S is given), we have

$$P = \frac{I}{rt}$$

$$= \frac{225}{3 \times 0.075}$$

$$= 1000$$

So, Joy's initial deposit into the account was $1 000.

EXAMPLE 13.10

James took out a loan today knowing that it was to be repaid in 3 months time with a future value of $2 639, which includes simple interest at 6%p.a. How large was the original loan?

Solution

Here we are given $S = 2639$, $t = \dfrac{3}{12}$, $r = 0.06$.

So, using an appropriate formula (noting that we have S rather than I), we have

$$P = \frac{S}{(1 + rt)}$$

$$= \frac{2639}{\left[1 + \left(\dfrac{3}{12} \times 0.06\right)\right]}$$

$$= \frac{2639}{1.015}$$

$$= \$2\,600$$

Note: Whenever an interest bearing transaction is involved, then the *present value*, P, is **always** smaller than the *future value*, S, due to the fact that money earns interest with time.

EXAMPLE 13.11

An amount of $34\,000 is due to be paid to an investor 18 months from today. If money is worth 5.5%p.a. simple interest, what is the present value of the investment? That is, how much is it worth today?

Solution

Here we are given $S = 34000$, $t = 1.5$, $r = 0.055$. So, using an appropriate formula, we have

$$P = \frac{S}{(1 + rt)}$$

$$= \frac{34000}{1 + (0.055 \times 1.5)}$$

$$= 31\,408.78$$

Therefore, the investment is worth $31\,408.78 today – the required present value.

YOUR TURN

13.8 What principal invested today at 10%p.a. simple interest will amount to $1\,058.33 in 9 months time?

13.9 Morris invested money for 18 months at 12%p.a. simple interest. If the amount of interest earned in that period was $750, find the original principal, or present value, of the investment.

13.10 Two years from today, the ABC company must pay off a debt with a total amount of $14\,000.

13.10 *Cont.*

Find the present value of this debt **today** if money can earn simple interest at
(a) 5%p.a.
(b) 15.5%p.a.
(c) 7.75%p.a.

13.11 If money earns 5%p.a. simple interest, find the amount of money (principal) you would need to invest **today** so that in 3 years time it will have
(a) earned $500 simple interest
(b) accumulated to a future value of $3\,000.

13.12 Fred borrowed some money and is required to repay the loan with 2 payments, one of $500 to be paid in 4 months time and the other of $750 to be paid in 6 months time. If money is worth 9%p.a. simple interest, find the single value of the loan *today*.

(*Hint:* Find the present value of each of the 2 payments and then add the present values together.)

13.2.3 The time between dates

In all of our simple interest problems so far, the length of the transaction has involved a given number of either years, months or weeks, all of which must be expressed in years for the purpose of calculation.

However, in many simple interest problems it is important to know the exact number of days involved in a transaction. This type of problem will occur whenever we are told the actual date that money is invested or borrowed and the maturity or termination date.

Provided both the starting and finishing dates are given it is, in fact, very simple to determine the length of the transaction – that is, the number of days. This is best accomplished using Table III in Appendix A. This table is essentially a calendar that gives the serial numbers of the days in the year. It assumes that the day the transaction starts is included whilst the termination date is excluded. Then the number of days concerned (the length of the transaction) is just the difference between the serial numbers of the given dates.

(*Note:* In leap years, the serial numbers of the days is increased by 1 for all dates after February 28.)

Once the exact number of days has been determined, the length of the transaction, (in *years*), t, may be found by dividing by 365, (or 366 in a leap year).

EXAMPLE 13.12

Using Table III in Appendix A, find the number of days between

(a) April 18 and November 3

(b) October 2 and June 15.

Solution

(a) From the table,

April 18 is the 108th day of the year and

November 3 is the 307th day of the year.

Therefore, the number of days is

$$307 - 108 = 199 \text{ days.}$$

(b) In this example, October 2 is in one year and June 15 is in the following year. Therefore, we really need to determine the number of days between October 2 and the end of the year (December 31), and add this to the tabulated serial number for June 15 (which assumes the starting point is January 1).

So, from Table III, the number of days from October 2 to December 31 is $365 - 275 = 90$, while the number of days to June 15 of the following year (from the start of the year) $= 166$.

Therefore, the total number of days from October 2 to June 15 is $90 + 166 = 256$ days.

EXAMPLE 13.13

Susan opened a bank account on May 18, 1989 with a deposit of $2 000. She closed the account on April 8, 1990. If the money in her account earned simple interest at a rate of 9.5%p.a., how much was in the account at the time of closure?

Solution

Here we are given $P = 2000$, $r = 0.095$, $t = \dfrac{\text{days}}{365}$.
To find t we have, using Table III,

(i) from May 18 to the end of the year is $365 - 138 = 227$ days, and

(ii) from the start of 1990 to April 8 is 98 days.

Therefore, total number of days from May 18, 1989 to April 8, 1990 is $227 + 98 = 325$ days.

So, $\quad t = \dfrac{325}{365}$ (years).

Now, using an appropriate formula, we have

$$
\begin{aligned}
S &= P(1 + rt) \\
&= 2000\left(1 + 0.095 \times \frac{325}{365}\right) \\
&= 2000\,(1 + 0.084589) \\
&= 2000 \times 1.084589 \\
&= 2169.18
\end{aligned}
$$

Therefore, on April 8, 1990, there should be a total of $2 169.18 in Susan's account.

13.13 Using Table III (Appendix A), find the number of days between
(a) March 25 and August 18
(b) September 28, 1986 and July 13, 1987
(c) November 16, 1989 and May 24, 1991.

13.14 On April 7, 1986, a woman borrows $1 000 at 8%p.a. simple interest. She agrees to repay the debt plus interest on November 22, 1986. How much interest will she have to pay?

13.15 On November 1, Pamela invested $12 000 in an account earning $8\frac{3}{4}$%p.a. On June 30 of the following year she decided to close the account. What total amount, including interest, should be in the account at the time of maturity?

13.16 On March 31, you agree to lend a friend some money provided she will repay the loan with interest on October 1. If interest is charged at 13%p.a. simple interest, and if the full repayment amount is $500, how much did you lend your friend?

13.3 Compound interest

Remember that, with **simple interest**, it is only the original principal that earns interest for the full length of the transaction, and this interest is added on to the principal at the end of the term to give us the future or accumulated value of the transaction.

With **compound interest**, we find that the interest earned during any given time period is **added** to the principal, and then it is this new amount (principal plus interest) that earns interest during the next time period, and so on.

When interest is added to the principal, it is often said to be compounded or converted into principal (hence the name), and thereafter also earns interest. Therefore, the principal increases periodically and the interest compounded into principal also increases periodically throughout the term of the transaction.

> When interest is earned (or charged) in the above manner, then the difference between the amount due at the end of the transaction and the original principal is called the **compound interest.**

Before introducing formulae for calculating quantities involving compound interest, here is an example showing how interest is compounded step by step, resulting in a new amount each time.

EXAMPLE 13.14

Find the accumulated amount after 4 years if a principal of $2 000 is invested today at 7%p.a. Assume that interest is compounded annually.

Solution

With **compound interest** we have the following steps:

(i) With the original $P = 2000$, the new amount at the end of the **first** year, with one year's interest, is

$$2000 (1 + 0.07 \times 1) = 2000 \times 1.07$$
$$= \$2\ 140$$

(ii) With the new $P = 2140$, the new amount at the end of the **second** year, adding one year's interest, is

$$2140 (1 + 0.07 \times 1) = 2140 \times 1.07$$
$$= \$2\ 289.80$$

(iii) With the new $P = 2289.80$, the new amount at the end of the **third** year, adding one year's interest, is

$$2289.80 (1 + 0.07 \times 1) = 2289.80 \times 1.07$$
$$= \$2\ 450.09$$

(iv) With the new $P = 2450.09$, the new amount at the end of the **fourth** year, adding one year's interest, is

$$2450.09 (1 + 0.07 \times 1) = 2450.09 \times 1.07$$
$$= \$2\ 621.59$$

Therefore, the compound amount after 4 years is $2 621.59.

So, the **compound interest** earned during the four years is $2 621.59 – $2 000 = $621.59.

As expected, more interest is earned with compound interest since the interest earned in one time period itself earns interest in all following time periods.

Clearly the above method for calculating compound interest is not very efficient. However, careful examination of the steps involved will show that the above process can be summarised by a single formula that will be presented in the next sub-section.

Note: One final point before moving on. With compound interest, interest is earned during a specified interest period, and compounded to the principal at the end of that period. Remembering that N = number of interest periods per year, we generally find that interest is compounded:

(i) annually, $N = 1$.

(ii) semi-annually, $N = 2$.

(iii) quarterly, $N = 4$.

(iv) monthly, $N = 12$.

(v) weekly, $N = 52$.

(vi) daily, $N = 365$.

13.3.1 Future value at compound interest

If money is borrowed or invested at **compound** interest, then at the end of the term, the accumulated amount, comprising principal plus compound interest, is called the **compound amount** or **future value** of the transaction, and is denoted S.

Appropriate formula

The future amount, S, of a transaction earning compound interest at a rate i per interest period for a total of n periods, covering the full length of the term, is given by:

$$S = P(1 + i)^n$$

where S = future or compound value at the end of the transaction

P = principal or present value at the beginning of the transaction

$i = \dfrac{r}{N} = \dfrac{\text{per annum rate}}{\text{no. of periods per year}}$

= interest rate **per period** (as a decimal)

$n = Nt$ = no. of periods per year x length of term.

Hence, n = no. of interest periods throughout the full term.

Note: (for interest only). If you examine the steps in example 13.14 very carefully, you may notice that, with $P = 2000$ and $i = 0.07$ (since $N = 1$),

Step (i) gives $P(1 + i)$

Step (ii) gives $P(1 + i)(1 + i) = P(1 + i)^2$

Step (iii) gives $P(1 + i)^2(1 + i) = P(1 + i)^3$

The successive amounts $P(1 + i)$, $P(1 + i)^2$,

$P(1 + i)^3$ ... form a **geometric** progression whose

n^{th} term is $S = P(1 + i)^n$.

EXAMPLE 13.15

Let us re-do example 13.14, solving the problem here using the compound interest formula.

Remember that the question asked for the future amount after 4 years if a principal of $2 000 is invested today at 7%p.a. compounded annually.

Solution

Here we are given:
principal, $P = 2000$
per annum interest rate, $r = 0.07$
length of the term, $t = 4$ years
no. of interest periods in 1 year, $N = 1$ (since interest is compounded annually).
So, using the appropriate formula, we have

$$S = P(1 + i)^n$$

where i = interest rate **per period**

$$= \frac{r}{N} = \frac{0.07}{1} = 0.07$$

n = no. of interest periods altogether
$$= Nt = 1 \times 4 = 4.$$

Hence, $S = 2000(1 + 0.07)^4$

$$= 2000(1.07)^4$$
$$= 2000(1.310796)$$
$$= \$2\,621.59, \text{ as before.}$$

Therefore, the future amount after 4 years is $2 621.59, implying that $621.59 was earned in compound interest throughout the term.

Note: The most efficient method of calculating $(1.07)^4$ is to use the $\boxed{x^y}$ function key on your calculator.
Thus, $(1.07)^4$ is

$$1.07 \boxed{x^y} \; 4 \; \boxed{=} \quad \boxed{1.310796}$$

EXAMPLE 13.16
A deposit of $400 is placed in an account earning interest at 10%p.a. compounded quarterly. Find the future amount in the account after 3 years. How much compound interest was earned?

Solution
Here we are given $P = \$400$, $r = 0.10$, $t = 3$ and $N = 4$, since interest is compounded quarterly (that is, there are 4 quarterly interest periods in 1 year). So, using an appropriate formula, we have

$$S = P(1 + i)^n$$

where $i = \frac{r}{N} = \frac{0.10}{4} = 0.025$

$$= \text{ interest rate \textbf{per quarter}}$$
$$n = Nt = 4 \times 3 = 12$$
$$= \text{ no. of quarters in 3 yrs}$$

Hence, $S = 400(1 + 0.025)^{12}$

$$= 400(1.025)^{12}$$
$$= 400 \times 1.3448888$$
$$= 537.9555$$

Therefore, the future amount in the account after 3 years is $537.96.

The amount of compound interest earned is just the difference between the amount at the end of the term and the original principal.
So, compound interest $= S - P$

$$= \$537.96 - \$400$$
$$= \$137.96$$

EXAMPLE 13.17
Mary borrows $350 and agrees to repay the loan with interest in 9 months time. If money is worth 18%p.a. compounded monthly, find

(a) total amount due in 9 months, and

(b) the amount of compound interest earned during the 9 month period.

Solution
Here we are given $P = 350$, $r = 0.18$, $t = 0.75$ years and $N = 12$ (since there are 12 monthly periods in 1 year).

Therefore, $i = \frac{0.18}{12} = 0.015$ is the interest rate per month

$n = 9$, since the full term covers only 9 months, or 9 interest periods.

(a) Using an appropriate formula, we have

$$S = P(1 + i)^n$$
$$= 350(1 + 0.015)^9$$
$$= 350(1.015)^9$$
$$= 350 \times 1.14339$$
$$= \$400.19$$

(b) Compound interest $= S - P$
$$= (400.19 - 350)$$
$$= \$50.19.$$

Therefore, in 9 months time, Mary must settle her debt with a payment of $400.19. This amount is the original loan plus compound interest of $50.19.

13.17 Find
  (i) i, the rate of interest per interest period, as a decimal,
  (ii) n, the total number of interest periods during the entire transaction,

  when a given principal, P, is invested:
  (a) for 2 years at 7%p.a. compounded quarterly
  (b) for 6 months at 13%p.a. compounded weekly
  (c) for 5 years 4 months at 15%p.a. compounded monthly
  (d) from January 1, 1985 to July 1, 1992 at 6%p.a. compounded semi-annually.

13.18 A principal of $700 is to be placed in an interest-bearing account. How much is in the account after 7 years, if interest is earned at
  (a) 6%p.a. compounded annually?
  (b) 6%p.a. compounded semi-annually?
  (c) 6%p.a. compounded quarterly?
  (d) 6%p.a. compounded monthly?

13.19 A building society pays interest at 10%p.a. If I deposit $500 today, how much will be in the account after 6 years? Assume the interest is compounded
  (i) quarterly
  (ii) semi-annually
  (iii) annually.

13.20 Find the future accumulated amount, S, if a principal of $2 500 is invested at 8%p.a. compounded semi-annually for
  (a) 5 years
  (b) 9 years
  (c) 15 years.

  When will the compound amount be approximately twice the original principal?

13.21 An endowment policy for $15 000 which matured on June 1, 1980 was left with the insurance company at 5.5% compounded annually. How much compound interest was earned between June 1, 1980 and June 1, 1990?

13.3.2 Present value at compound interest

Here the problem presented in 13.3.1 is reversed – given an amount in the future, we wish to find the sum of money that, if deposited **today**, would accumulate to the given future amount. The sum of money to be deposited today is called the **present value** of a given future amount, and is denoted P.

Appropriate formula

The present value, P, of a given future amount, S, due after n interest periods at an interest rate i per interest period (or equivalently, the present value P, which, if invested **today** at a rate i per interest period, would amount to a given S after n interest periods), is given by

$$P = \frac{S}{(1+i)^n}$$

$$= S(1+i)^{-n}$$

where S = given future amount
  i = interest rate **per period** (as a decimal)
  n = no. of interest periods in the full term.

Note:

(i) The formula $P = \dfrac{S}{(1+i)^n}$ is found by taking the formula $S = P(1+i)^n$ (from 13.3.1) and making P the subject.

(ii) We already know that $\dfrac{a}{b^x} = a(b)^{-x}$ and that, with modern calculators, the use of $a(b)^{-x}$ is preferable.

EXAMPLE 13.18
Stewart needs to have $23 000 four and a half years from today. How much must he invest today at 9%p.a. compounded monthly to reach the target?

Solution
Here we are given:
  S = future amount = $23 000
  r = per annum interest rate = 0.09
  t = length of the term = 4.5 years
  N = no. of interest periods in 1 year = 12.

From this, we know that
i = interest rate per interest period (month)
  $= \dfrac{0.09}{12} = 0.0075$

n = no. of months in 4.5 years = 12 × 4.5 = 54.

So, using the appropriate formula, we have

$$P = S(1+i)^{-n}$$
$$= 23000(1+0.0075)^{-54}$$
$$= 23000(1.0075)^{-54}$$
$$= 23000 \times 0.6679855$$
$$= 15363.667$$

Therefore, in order to reach the desired target, Stewart must invest $15 363.67 **today**.

Note:
The simplest and most efficient method of calculating $(1.0075)^{-54}$ is to use your calculator as follows:

1.0075 $\boxed{x^y}$ 54 $\boxed{^+/_-}$ $\boxed{=}$ $\boxed{0.6679855}$

EXAMPLE 13.19
Mr Smith bought a block of land in the country today under the following arrangement. He paid $15 000 cash immediately and then agreed to pay a total of $16 500 in three years time. If money is worth 7%p.a. compounded semi-annually, find the full cash value of the land **today**.

Solution
Remember: The full cash value today will be the $15 000 cash already paid **plus** the present value of the amount $16 500 to be paid in 3 years.
Here we are given: $S = 16500$, $r = 0.07$, $t = 3$, $N = 2$ (semi-annual).
Hence, we have

$$i = \frac{0.07}{2} = 0.035 \text{ per 6 month period, and}$$

$n = 3 \times 2 = 6$ semi-annual periods in 3 years.

So, using an appropriate formula, we have

$$
\begin{aligned}
P &= S(1+i)^{-n} \\
&= 16500(1+0.035)^{-6} \\
&= 16500(1.035)^{-6} \\
&= 16500 \times 0.8135006 \\
&= 13422.76
\end{aligned}
$$

Therefore, the full cash value of the land **today** is $15 000 + $13 422.76 = $28 422.76

YOUR TURN

13.22 Find the present value of a future amount of $60 000 due to be paid in 10 years time if money can earn interest at 6%p.a. compounded:
  (a) once a year
  (b) quarterly
  (c) semi-annually.

13.23 The Jack Sprat company needs to have an accumulated amount of $5 000 at the end of an 18-month period.

  What lump sum must the company place in an account at the beginning of the 18-month term if interest is earned at:
  (a) 12%p.a. compounded monthly?
  (b) 9%p.a. compounded quarterly?

13.24 At the birth of a child, a father wishes to invest sufficient, at 5%p.a. compounded semi-annually, so that when the child turns 21 there will be $10 000 in the account. How much must he invest at the time of the birth?

13.25 You have a debt that you may discharge in one of two ways:
  (a) by paying $6 000 now; or
  (b) by paying $8 200 six years from now.

  If money is worth 6%p.a. compounded quarterly, would you accept **(a)** or **(b)**? Show your reasoning.

  (*Hint:* Find the present value of $8 200 and compare with **(a)**.)

13.3.3 Finding the interest rate
When S, P and n are given, we can substitute the given values into the compound interest formula $S = P(1+i)^n$ and solve it for the unknown interest rate, i.
Once we have found i, the interest rate per interest period, it is a simple matter to convert this to r, the **per annum** rate of compound interest.

Appropriate formula
If we are given the values for the future amount, S, the present value, P, and the total number of interest periods in the full term, n, then we can find the unknown value of i, the interest rate per interest period, by re-arranging the formula

$$S = P(1+i)^n .$$

Remembering that $S = P(1+i)^n$ can be written $P(1+i)^n = S$, we need only re-arrange the latter to make i the subject.

Hence,
$$
\begin{aligned}
P(1+i)^n &= S \\
(1+i)^n &= \frac{S}{P} \\
(1+i) &= \left(\frac{S}{P}\right)^{\frac{1}{n}}
\end{aligned}
$$

Thus, $\quad i = \left(\frac{S}{P}\right)^{\frac{1}{n}} - 1 .$

To find the per annum rate of interest we must first find r, the per annum rate as a decimal. This is most easily accomplished by multiplying i by N, the number of interest periods in *one* year. Finally, the per annum rate is simply $r \times 100$. Hence, the per annum rate of compound interest, as a percentage, is clearly $i \times N \times 100 = r \times 100$,

where i = rate per period (decimal)
N = number of periods in 1 year
r = rate per year (decimal).

EXAMPLE 13.20

At what per annum rate of interest compounded quarterly will $1 250 invested today amount to $1 900 in 10 years?

Solution

Here we are given $P = 1250$, $S = 1900$, $t = 10$ and $N = 4$ (since there are 4 quarterly periods in 1 year). Now, with $t = 10$ and $N = 4$, the total number of quarterly periods in the full term is $n = 4 \times 10 = 40$. Before we can find the per annum rate of compound interest, we must first determine i, the rate per period. Using the appropriate formula, we have

$$i = \left(\frac{S}{P}\right)^{\frac{1}{n}} - 1$$

$$= \left(\frac{1900}{1250}\right)^{\frac{1}{40}} - 1$$

$$= (1.52)^{0.025} - 1$$

$$= 1.0105227 - 1 \quad \text{(using the } x^y \text{ button)}$$

$$= 0.010523$$

So, the interest rate *per quarter* is $i = 0.0105$ or 1.05%.

Therefore, the per annum rate of interest compounded quarterly is $0.0105 \times 4 \times 100 = 4.2\%$p.a.

Note:

Using your calculator, the value of $\frac{1}{n}$ may be calculated in two ways.

For example: $\frac{1}{40}$

a. 1 $\boxed{\div}$ 40 $\boxed{=}$ $\boxed{0.025}$

b. 40 $\boxed{1/x}$ $\boxed{0.025}$

Remember that $\frac{1}{x}$ is the reciprocal function.

YOUR TURN

13.26 For each of the following cases, calculate the per annum rate of compound interest.
(a) $P = \$2\,000$, $S = \$3\,000$, time = 3 years 9 months, interest is compounded quarterly.

13.26 *Cont.*
(b) $P = \$100$, $S = \$150$, time = 4 years 7 months, interest is compounded monthly.
(c) $P = \$1\,000$, $S = \$1\,581$, time = 3 years 6 months, interest is compounded semi-annually.

13.27 If an investment increased from $4 000 to $6 000 in 3 years, what is the per annum rate of interest being earned, assuming that interest is compounded monthly?

13.28 You borrow $2 500 today and agree to pay an amount of $3 250, comprising the debt plus interest, in 4 years time. What per annum rate of interest compounded semi-annually is being charged?

Comment

When discussing compound interest, there is one further quantity that may be calculated – namely, **the length of the term**, t. However, since this is slightly more complicated we present only the relevant formula, leaving further exploration to the reader.

The length of the term, t, in years, is

$$t = \frac{n}{N}$$

where N = no. of interest periods in *one* year
n = total number of interest periods

$$= \frac{\log\left(\frac{S}{P}\right)}{\log(1 + i)}$$

Note: The above formula for n is obtained by re-writing the basic formula for compound interest.

13.3.4 Nominal and effective rates of compound interest

When interest is compounded **more than** once a year (e.g. compounded quarterly), the given annual rate is called the **nominal rate** or **nominal annual rate.**

For example: If money is worth 8%p.a. compounded monthly, then 8%p.a. is the **nominal** interest rate.

We have already seen that a given nominal rate pays more as the frequency of compounding grows.

For example: A nominal 10%p.a. compounded semi-annually turns $100 into $110.25. If compounded quarterly, $100 becomes $110.38

while if compounded monthly, the $100 would become $110.47.

Yet all of these are called a **nominal** 10%p.a. even though they all result in a different amount of interest being earned in one year. Clearly, we need some standardised way of comparing differing nominal rates at varying compounding frequencies, to provide a fair basis for comparison.

This brings us to the **effective** rate of interest.

> The rate of interest actually earned in **one year** – that is, the annual rate when interest is compounded **annually** – is called the **effective rate** or **effective annual rate**.

So, the **effective** rate for a given **nominal** rate would be that per annum rate which produces the same result (i.e. the same annual interest) if it were compounded once a year.

For example: Since the nominal rate of 10%p.a. compounded quarterly turns $100 into $110.25 after **one** year, the corresponding effective annual rate would be 10.25%p.a., because 10.25% compounded annually turns $100 into $110.25.

Similarly, since the nominal rate of 10%p.a. compounded monthly turns $100 into $110.47 after **one** year, the corresponding effective annual rate would be 10.47%p.a., because 10.47%p.a. compounded annually turns $100 into $110.47.

As this subsection was just by way of explanation, there are no worked examples or '**your turn**' exercises.

13.3.5 Compound interest versus simple interest

Remember from the previous section that, with simple interest rates, interest is paid on the full amount borrowed even after part of the capital is repaid.

Therefore, as the simple (or sometimes called '**flat**') rate of interest applies to the original loan for the whole term for all simple (flat) interest rate loans, we **cannot** directly compare the rate of interest quoted for these loans with the rate used by a compound interest loan.

In fact, the *real* rate (that is, the compound rate) of interest being charged is somewhat **higher** than the quoted flat (simple) rate.

As a guide, it can be shown that the annual effective rate of compound interest is approximately **double** the simple interest rate minus about one or two percent.

For example: 14.5% simple interest over a two year term is roughly equivalent to 26%p.a. compounded monthly. Similarly, 7% flat over five years is roughly equivalent to 12.5%p.a.

compounded monthly. This approximate relationship applies to all simple (flat) interest rate loans.

So, although flat rate loans may 'appear' cheaper than compound reducible loans, this is, in fact, generally not the case. For this reason, many countries insist that loans at flat (simple) interest rates also show their effective rates of compound interest.

In the following exercises remember that the real or effective annual compound rate ≈ 2(simple or flat rate) − (1 or 2)%.

13.29 Would you rather borrow money at
  (a) 10%p.a. flat or 10%p.a. compounded monthly?
  (b) 13%p.a. flat or 28%p.a. compounded monthly?
  (c) 6.5%p.a. flat or 9%p.a. compounded monthly?

13.30 Would you rather borrow money over a 2-year term at
  (a) 18%p.a. compounded monthly or 18%p.a. flat?
  (b) 7%p.a. flat or 16.2%p.a. compounded quarterly?

13.4 Simple annuities

Most borrowings or loans do not usually involve repayment of the entire loan, with interest, at one single or several irregular future dates, but rather they involve repayment in equal stages.

For example: A loan of $2 000 might be repaid over 2 years by 24 monthly payments of $100 each. The total number of dollars repaid is always greater than the loan because of the interest being charged.

As already mentioned (in *section 1*), an annuity is a sequence of equal payments made at equal intervals of time. The loan repayment described above is an annuity because the 24 payments are all of equal size and are all equally spaced in time.

Some other examples of annuities are rent payments, mortgage payments, fixed deposits in a savings account, salaries, proceeds to life insurance policies, instalment payments on purchases etc.

Note: Any given annuity may be classified as one of two types – namely, a **simple** annuity or a **general** annuity.

A **simple annuity** is one where the payment interval and the compounding interest period coincide, whereas a **general annuity** is one where the payment interval is different from the compounding interest period.

For example: An annuity that involves *monthly* repayments at an interest rate of 12%p.a. compounded *monthly* is a **simple** annuity, whereas an annuity that involves *monthly* repayments at an interest rate of 16%p.a. compounded *quarterly* is a **general** annuity.

In this text we discuss **simple annuities** only and begin by defining the two most common cases, each of which are explored further in the following subsections.

Case 1
An **ordinary annuity** is one in which the periodic payments are made **at the end** of each payment interval. This is often called 'payment in **arrears**', e.g. salaries.

Case 2
An **annuity due** is one in which the periodic payments are made **at the beginning** of each payment interval. This is often called 'payment **in advance**', e.g. rent.

Note: It is very important when solving an annuity problem to decide initially whether you have an **ordinary** annuity or an annuity **due**, as slightly different computational formulae are involved.

13.4.1 The future value (amount), S, of an annuity

General formulae
Case 1:
The **future amount, S**, of an **ordinary** annuity is given by

$$S = R\left[\frac{(1+i)^n - 1}{i}\right]$$

Case 2:
The **future amount, S**, of an annuity **due** is given by

$$S = R\left[\left(\frac{(1+i)^{n+1} - 1}{i}\right) - 1\right]$$

where r = per annum nominal interest rate (as a decimal)

N = number of interest periods in **one** year

t = the term of the annuity

R = periodic payment of an annuity

$i = \dfrac{r}{N}$ = interest rate per interest period (as a decimal)

$n = Nt$ = number of periodic payments to be made during the whole term.

$\dfrac{(1+i)^n - 1}{i}$ = **future** amount of an annuity of $1 per period for n periods at interest rate, i.

Abbreviation
By convention, the term $\dfrac{(1+i)^n - 1}{i}$ is abbreviated to $s_{n\rceil i}$, (read as 's angle n at i'). Therefore,

(i)   the **future amount** of an **ordinary** annuity is
$$S = Rs_{n\rceil i}$$

where $s_{n\rceil i} = \dfrac{(1+i)^n - 1}{i}$

(ii)   the **future amount** of an annuity **due** is
$$S = R(s_{n+1\rceil i} - 1)$$

where $s_{n+1\rceil i} = \dfrac{(1+i)^{n+1} - 1}{i}$.

Important note:
For many values of i and n, the value of $s_{n\rceil i}$ may be found in Table I Appendix A. Use this table whenever possible so as to reduce calculation errors.

EXAMPLE 13.21
A business is due to receive $1000 at the *end* of each three months for the next year. The business considers money to be worth 12%p.a. compounded quarterly. Calculate the future amount of this annuity.

Solution
Since payments are made at the *end* of each period, we are dealing with an **ordinary** annuity.
So, the appropriate formula for finding the future amount, S, is $S = Rs_{n\rceil i}$.
For our problem, with $r = 0.12$, $N = 4$ and $t = 1$, we have

$i = \dfrac{r}{N} = \dfrac{0.12}{4} = 0.03$ (interest rate per quarter)

$n = Nt = 4 \times 1 = 4$ (total no. of payments).

So, with $R = 1000$ (periodic payment), we have

$S = Rs_{n\rceil i}$

$= 1000 \times s_{4\rceil 0.03}$

From Table I (Appendix A), $s_{4\rceil 0.03} = 4.183627$

Hence,
$$S = 1000 \times 4.183627$$
$$= 4183.627$$

Therefore, the future amount of the given ordinary annuity is \$4 183.63.

EXAMPLE 13.22

John decides to invest \$500, at the *beginning* of each 6-month period, in a company paying interest at 10%p.a. compounded semi-annually.
How much will John's investment be worth at the end of three and a half years?

Solution

Since payments are made at the *beginning* of each six-month period, we are dealing with an annuity **due**.
The appropriate formula for finding the future amount, S, is $S = R[s_{n+1\rceil i} - 1]$.

For this problem, with $r = 0.10$, $N = 2$ and $t = 3.5$, we have

$$i = \frac{r}{N} = \frac{0.1}{2} = 0.05 \text{ (interest rate per 6-months)}$$

$$n = Nt = 2 \times 3.5 = 7 \text{ (total no. of 6-month periods)}.$$

So, with $R = 500$ (periodic payment), we have
$$S = R[s_{n+1\rceil i} - 1]$$
$$= 500[s_{8\rceil 0.05} - 1]$$

From Table I, Appendix A, $s_{8\rceil 0.05} = 9.549109$

So,
$$S = 500(9.549109 - 1)$$
$$= 4274.5545$$

Therefore, the future amount of the given annuity due is \$4 274.55.

Note:

Clearly, Table I in Appendix A cannot cover every possible value of n and i. Whenever the tables do **not** suffice, you must perform all the necessary calculations yourself, using the formula

$$s_{n\rceil i} = \frac{(1+i)^n - 1}{i}.$$

EXAMPLE 13.23

On the birth of his son, James opened an account at the local building society. He intends to deposit \$15 each month until his son's eighteenth birthday, at which time the account will be closed and the full amount in the account given to his son.
If the building society is paying 12%p.a.

compounded monthly, how much will be in the account on his son's eighteenth birthday, if deposits are made

(a) at the *end* of each month?

(b) at the *beginning* of each month?

Solution

(a) This is an **ordinary** annuity problem (payments in arrears, at the *end* of each month).

Here we are given $r = 0.12$, $N = 12$ and $t = 18$.
So, $i = r/N = 0.01$ (interest rate per month)
$$n = Nt = 12 \times 18 = 216 \text{ (total no. of payment periods)}.$$

Since Table I in Appendix A does *not* contain $n = 216$, we must compute the future value, S, by hand.

So, with $R = 15$, we have

$$S = R\left[\frac{(1+i)^n - 1}{i}\right]$$
$$= 15\left[\frac{(1.01)^{216} - 1}{0.01}\right]$$
$$= 15(757.86063)$$
$$= 11367.909$$

Therefore, when his son turns 18, there should be \$11 367.91 in the account (payments in arrears).

(b) This is an annuity **due** problem (payments in advance, at the *beginning* of each month).

From (a), we know that $R = 15$, $i = 0.01$ and $n = 216$.

Since we cannot use Table I, we must find the future value, S, of an annuity due, by hand.
This gives

$$S = R\left[\left(\frac{(1+i)^{n+1} - 1}{i}\right) - 1\right]$$
$$= 15\left[\frac{(1.01)^{217} - 1}{0.01} - 1\right]$$
$$= 15[766.439236 - 1]$$
$$= 15 \times 765.439236$$
$$= 11481.588$$

Therefore, when his son turns 18 there should be \$11 481.59 in the account (payment in advance).

13.31 How much will you have at the end of 5 years, if you deposit $400 at the **end** of each year for the 5 years and if you earn interest at 6%p.a. compounded annually?

13.32 How much will you have at the end of 6 years, if you deposit $450 at the **beginning** of each 3-month period for the 6 years and if you earn interest at 6%p.a. compounded quarterly?

13.33 A family expects to buy a new car 5 years from now. They decide to put away $50 at the beginning of each month. If their money earns interest at 6%p.a. compounded monthly, how much will they have at the end of 5 years?

13.34 A person decides to save money for retirement. $250 is invested at the end of every three months at 8%p.a. compounded quarterly. How much will be in the retirement fund at the end of 40 years?

13.35 An investor intends to deposit $6 000 each six months into an account earning 16%p.a. compounded semi-annually. How much will be in this account at the end of 5 years, if deposits are made
   (a) at the end of each six-month period?
   (b) at the beginning of each six-month period?

13.4.2 The periodic payment, R, necessary to attain a given future amount, S

Appropriate formulae

The simplest method of finding R given S is to take the formulae from 13.4.1 and re-write them making R the subject of the equation, as follows:

Case 1:

The **periodic payment**, R, of an **ordinary** annuity needed to attain a given future amount, S, is calculated using

$$R = \frac{S}{s_{n\rceil i}}$$

Case 2:

The **periodic payment**, R, of an annuity **due** needed to attain a given future amount, S, is calculated using

$$R = \frac{S}{(s_{n+1\rceil i} - 1)}$$

where S = given future amount

i = interest rate for period (as a decimal)

n = total number of periodic payments to be made.

If the values $s_{n\rceil i}$ or $s_{n+1\rceil i}$ are not tabulated for given n and i (see Table I), then use the formula

(i) $\quad s_{n\rceil i} = \dfrac{(1+i)^n - 1}{i}$

or

(ii) $\quad s_{n+1\rceil i} = \dfrac{(1+i)^{n+1} - 1}{i}$.

EXAMPLE 13.24

Following the birth of a child, the parents wish to make a deposit on each of the child's subsequent birthdays so that $10 000 will have accumulated by the child's 20th birthday.
If interest is compounded annually at 5%p.a., find the value of each deposit.

Solution

Since deposits will be made at the *end* of each year, we are dealing with an **ordinary** annuity.
Here we are given

$\quad S = 10000$ (future amount)

$\quad i = r/N = 0.05$ (interest rate per period – here, annually)

$\quad n = 20$ (total no. of periods).

So, the required periodic payment, R, is calculated from $\quad R = \dfrac{S}{s_{n\rceil i}}$.

Now, $s_{n\rceil i} = s_{20\rceil 0.05} = 33.065954$, from Table I of

Appendix A.

Thus, $\quad R = \dfrac{S}{s_{n\rceil i}}$

$\qquad = \dfrac{10000}{33.065954}$

$\qquad = 302.42587$

Therefore, if the parents deposit $302.43 each birthday, there will be $10 000 in the account when the child reaches 20.

EXAMPLE 13.25

Suppose I know that in 12 months time I will have to pay the Taxation Department an amount of $2 500. How much should I invest at the beginning of each 3-month period in order to accumulate the required tax at the end of 12 months? Assume that my invested money earns interest at 12%p.a. compounded quarterly.

Solution

Since deposits will be made at the *beginning* of each quarter, we are dealing with an annuity **due**. Here we are given

$S = 2500$ (future amount)

$i = .12/4 = 0.03$ (interest rate per quarter)

$n = 4$ (total no. of payment periods).

So, the required periodic payment, R, is given by

$$R = \frac{S}{[s_{n+1]i} - 1]}$$

Now, $s_{n+1]i} = s_{5]0.03} = 5.309136$, from Table I of Appendix A.

Thus, $R = \frac{S}{[s_{n+1]i} - 1]}$

$= \frac{2500}{5.309136 - 1}$

$= \frac{2500}{4.309136}$

$= 580.16271$

Therefore, if I deposit $580.16 at the start of each quarter, I will reach my goal of $2 500 after one year.

YOUR TURN

13.36 A certain company must accumulate $12 000 during the next 10 years so that some of its machines can be replaced. What sum must it invest, at the end of each year, in a fund paying 3%p.a. compounded annually, in order to accumulate the required amount?

13.37 A young couple want to have a down-payment of $8 000 to buy a new house in 5 years time. They decide to make deposits at the beginning of each 6-month period in a fund earning 10%p.a. compounded semi-annually. Find the value of each deposit so that they will attain their goal.

13.38 In order to have $5 000 available eighteen months from today, equal monthly deposits are made into a fund earning 18%p.a. compounded monthly.

Find the deposit required if payments are made
(a) at the beginning of each month
(b) at the end of each month.

13.4.3 The present value, A, of an annuity

General formulae

Case 1:

The **present value**, A, of an **ordinary** annuity is

given by $A = R\left[\dfrac{1 - (1+i)^{-n}}{i}\right].$

Case 2:

The **present value**, A, of an annuity **due** is given

by $A = R\left[1 + \left(\dfrac{1 - (1+i)^{-(n-1)}}{i}\right)\right].$

where R = periodic payment of the annuity

i = interest rate per interest period (as a decimal)

n = total number of periodic payments to be made.

$\dfrac{1 - (1+i)^{-n}}{i}$ = **present** value of an annuity of $1 per period for n periods at interest rate, i.

Abbreviation

By convention, the term $\dfrac{1 - (1+i)^{-n}}{i}$ is abbreviated to $a_{n]i}$, (read as '*a* angle *n* at *i*'). Thus,

(i) the **present value** of an **ordinary annuity** is

$$A = Ra_{n]i}$$

where $a_{n]i} = \dfrac{1 - (1+i)^{-n}}{i}$

(ii) the **present value** of an annuity **due** is

$$A = R[1 + a_{n-1]i}]$$

where $a_{n-1]i} = \dfrac{1 - (1+i)^{-(n-1)}}{i}.$

Important note:

For many n and i, the value of $a_{n]i}$ may be found in Table II, Appendix A. Always use this table where possible and be careful not to confuse with $s_{n]i}$ in Table I.

EXAMPLE 13.26

Peter purchased a yacht, paying $5000 deposit and promising to pay $300 at the end of every quarter for the next 6 years. If money is worth 8%p.a. compounded quarterly, what is the cash value of the yacht **today**?

Solution

Since payments are made at the *end* of every quarter, we have an **ordinary** annuity.

So, the appropriate formula for finding the present value, A, is $A = Ra_{n]i}$.

In this problem, with $r = 0.08$, $N = 4$ and $t = 6$, we have $i = r/N = 0.02$ (interest rate per quarter)

$n = Nt = 24$ (total no. of payments).

So, with $R = 300$ (periodic payment), we have

$A = Ra_{n]i}$

$= 300 \times a_{24]0.02}$

From Table II, Appendix A, $a_{24 \rceil 0.02} = 18.913925$.

Hence, $A = 300 \times 18.913925$

$= 5674.1775$

Therefore, the present value of the 24 quarterly payments is $5 674.18.

Hence, the cash value of the yacht must be

$$C.V. = 5674.18 + \text{deposit}$$
$$= 5674.18 + 5000$$
$$= \$10\ 674.18$$

EXAMPLE 13.27

A company offers to sell a piece of machinery for equal payments of $150 to be paid at the beginning of each 6-month period for the next 2 years (the first payment being made today).

If interest is charged at 9%p.a. compounded semi-annually, find the **present** cash value of the machine.

Solution

Since payments are made at the *beginning* of each 6-month period, we are dealing with an annuity **due**.

The appropriate formula for finding the present value, A, is $A = R[1 + a_{n-1 \rceil i}]$.

In our problem, with $r = 0.09$, $N = 2$ and $t = 2$, we have

$$i = \frac{r}{N} = 0.045 \text{ (interest rate per 6-month period)}$$

$n = Nt = 4$ (total no. of payments).

So, with $R = \$150$ (periodic payment), we have

$$A = R[1 + a_{n-1 \rceil i}]$$
$$= 150[1 + a_{3 \rceil 0.045}]$$

From Table II, Appendix A, $a_{3 \rceil 0.045} = 2.748964$.

So, $A = 150[1 + 2.748964]$

$= 150 \times 3.748964$

$= 562.3446$

Therefore, the cash value of the machine today is $562.34.

Note: As we saw in Table I for values of $s_{n \rceil i}$, Table II in Appendix A cannot cover every possible value of n and i. Whenever the tables do **not** suffice, you must perform all the necessary calculations yourself, using the formula

$$a_{n \rceil i} = \frac{1 - (1+i)^{-n}}{i} .$$

EXAMPLE 13.28

Penny wishes to buy a small cottage by the sea.

She is able to pay $250 each month if she borrows money from the bank and is willing to continue doing so for 15 year. If interest is currently at 12%p.a. compounded monthly, how much could Penny borrow on these terms?

You may assume that monthly payments are made

(a) at the *end* of each month

(b) at the *beginning* of each month.

Solution

(a) This is an **ordinary** annuity problem (payment in arrears, at the *end* of each month).

Here we are given

$r = 0.12$, $N = 12$ and $t = 15$.

So, $i = r/N = 0.01$ (interest rate per month)

$n = Nt = 180$ (total payments).

Since Table II, Appendix A, does **not** contain $n = 180$, we must compute the present value, A, by hand.

So, with $R = 250$, we have

$$A = R\left[\frac{1 - (1+i)^{-n}}{i}\right]$$

$$= 250\left[\frac{1 - (1.01)^{-180}}{0.01}\right]$$

$$= 250 \times 83.321664$$

$$= 20830.416$$

Therefore, with the given terms, Penny could borrow $20 830.42 from the bank (with repayments in arrears).

(b) This is an annuity **due** problem (payment in advance, at the *beginning* of each month).

From **(a)** we know that $R = 250$, $i = 0.01$ and $n = 180$.

Since $n = 180$ is not given in Table II, we must find the present value, A, of an annuity due, by hand. This gives

$$A = R\left[1 + \left(\frac{1 - (1+i)^{-(n-1)}}{i}\right)\right]$$

$$= 250\left[1 + \left(\frac{1 - (1.01)^{-179}}{0.01}\right)\right]$$

$$= 250\ (1 + 83.154881)$$

$$= 250 \times 84.154881$$

$$= 21038.72$$

Therefore, with the given terms, Penny could borrow $21 038.72 from the bank (with repayments in advance).

13.39 What lump sum would have to be deposited now at 6%p.a. compounded monthly, so that withdrawals of $30 can be made at the end of every month for the next 3 years?

13.40 In settlement of a certain debt, Pamela agrees to pay $200 at the beginning of every six months (starting today) for 5 years. If interest is charged at 9%p.a. compounded semi-annually, find the present value of the debt.

13.41 Instead of paying $1 000 rent at the beginning of each quarter for the next 8 years, Mr Jones decides to buy a house. If money is worth 10%p.a. compounded quarterly, what is the cash equivalent *today* of the 8 years of rent?

13.42 Find the cash value *today* of $75 payments made each week for the next 4 years. Assume that money is worth 13%p.a. compounded weekly and that payments are made
  (a) at the beginning of each week
  (b) at the end of each week.

13.4.4 The periodic payment, R, given the present value, A, of an annuity

Appropriate formulae

As we saw when finding R given S, the simplest method of determining R given A is to take the formulae from 13.4.3 and re-write them making R the subject of the equation, as follows.

Case 1:

The **periodic payment**, R, of an **ordinary** annuity when given the present value, A, is calculated

using $R = \dfrac{A}{a_{n\rceil i}}$

Case 2:

The **periodic payment**, R, of an annuity **due** when given the present value, A, is calculated using

$$R = \frac{A}{(1 + a_{n-1\rceil i})}$$

where A = given present value

  i = interest rate per period (as a decimal)

  n = total number of periodic payments to be made.

If the values of $a_{n\rceil i}$ or $a_{n-1\rceil i}$ are not tabulated for given n and i (see Table II), then use the formula

(i) $a_{n\rceil i} = \dfrac{1 - (1 + i)^{-n}}{i}$, **or**

(ii) $a_{n-1\rceil i} = \dfrac{1 - (1 + i)^{-(n-1)}}{i}$

EXAMPLE 13.29

A person borrows $6 000 from a bank at an interest rate of 12%p.a. compounded monthly. The loan is to be paid off by equal monthly payments at the end of each month, for the next 3 years.
Find the cost of each payment.

Solution

Since payments will be made at the *end* of each month, we are dealing with an **ordinary** annuity. Here we are given:

  A = 6000 (present value)

  $i = r/N = 0.01$ (interest rate per month)

  $n = Nt = 36$ (total no. of payments).

So, the required periodic payment, R, is calculated

from $R = \dfrac{A}{a_{n\rceil i}}$.

Now, from Table II, Appendix A,

$$a_{n\rceil i} = a_{36\rceil 0.01} = 30.107505 .$$

Thus, $\quad R = \dfrac{A}{a_{n\rceil i}}$

$\qquad = \dfrac{6000}{30.107505}$

$\qquad = 199.28586$

So, if $6 000 is borrowed, the cost of each monthly repayment if $199.29.

EXAMPLE 13.30

A television set worth $780 may be purchased by paying $80 down and the balance in quarterly instalments, payable at the *beginning* of each quarter, for 2 years. Find the quarterly instalment if the dealer charges 16%p.a. compounded quarterly, and the first instalment is due today.

Solution

Since instalments will be made at the *beginning* of each quarter, we are dealing with an annuity **due**. Here we are given

  $A = 780 - 80 = 700$ (amount borrowed is the cash value – deposit)

  $i = .16/4 = .04$ (interest rate per quarter)

  $n = 4 \times 2 = 8$ (total no. of payments to be made).

So, the required periodic payment, R, is calculated from

$$R = \frac{A}{(1 + a_{n-1\rceil i})}.$$

Now, $a_{n-1\rceil i} = a_{7\rceil 0.04} = 6.002055$, from Table II.

Thus, $R = \dfrac{A}{(1 + a_{n-1\rceil i})}$

$\qquad = \dfrac{700}{1 + 6.002055}$

$\qquad = \dfrac{700}{7.002055}$

$\qquad = 99.97065$

Therefore, if a debt of \$700 is to be paid off in 2 years with quarterly instalments paid in advance, each instalment would be \$99.97.

YOUR TURN

13.43 A family borrows \$9 000 to cover the cost of extensions to their home. The loan is to be paid off by equal quarterly payments, at the end of each quarter, for the next 3 years.

If interest is charged at 12%p.a. compounded quarterly, find the cost of each payment.

13.44 Lease payments are usually made in advance. A \$340 000 machine is leased over a six-year period at 16%p.a. compounded semi-annually. What is the cost of each six-monthly payment?

(*Note:* \$340 000 is the cash value of the machine today.)

13.45 A personal loan of \$8 000, with interest at 26%p.a. compounded weekly, is to be paid off in equal weekly instalments over a 5 year period.

Find the cost of each weekly instalment, if payment is made
  (a) at the *beginning* of each week
  (b) at the *end* of each week.

Comment:

When discussing simple annuities there are two additional quantities that may be calculated – namely, the per annum interest rate, r, and the length of the term, t.

Because the exact values of r and t are rather difficult to determine, we instead present in the following table, formulae that allow approximate values to be determined.

Remember that it is first necessary to find values for i, the *per period* interest rate and n, the total number of periodic payments.

	Ordinary annuity	Annuity due
For given future amount, S, (i) approximate interest rate per period, i; or (ii) approximate number of periodic payments, n	$s_{n\rceil i} = \dfrac{S}{R}$	$s_{n+1\rceil i} = \left(\dfrac{S}{R}\right) + 1$
	Use Table I (Appendix A) to find (i) approx. i given n, or (ii) approx. n given i. Then, (i) $r = i \times N$ = annual interest rate, or (ii) $t = n/N$ = length of the term (in years).	
For given present value, A, (i) approximate interest rate per period, i; or (ii) approximate number of periodic payments, n	$a_{n\rceil i} = \dfrac{A}{R}$	$a_{n-1\rceil i} = \left(\dfrac{A}{R}\right) - 1$
	Use Table II (Appendix A) to find (i) approx. i given n, or (ii) approx. n given i. Then, (i) $r = i \times N$ = annual interest rate, or (ii) $t = n/N$ = length of term (in years).	

Note:

The purpose of the above table is two-fold; for interest and for the sake of 'completeness' of our discussion of simple annuities. However, as the concepts and calculations involved are not difficult, further exploration will be left to the reader.

Summary

In this chapter we have discussed in some detail, the concepts of simple interest, compound interest and simple annuities and have examined the mechanics of a number of calculation techniques by solving many practical problems.

To sum up, let us take each of the three main topics in turn.

A.   Simple interest

Here we discussed how to calculate

(i)   the amount of interest, the interest rate or the length of the term associated with a particular transaction

(ii) the future value of a transaction at simple interest

(iii) the present value or principal of a transaction at simple interest

(iv) the time between actual dates when simple interest is involved.

Appropriate formulae for calculating (i) to (iv) above

Assuming that

S = future or accumulated value of a transaction

P = present value or original principal

I = simple interest earned

r = per annum simple interest rate (as a decimal)

t = time or length of the term (in years),

we have

a. amount of interest: $I = Prt$

b. length of the term: $t = \dfrac{I}{Pr}$

c. rate of interest: $r = \dfrac{I}{Pt}$

d. future value: $S = P(1 + rt)$

e. principal: $P = \dfrac{I}{rt}$

f. present value: $P = \dfrac{S}{(1 + rt)}$

B.   Compound interest

Here we discussed how to calculate

(i) the future value of a transaction at compound interest

(ii) the present value or principal of a transaction at compound interest

(iii) the amount of compound interest earned

(iv) the per annum rate of compound interest

and mentioned briefly

(v) the length of the term of the transaction

(vi) the equivalence between nominal and effective annual rates of interest

(vii) the relationship between simple (or flat) interest rates and compound interest rates.

Appropriate formulae for calculating (i) to (v) above

Assuming that

S = future or accumulated value of a transaction

P = principal or present value of a transaction

r = nominal annual rate of interest (in decimal form)

t = length of the transaction (in years)

N = no. of interest periods in **one** year

$i = r/N$ = interest rate per interest period (in decimal form)

$n = Nt$ = no. of interest periods in the *whole* transaction,

we have

a. future amount: $S = P(1 + i)^{n}$

b. present value: $P = S(1 + i)^{-n}$

c. interest rate per period: $i = \left(\dfrac{S}{P}\right)^{\frac{1}{n}} - 1$

d. percentage nominal annual rate: $i \times N \times 100 = r \times 100$

e. total number of interest periods: $n = \dfrac{\log\left(\dfrac{S}{P}\right)}{\log(1 + i)}$

f. length of the term, in years: $t = \dfrac{n}{N}$

C.   Simple annuities

Here we discussed how to calculate the following quantities for each of the two types of simple annuities – namely, an **ordinary** annuity and an annuity **due**,

(i) the future value (amount) of an annuity

(ii) the present value of an annuity

(iii) the value of the periodic payment.

We also mentioned briefly (with no examples) how to determine

(iv) the *approximate* nominal rate of interest

(v) the *approximate* term of the annuity.

Appropriate formulae for calculating (i) to (iii) above

Assuming that

S = future amount of an annuity

A = present value of an annuity

R = periodic payment of an annuity

r = per annum nominal rate of compound interest (as a decimal)

N = number of interest or payment periods **per year**

t = length of the term of an annuity (in years)

$i = r/N$ = interest rate per interest (payment) period (as a decimal)

$n = Nt$ = total number of periodic payments to be made

$s_{\overline{n}|i}$ = *future* amount of an annuity of $1 per payment period for n periods at interest rate i per period, and

$$s_{\overline{n}|i} = \dfrac{(1 + i)^{n} - 1}{i}$$

$a_{n\rceil i}$ = *present* value of an annuity of \$1 per payment period for n periods at interest rate i per period, and

$$a_{n\rceil i} = \frac{1 - (1+i)^{-n}}{i}.$$

The following table best summarises the basic formulae necessary for solving the annuity problems discussed in *section 4*.

Quantity to be calculated / Type of annuity	Ordinary annuity – payment made at the end of each period	Annuity due – payment made at the beginning of each period
1. Future amount, S	$S = Rs_{n\rceil i}$	$S = R\left[s_{n+1\rceil i} - 1\right]$
2. Payment, R, given future amount, S	$R = \dfrac{S}{s_{n\rceil i}}$	$R = \dfrac{S}{\left[s_{n+1\rceil i} - 1\right]}$
3. Present value, A	$A = Ra_{n\rceil i}$	$A = R\left[1 + a_{n-1\rceil i}\right]$
4. Payment, R, given present value, A	$R = \dfrac{A}{a_{n\rceil i}}$	$R = \dfrac{A}{\left[1 + a_{n-1\rceil i}\right]}$
5. Approx. interest rate per period, i, given future amount, S	$s_{n\rceil i} = \dfrac{S}{R}$	$s_{n+1\rceil i} = \left(\dfrac{S}{R}\right) + 1$
	Use **Table I** (Appendix A) to find approx. i given n. Then, $r = i \times N$ = annual interest rate.	
6. Approx. interest rate per period, i, given present value, A	$a_{n\rceil i} = \dfrac{A}{R}$	$a_{n-1\rceil i} = \left(\dfrac{A}{R}\right) - 1$
	Use **Table II** (Appendix A) to find approx. i given n. Then, $r = i \times N$ = annual interest rate.	
7. Approx. no. of periodic payments, n, given future amount, S	$s_{n\rceil i} = \dfrac{S}{R}$	$s_{n+1\rceil i} = \left(\dfrac{S}{R}\right) + 1$
	Use **Table I** (Appendix A) to find approx. n given i. Then, $t = n/N$ = length of the term (in years).	
8. Approx. no. of periodic payments, n, given present value, A	$a_{n\rceil i} = \dfrac{A}{R}$	$a_{n-1\rceil i} = \left(\dfrac{A}{R}\right) - 1$
	Use **Table II** (Appendix A) to find approx. n given i. Then, $t = n/N$ = length of the term (in years).	

13.1 Mr Young invested $15 400 in a fund earning 12%p.a. simple interest for a period of 5 months. How much interest was earned during that time?

13.2 Find how long Crown Furniture must invest $14 000 of spare cash at 18%p.a. simple interest to earn $315 interest.

13.3 To an account that was opened with $5 900, a credit union credited $520 simple interest for an 18 month period. Find the per annum rate of simple interest paid by the credit union.

13.4 What is the maturity value of $2 500 invested today at 6%p.a. simple interest for 4 months?

13.5 What sum invested today will amount to $1 000 in 8 months time at 5%p.a. simple interest?

13.6 How much was deposited in an account on February 1 if, when the account was closed on November 25, it contained $1 820? Assume the money was earning 11.5%p.a. simple interest.

13.7 Henry borrowed $1 500 on June 5, 1993 and agreed to repay the debt plus interest at 10%p.a. simple interest on November 20, 1994. What is the full amount that Henry was required to pay on November 20?

13.8 Answer true (T) or false (F) to each of the following statements, (be sure that you understand why).
  (a) The present value of an interest bearing loan is usually greater than the future value.
  (b) The present value of a future amount of $2 000 due 5 years from today is greater at 7%p.a. compounded monthly than at 8%p.a. compounded monthly.
  (c) Money loaned at 10%p.a. compounded quarterly earns more interest than money loaned at 10%p.a. compounded semi-annually.
  (d) A nominal rate of 13%p.a. compounded quarterly is equivalent to an effective rate of **less** than 13%p.a.

13.9 Jane borrows $850, agreeing to repay the principal with interest at 8%p.a. compounded semi-annually. What is the debt at the end of 6 years?

3.10 Josephine bought a lounge suite worth $2 000. She paid 20% deposit and agreed to pay the balance with interest as a lump sum in 2 years time. If interest is charged at 12%p.a. compounded monthly, what is the accumulated amount she must pay in 2 years?

13.11 If money is worth 14%p.a. compounded quarterly, how much could I borrow today if I know I can repay $3 500 four years from today?

13.12 Henry borrowed $3 200 today at 10%p.a. compounded quarterly, and settled his full debt with a payment of $4 800. When was this payment made?

13.13 John purchased a car at an agreed value of $8 000. He paid $2 000 down and agreed to pay the balance as a single payment of $6 800 at the end of 2 years. If interest is compounded quarterly, what is the nominal annual rate being charged?

13.14 How much money will be required on 31 December, 1995 to repay a loan of $2 000 made on 31 December, 1992, if interest is charged at 12%p.a. compounded quarterly.

13.15 An obligation of $3 000 was due on 31 December, 1993. What was the present value of this obligation on 30 June, 1989, if interest is charged at 13%p.a. compounded semi-annually?

13.16 How long will it take to double your deposit of $1 000 in a savings account, if interest is earned at
  (a) 18%p.a. compounded monthly?
  (b) 9%p.a. compounded semi-annually?

13.17 I need to have saved $5 000 at the end of 2 years. I intend to deposit $R at the end of each month into an account earning 12%p.a. compounded monthly. Find the required value of R.

13.18 Find the future value, S, of 16 quarterly payments of $564, if deposits earn interest at 14%p.a. compounded quarterly. Assume that deposits are made
  (a) at the beginning of each quarter
  (b) at the end of each quarter.

*13.19 Mary has just decided to save for an extended holiday trip in 2 years time. She intends to save $8000 for this by making regular weekly deposits into a credit union savings account that earns 13%p.a. compounded weekly.

How much should she set aside each week if deposits are to be made
  (a) at the end of each week?
  (b) at the beginning of each week?

13.20 An heir is to receive an inheritance of $1 000 each half-year for 10 years. If money is worth 10%p.a. compounded semi-annually, what is the cash value of this inheritance **today** if each payment is received

(a) at the beginning of each 6-month period?

(b) at the end of each 6-month period?

13.21 A car selling for $8 500 may be purchased for $1 500 down and the balance in equal monthly instalments for 3 years. If money earns 18%p.a. compounded monthly, find the cost of each monthly payment, assuming payments are made

(a) at the end of each month

(b) at the beginning of each month.

13.22 At what nominal rate of interest would an annuity of $900 per quarter, paid at the *end* of each quarter, grow to a total of $60 662.30 over a period of 10 years?

13.23 A man deposits $500 at the end of each quarter with a credit union that pays interest at the rate of 9.5%p.a. compounded quarterly. How much does he have to his credit after 4 years?

*13.24 A certain fund earns 12%p.a. compounded semi-annually. How long will it take to accumulate a total future amount of $8 000 in the fund if $2 000 is deposited

(a) at the end of each 6-month period?

(b) at the beginning of each 6-month period?

*13.25 A loan of $6 000 taken today is to be repaid with equal monthly payments of $230 (paid at the beginning of each month) over a 2.5 year period. Find the approximate nominal annual rate of interest being charged on the loan.

14 Extras – Trigonometry revisited

Introduction

This is the second of two chapters that deal with an important branch of mathematics called **trigonometry.** In Chapter 10 we discussed the origin of the word *trigonometry* and then, after listing the new terminology and notation, we proceeded to examine

(i) trigonometric ratios

(ii) reciprocal ratios

(iii) the use of both the sine and cosine rules, and

(iv) Pythagorean identities,

all of which are useful in solving certain types of mathematical problems.

In this chapter we concentrate on the study of **trigonometric functions** involving, at the outset, the introduction of an alternative measuring system. Just as decimal currency was introduced to simplify monetary problems and the metric system was introduced to simplify measurement problems, so too is **radian measure** introduced here to simplify problems involving *trigonometric functions.*

In our earlier chapter on trigonometry we used the more common 'degree' or 'sexagesimal' system (since $1° = 60'$ and $1' = 60''$) to measure angles. In this chapter, a more appropriate unit for measuring angles, in the context of trigonometric functions, is the **radian.**

We begin, in *Section 2*, with a definition of radian measure followed by a discussion of how to convert

(i) degrees to radians, and

(ii) radians to degrees.

Section 3 presents some specific *applications* of radian measure.

Since the main purpose of this chapter is to examine trigonometric functions, *Section 4* discusses the **graphs** of the more common trigonometric functions – namely, the sine, cosine and tangent functions, while *Section 5* briefly looks at calculus and the trigonometric functions.

As with all the later chapters in this text, the material on trigonometric functions presented here is certainly not particularly relevant to *all* disciplines. However, it should be of interest and benefit to those readers intending to pursue studies in physics, engineering and mathematics.

14.1 Terminology and notation

> An **arc** is a portion of the circumference of a circle (usually the smaller portion).

Diagrammatically, an arc is labelled in one of two ways:

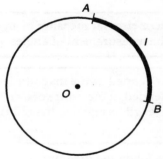

(i) with capital letters marking the endpoints, or

(ii) with the small letter, *l*, denoting the length.

For example: In the diagram above, displaying a circle with centre *O*, the smaller (bold) portion between the two points *A* and *B*, located on the circumference, is called the **arc** *AB* or the **arc** *l*.

> The word **unit**, when used as an adjective to describe a particular shape or measurement, means **one.**

For example:

(i) unit circle = circle with a radius of 1.

(ii) unit length = length of 1.

(iii) unit radius = radius of unit length

= radius of length 1.

In geometry and trigonometry, the word **subtend**, when discussing lines, points and angles, means **to be opposite**. Thus two points, especially if they are on a circle, are said to subtend an angle at a third point when they are joined to that point.

For example:

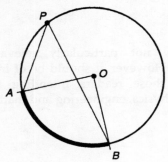

In the diagram above, the points A and B subtend $\angle AOB$ at O, the centre of the circle, and also at P on the circumference of the circle.

14.2 Radian measure

14.2.1 Definitions

The **radian** or **circular measure** of angles involves the measurement of angles in **radians**.

A **radian** is defined as the magnitude of an angle subtended at the centre of a circle by an arc whose length is equal to the radius of the circle.

Hence, in the following diagram,

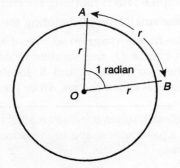

(i) the radius is of length r

(ii) the arc AB is of length r.

Therefore, $\angle AOB = 1$ radian.

Notation:
Although most text books do not use a special symbol to denote that an angle is measured in radians, this often leads to confusion. Thus, in this chapter, we will use the superscript c to denote radian measure. That is, 1 radian = 1^c.

Note:
Suppose we have a circle of unit radius (that is, radius = 1). Then, if an angle, say $\angle AOB$, has its vertex at the centre, O, then its radian measure is equal to the length of the arc on which it stands.

For example:

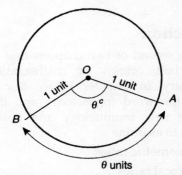

In the above diagram, since the circle has unit radius, we know that, if the arc AB is of length θ units then the angle $\angle AOB$ is θ radians or θ^c.

Similarly,

(i) if the length of arc AB = 2.4 units, then
$\angle AOB = 2.4^c$

(ii) if arc AB measures 0.61 units, then
$\angle AOB = 0.61^c$.

From the preceding discussion we may conclude that the length of any arc measured by a real number may be interpreted as the magnitude of an angle measured in radians.

Finally, it should also be noted that, although *radian measure* is always used when theoretical work is being carried out with trigonometric functions, *degree measure* is still very important in more practical situations such as carpentry or surveying.

14.2.2 Conversions between radians and degrees

Because both degree and radian measure are used in many geometric and trigonometric situations, it is important to be able to convert between degrees and radians.

In fact, although most scientific calculators can carry out the conversion automatically, equivalent values may always be found by means of a simple relationship.

We already know that the circumference of a circle is given by $C = 2\pi r$, where
$\qquad \pi$ is the irrational number with
$\qquad\qquad$ approximate value 3.14159,
$\qquad r$ is the radius of the circle.

Consider again the unit circle with diameter AB.

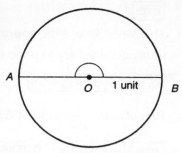

Clearly, since $C = 2\pi r$, we have that the circumference, $C = 2\pi$ when $r = 1$.

Therefore, the length of the semi-circular arc $AOB = \pi$ units (that is, half the circumference) and hence it follows that the straight angle $\angle AOB = \pi^c$.

> Since π is the length of the semi-circular arc AB in a unit circle, the straight angle AOB has magnitude π radians.

However, since we also know that there are 360 degrees in one full circular revolution, and that the straight angle AOB measures $180°$, it follows that

> **(i)** 2π radians $= 2\pi^c = 360°$.
>
> **(ii)** π radians $= \pi^c = 180°$.

From the above relationship we may deduce immediately that

> **(i)** 1 radian $= 1^c = \left(\dfrac{180}{\pi}\right)$ degrees
>
> $\qquad = \left(\dfrac{180}{\pi}\right)° \approx 57.2958°$.
>
> **(ii)** 1 degree $= 1° = \left(\dfrac{\pi}{180}\right)$ radians
>
> $\qquad = \left(\dfrac{\pi}{180}\right)^c \approx 0.01745$ radians.

We can now use the relationship given above to convert
(i) degrees to radians, and
(ii) radians to degrees.

EXAMPLE 14.1
Convert each of the following measures in degrees to an equivalent measure in radians.
(a) 30°   **(b)** 125°   **(c)** 270°   **(d)** 420°

Solution
(a) Since $1° = \left(\dfrac{\pi}{180}\right)^c$, we have

$$30° = \left(\dfrac{\pi}{180}\right)^c \times 30$$

$$= \left(\dfrac{\pi}{6}\right)^c$$

Hence, 30 degrees is equivalent to $\dfrac{\pi}{6}$ radians.

(b) Since $1° = \left(\dfrac{\pi}{180}\right)^c$, we have

$$125° = \left(\dfrac{\pi}{180}\right)^c \times 125$$

$$= \left(\dfrac{25\pi}{36}\right)^c \text{ or } 2.18^c$$

Hence, 125 degrees is equivalent to 2.18 radians.

(c) Since $1° = \left(\dfrac{\pi}{180}\right)^c$, we have

$$270° = \left(\dfrac{\pi}{180}\right)^c \times 270$$

$$= \left(\dfrac{3}{2}\pi\right)^c \text{ or } 4.71^c$$

Hence, 270 degrees is equivalent to 4.71 radians.

(d) Since $1° = \left(\dfrac{\pi}{180}\right)^c$, we have

$$420° = \left(\dfrac{\pi}{180}\right)^c \times 420$$

$$= \left(\dfrac{7}{3}\pi\right)^c \text{ or } 7.33 \text{ radians.}$$

EXAMPLE 14.2
Convert each of the following measures in radians to an equivalent measure in degrees.

(a) $\left(\dfrac{\pi}{6}\right)^c$

(b) $\left(\dfrac{3\pi}{4}\right)^c$

(c) $\left(\dfrac{\pi}{90}\right)^c$

(d) 4.5 radians.

Solution
(a) Since $1^c = \left(\dfrac{180}{\pi}\right)°$, we have

$$\left(\dfrac{\pi}{6}\right)^c = \left(\dfrac{180}{\pi}\right)° \times \dfrac{\pi}{6}$$

$$= 30°$$

Hence, $\dfrac{\pi}{6}$ radians is equivalent to 30 degrees.

(b) Since $1^c = \left(\dfrac{180}{\pi}\right)^\circ$, we have

$$\left(\dfrac{3\pi}{4}\right)^c = \left(\dfrac{180}{\pi}\right)^\circ \times \dfrac{3\pi}{4}$$

$$= 135^\circ$$

Hence, $\dfrac{3\pi}{4}$ radians is equivalent to 135 degrees.

(c) Since $1^c = \left(\dfrac{180}{\pi}\right)^\circ$, we have

$$\left(\dfrac{\pi}{90}\right)^c = \left(\dfrac{180}{\pi}\right)^\circ \times \dfrac{\pi}{90}$$

$$= 2^\circ$$

(d) Since $1^c = \left(\dfrac{180}{\pi}\right)^\circ$, we have

$$4.5^c = \left(\dfrac{180}{\pi}\right)^\circ \times 4.5$$

$$= \left(\dfrac{810}{\pi}\right)^\circ \approx 257.83^\circ$$

14.2.3 Using your calculator

For more complicated examples involving conversions and certainly for evaluating the trigonometric ratios of angles measured in radians, the calculator is extremely useful.

A. ⟦° ' "⟧

This key converts a **sexagesimal** number (one given in degrees, minutes and seconds) to **decimal** notation.

For example:

(i) 67° 24' as decimal is 67.4°.

67 ⟦° ' "⟧ 24 ⟦° ' "⟧   ⟦ 67.4 ⟧

(ii) 243° 18' 51" as a decimal is 243.3142°.

243 ⟦° ' "⟧ 18 ⟦° ' "⟧ 51 ⟦° ' "⟧ ⟦ 243.314167 ⟧

B. ⟦shift⟧ ⟦° ' "⟧

These keys convert degrees in **decimal** notation to **sexagesimal** notation (that is, degrees, minutes and seconds).

For example:

(i) 126.43° in ° ' " is 126° 25' 48".

126.43 ⟦shift⟧ ⟦° ' "⟧ ⟦ 126° 25° 48. ⟧

(ii) 16.84° in ° ' " is 16° 50' 24".

16.84 ⟦shift⟧ ⟦° ' "⟧ ⟦ 16° 50° 24. ⟧

C. ⟦exp⟧

This key displays the approximate value of π (it is **not** necessary to press the ⟦shift⟧ key).

So, ⟦exp⟧ ⟦ 3.141592654 ⟧

For example: $\dfrac{\pi}{180} = 0.174533$

⟦exp⟧ ⟦÷⟧ 180 ⟦=⟧ ⟦ 0.017453292 ⟧

D. To evaluate $\sin x^c$, $\cos x^c$ or $\tan x^c$ on your calculator, you *must* first change to 'radian mode' by pressing ⟦mode⟧ ⟦5⟧, on our Casio.

Notice that after pressing ⟦mode⟧ ⟦5⟧ the display will show RAD instead of DEG at the top.

EXAMPLE 14.3

Using your calculator,

(a) convert 241° 17' to radians.

(b) express in degrees, minutes and seconds an angle that measures 2.75 radians.

(c) give the equivalent radian measure of the angle 6° 25' 47".

(d) rewrite, the following angles measured in radians, in their sexagesimal form.

(i) $(0.04)^c$ **(ii)** $\left(\dfrac{17}{6}\pi\right)^c$

Solution

(a) Since $1^\circ = \left(\dfrac{\pi}{180}\right)^c$, we have

$$241° 17' = \left(\dfrac{\pi}{180}\right)^c \times 241° 17'$$

Using the calculator, we have

241 ⟦° ' "⟧ 17 ⟦° ' "⟧ ⟦x⟧ ⟦exp⟧ ⟦÷⟧ 180
⟦=⟧ ⟦ 4.211188597 ⟧

Hence, $241°17' \approx 4.2^c$.

(b) Since $1^c = \left(\dfrac{180}{\pi}\right)^\circ$, we have

$$2.75^c = \left(\dfrac{180}{\pi}\right)^\circ \times 2.75$$

Using the calculator, we have

180 ⟦÷⟧ ⟦exp⟧ ⟦x⟧ 2.75
⟦=⟧ ⟦shift⟧ ⟦° ' "⟧ ⟦ 157°33°48.2 ⟧

Hence, 2.75 radians = 157° 33' 48.2".

(c) Since $1^\circ = \left(\dfrac{\pi}{180}\right)^c$, we have

$$6°25'47" = \left(\dfrac{\pi}{180}\right)^c \times 6°25'47"$$

Using the calculator, we have

6 $\boxed{\circ\,'\,''}$ 25 $\boxed{\circ\,'\,''}$ 47 $\boxed{\circ\,'\,''}$ $\boxed{\text{x}}$ $\boxed{\text{exp}}$ $\boxed{\div}$ 180
$\boxed{=}$ $\boxed{0.112219822}$

Hence, $6° 25' 47" \approx 0.11222^c$.

(d) **(i)** Since $1^c = \left(\dfrac{180}{\pi}\right)^\circ$, we have

$$0.04^c = \left(\dfrac{180}{\pi}\right)^\circ \times 0.04$$

Using the calculator, we have

180 $\boxed{\div}$ $\boxed{\text{exp}}$ $\boxed{\text{x}}$ .04
$\boxed{=}$ $\boxed{\text{shift}}$ $\boxed{\circ\,'\,''}$ $\boxed{2°17°30.59}$

Hence, 0.04 radians = 2° 17' 30.59".

(ii) Since $1^c = \left(\dfrac{180}{\pi}\right)^\circ$, we have

$$\left(\dfrac{17}{6}\pi\right)^c = \left(\dfrac{180}{\pi}\right)^\circ \times \dfrac{17\pi}{6}$$

$$= \left(\dfrac{180 \times 17}{6}\right)^\circ$$

Using the calculator, we have

180 $\boxed{\text{x}}$ 17 $\boxed{\div}$ 6
$\boxed{=}$ $\boxed{\text{shift}}$ $\boxed{\circ\,'\,''}$ $\boxed{510°0°0.}$

Hence, $\dfrac{17}{6}\pi$ radians = 510°.

EXAMPLE 14.4

With your calculator set to *radian mode* find, correct to 3 decimal places:

(a) $\cos 2^c$

(b) $\sin \dfrac{3\pi}{10}$

(c) $\tan 3.7$

Solution

(a) Check that your calculator is in *radian mode*.

Then $\cos 2^c$ is

2 $\boxed{\cos}$ $\boxed{-0.416146836}$

Hence, $\cos 2^c \approx -0.416$.

(b) Using the calculator, in *radian mode*,

$\sin \dfrac{3\pi}{10}$ is

3 $\boxed{\text{x}}$ $\boxed{\text{exp}}$ $\boxed{\div}$ 10 $\boxed{=}$ $\boxed{\sin}$ $\boxed{0.809016994}$

Hence, $\sin \dfrac{3\pi}{10} \approx 0.809$.

(c) Using your calculator, tan 3.7 is

3.7 $\boxed{\tan}$ $\boxed{0.624733075}$

Hence, $\tan 3.7 \approx 0.625$.

Note 1:
In Example 14.1 (c) we found that

$$270^\circ = \left(\dfrac{\pi}{180}\right)^c \times 270 = 1.5\pi^c.$$

Using the calculator, we have

270 $\boxed{\text{x}}$ $\boxed{\text{exp}}$ $\boxed{\div}$ 180 $\boxed{=}$ $\boxed{4.71238898}$.

The difference between the answer obtained using the calculator (4.71238898) and the answer $\left(\dfrac{\pi}{6}\right)^c$ (calculated by hand from the formula), is best explained by remembering that the answer given by the calculator is only an *approximation*, while the answer $\left(\dfrac{\pi}{6}\right)^c$ is an *exact* value.

Note 2: The approximation is due to the fact that π can *never* be evaluated *precisely*.

YOUR TURN

14.1 Without using your calculator, express each of the following angles in radians, leaving your answer in terms of π.
(a) 120°
(b) 18°
(c) 300°
(d) 22.5°

14.2 State in degrees, using decimal notation, the following angles measured in radians.
(a) 3π
(b) $\dfrac{7\pi}{18}$
(c) $\dfrac{\pi}{12}$
(d) $\dfrac{5}{3}\pi$

14.3 Using your calculator, express the following angles in radians, correct to 4 decimal places.
(a) 61° 38'
(b) 266° 13' 58"
(c) 00° 01'

14.3 Applications of radian measure

One of the main reasons for introducing a different measuring system is to simplify the mathematical calculations required in given practical situations.

In the preceding section we defined the term 'radian' and learned how to convert from radians to degrees and vice versa, when measuring angles.

In this section we now present two important applications of radian measure.

14.3.1 The length of a circular arc

> The length, l, of an arc AB, measured in radians, that subtends an angle, θ, at the centre of a circle of radius, r, is given by
> $$l = r\theta.$$

Diagrammatically, we have

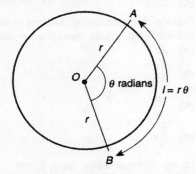

From the above formula for arc length, it follows that the measure of an angle θ, in radians, is given by

$$\theta \text{ radians} = \frac{\text{arc length}}{\text{length of radius}}$$

Hence, $\theta = \dfrac{l}{r}$ radians.

Note:
The formula for arc length, $l = r\theta$, is correct only when θ is measured in radians.

EXAMPLE 14.5
Find the length of the arc of a circle of radius 9 cm, subtending an angle at the centre of

(a) $\left(\dfrac{\pi}{3}\right)^c$

(b) 0.86^c

(c) $300°$

Solution

(a) Since $l = r\theta$, given $r = 9$ and $\theta = \left(\dfrac{\pi}{3}\right)^c$, we have

$$\begin{aligned} l &= 9 \times \frac{\pi}{3} \\ &= 3\pi \\ &\approx 9.425 \end{aligned}$$

Hence, arc length ≈ 9.425 cm (correct to 3 decimal places).

(b) Given $r = 9$ and $\theta = 0.86^c$, we have
$$\begin{aligned} l &= r\theta \\ &= 9 \times 0.86 \\ &= 7.74 \end{aligned}$$

Hence, arc length ≈ 7.74 cm.

(c) Given $r = 9$ and $\theta = 300°$, we must first convert $300°$ to radians before applying the formula to determine arc length.

Since $1° = \left(\dfrac{\pi}{180}\right)^c$,

$$\begin{aligned} 300° &= \left(\frac{\pi}{180}\right)^c \times 300 \\ &= \left(\frac{5\pi}{3}\right)^c \end{aligned}$$

Thus, $\begin{aligned}[t] l &= r\theta \\ &= 9 \times \frac{5\pi}{3} \\ &= 15\pi \\ &\approx 47.124 \text{ cm} \end{aligned}$

Hence, arc length ≈ 47.124 cm (correct to 3 decimal places).

EXAMPLE 14.6

For a circle with radius, r, central angle, θ, and arc length, l, find the unknown value in each of the following:

(a) $r = 7.8$ cm, $l = 12.5$ cm, find θ (in radians).

(b) $\theta = 120°$, $l = 8.4$ cm, find r (in cms).

(c) $r = 4.6$ cm, $l = 2.9$ cm, find θ (in degrees, minutes and seconds).

Solution

(a) Given $r = 7.8$ and $l = 12.5$, we have

$$\theta = \left(\frac{l}{r}\right)^c$$

$$= \left(\frac{12.5}{7.8}\right)^c$$

$$\approx 1.603^c$$

Hence, the central angle $\theta = 1.603$ radians.

(b) Given $\theta = 120°$ and $l = 8.4$, we must first convert $\theta = 120°$ to radian measure before calculating the value of r.

Since $1° = \left(\frac{\pi}{180}\right)^c$,

$$120° = \left(\frac{\pi}{180}\right)^c \times 120$$

$$= \left(\frac{2\pi}{3}\right)^c$$

Now, from the relationship $l = r\theta$ we see that

$$r = \frac{l}{\theta}.$$

Thus, with $l = 8.4$ and $\theta = \left(\frac{2\pi}{3}\right)^c$, we have

$$r = \frac{l}{\theta}$$

$$= \frac{8.4}{(2\pi)/3}$$

$$= \frac{8.4 \times 3}{2\pi}$$

$$\approx 4.011 \text{ cm (using the calculator)}$$

Hence, the radius ≈ 4.011 (correct to 3 decimal places).

(c) Given $r = 4.6$ and $l = 2.9$, we have

$$\theta = \left(\frac{l}{r}\right)^c$$

$$= \left(\frac{2.9}{4.6}\right)^c$$

$$= 0.63 \text{ radians} \quad \text{(correct to 2 decimal places)}.$$

To convert this to degrees, we have

$$1° = \left(\frac{180}{\pi}\right)°$$

Hence, $0.63^c = \left(\frac{180}{\pi}\right)° \times 0.63$

$$= 36.09634109$$

$$= 36° \, 5' \, 46.83''.$$

Using the calculator keys,

180 [x] .63 [÷] [exp]
[=] [shift] [° ' "] [36°5°46.83]

Hence, the central angle $\theta = 36° \, 5' \, 46.83''$.

14.6 For a circle with radius r, central angle θ, and arc length l, find the unknown value in each of the following.

(a) $r = 10.5$ cm, $\theta = \left(\frac{4\pi}{3}\right)^c$, $l = ?$

(b) $r = 6.5$ cm, $\theta = 1.8$ radians, $l = ?$

(c) $\theta = 84°$, $r = 9.2$ cm, $l = ?$

(d) $r = 18.4$ cm, $l = 25.2$ cm, $\theta = ?$ radians.

(e) $l = 4.7$ cm, $r = 7.9$ cm, $\theta = ?$ (° ' ")

(f) $l = 8.3$ cm, $\theta = \left(\frac{3}{4}\pi\right)^c$, $r = ?$

(g) $l = 1.8$ cm, $\theta = 2.4$ radians, $r = ?$

(h) $l = 3.7$ cm, $\theta = 64° \, 32' \, 12''$, $r = ?$

14.3.2 The area of a sector

In the diagram, the region that is shaded is called a **sector** of the circle of radius, r, with a central angle of θ radians.

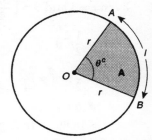

To find the area, A, of a sector, note that the area is, in fact, proportional to the angle at the centre. So, as the angle at the centre increases, so does the area of the sector.

In fact, when $\theta = 2\pi$ radians (a complete revolution), A (sector area) becomes the area of the whole circle (πr^2).

Hence, $\dfrac{\text{area of sector } AOB}{\text{area of circle}} = \dfrac{\theta \text{ radians}}{2\pi \text{ radians}}$.

Symbolically, with A = area of sector, we have

$$\frac{A}{\pi r^2} = \frac{\theta}{2\pi}$$

Hence, $A = \dfrac{\theta \pi r^2}{2\pi}$

$\qquad = \dfrac{1}{2}\theta r^2$

Therefore,

> Given a circle of radius, r, the area, A, of a sector containing an angle of θ radians at the centre, is given by
>
> $$A = \dfrac{1}{2}r^2\theta \quad \text{square units.}$$

EXAMPLE 14.7
Find the area of the sector of a circle, given
(a) radius, $r = 3$ cm and central angle, $\theta = 4$ radians.
(b) radius, $r = 4$ cm and central angle $\theta = \dfrac{3\pi}{4}$ radians.
(c) radius, $r = 6.5$ cm and central angle, $\theta = 150°$.

Solution
(a) Given $r = 3$ and $\theta = 4^c$, we have

$A = \dfrac{1}{2}r^2\theta$

$\quad = \dfrac{1}{2} \times 3^2 \times 4$

$\quad = \dfrac{1}{2} \times 9 \times 4$

$\quad = 18$

Hence, the area of the sector is 18 square cm (or cm^2).

(b) Given $r = 4$ and $\theta = \left(\dfrac{3\pi}{4}\right)^c$, we have

$A = \dfrac{1}{2}r^2\theta$

$\quad = \dfrac{1}{2} \times 4^2 \times \dfrac{3\pi}{4}$

$\quad = 6\pi$

$\quad \approx 18.85 \quad$ (using the $\boxed{\text{exp}}$ button on your calculator).

Hence, the area of the sector is 18.85 square cm (correct to 2 decimal places).

(c) Given $r = 6.5$ and $\theta = 150°$, we must convert $150°$ to radian measure before evaluating the required area.

Since $1° = \left(\dfrac{\pi}{180}\right)^c$,

$150° = \left(\dfrac{\pi}{180}\right)^c \times 150$

$\qquad = \left(\dfrac{5\pi}{6}\right)^c$

We can now find the area of the sector, using

$A = \dfrac{1}{2}r^2\theta$

$\quad = \dfrac{1}{2} \times 6.5^2 \times \dfrac{5\pi}{6}$

$\quad = \dfrac{211.25\pi}{12}$

$\quad \approx 55.305 \quad$ (using the $\boxed{\text{exp}}$ button)

Hence, the area of the sector is 55.305 cm^2 (correct to 3 decimal places).

EXAMPLE 14.8
In the diagram, find the area of the *shaded* region, giving your answer correct to 1 decimal place.

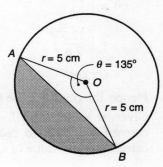

Solution
Looking carefully at the diagram we see that:
Area of shaded region = area of sector AOB
$\qquad\qquad\qquad\qquad$ − area of $\triangle AOB$

$\qquad = \dfrac{1}{2}r^2\theta - \dfrac{1}{2}ab\sin\theta$

$\qquad = \dfrac{1}{2}r^2\theta - \dfrac{1}{2}r^2\sin\theta$ (since $a = b = r$)

$\qquad = \dfrac{1}{2}r^2(\theta - \sin\theta)$

Since $\theta = 135°$ *must* be in **radians**, we have

$1° = \left(\dfrac{\pi}{180}\right)^c$.

Thus, $135° = \left(\dfrac{\pi}{180}\right)^c \times 135$

$\qquad\qquad = \left(\dfrac{3}{4}\pi\right)^c$.

So, with $r = 5$ and $\theta = \dfrac{3}{4}\pi$ radians, we have:

Shaded area $= \dfrac{1}{2}r^2(\theta - \sin\theta)$

$\qquad = \dfrac{1}{2} \times 5^2 \times \left(\dfrac{3}{4}\pi - \sin\dfrac{3}{4}\pi\right)$

$\qquad = \dfrac{1}{2} \times 25 \times (2.35619449 - 0.70710678)$

$\qquad = \dfrac{1}{2} \times 25 \times 1.64908771$

$\qquad = 20.61359638$

Hence, the shaded area ≈ 20.6 square cm (correct to 1 decimal place).

Note:

In this example be sure to set your calculator to **radian** mode, so that the correct value for $\sin\theta$ may be found.

14.7 Find the area of the sector of a circle, given

    (a) $r = 103$ mm, $\theta = \dfrac{5\pi}{6}$.

    (b) $r = 8.2$ cm, $\theta = 3.4$ radians.

    (c) $r = 2.3$ cm, $\theta = 56° 12'$.

14.8 A pendulum of length 1.5 metres swings through an angle of $\dfrac{\pi}{6}$ radians. What is

    (a) the length of the arc traced out by the end of the pendulum?

    (b) the area of the sector formed by the swing of the pendulum?

14.9 In the given diagram, find the area of the shaded region, given

    (i) $\theta = 105°$

    (ii) $r = 11.8$ cm

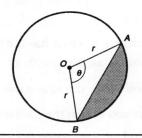

14.4 Graphs of trigonometric functions

The three most important trigonometric functions are

(i) $y = f(x) = \sin x$

(ii) $y = f(x) = \cos x$

(iii) $y = f(x) = \tan x$

where $\sin x$, $\cos x$ and $\tan x$ are defined as the sin, cos or tan of an angle of size x radians.
Hence,

$$\sin x = \sin x^c$$

$$\cos x = \cos x^c$$

$$\tan x = \tan x^c$$

Note:

For every value of x, there is one and only one value of $\sin x$, $\cos x$ or $\tan x$.

As you will see as you read on, the graphs of these trigonometric functions do have a common feature – the same pattern is repeated over and over again.

In fact, any function whose graph repeats exactly the same pattern is called a **periodic function.**

> The **period** of a periodic function is the smallest number, k, such that
> $$f(x + k) = f(x), \text{ for all values of } x.$$

> The **amplitude** of a periodic function is half the difference between its maximum and minimum values.

14.4.1 The graph of the sine function

The graph of $y = \sin x$, for $-2\pi \le x \le 2\pi$.

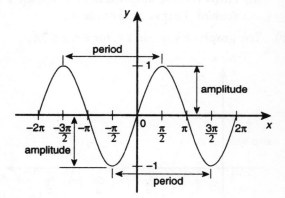

Note:

(i) The graph of $y = \sin x$ is periodic and resembles a wave motion, with the distance between two successive troughs or crests being called the **period**. Here the period is 2π, implying that the graph repeats itself at intervals of width 2π.

Hence, $\sin x = \sin(x + 2\pi)$.

(ii) The **amplitude** of $y = \sin x$ is 1, being, by definition, $\dfrac{1}{2}$ (maximum – minimum).

EXAMPLE 14.9
Draw the graph of

(a) $y = 2\sin x$, for $0 \le x \le 2\pi$

(b) $y = \sin 2x$, for $0 \le x \le 2\pi$,
and hence, write down

    (i) the period

    (ii) the amplitude
of each function.

Solution

(a) The graph of $y = 2\sin x$, for $0 \le x \le 2\pi$.

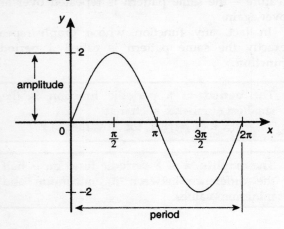

(i) From the graph we see that the **period** of this function is again 2π (as for $y = \sin x$).

(ii) However, the **amplitude** of $y = 2\sin x$ is *doubled*. That is, amplitude = 2.

(b) The graph of $y = \sin 2x$, for $0 \le x \le 2\pi$.

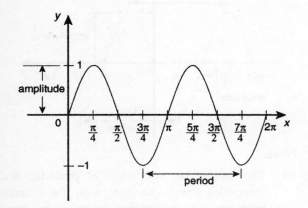

(i) From the graph we see that the **period** of this function is π (half that of $y = \sin x$), with two full waves between 0 and 2π.

(ii) However, the **amplitude** of $y = \sin 2x$ remains the same as for $y = \sin x$. That is, amplitude = 1.

YOUR TURN

14.10 Draw sketch graphs of each of the following periodic functions, and hence state (i) the period, and (ii) the amplitude.

(a) $y = 3\sin x$

(b) $y = \sin 4x$

(c) $y = 2\sin\left(\dfrac{\pi}{2} + x\right)$

Note:
As you have probably already discovered:

> 1. Periodic functions of the form $y = a\sin x$ have *period* 2π and *amplitude* = a.
> 2. Periodic functions of the form $y = \sin bx$ have *period* = $\dfrac{2\pi}{b}$ and *amplitude* = 1.

14.4.2 The graph of the cosine function

The graph of $y = \cos x$, for $-2\pi \le x \le 2\pi$.

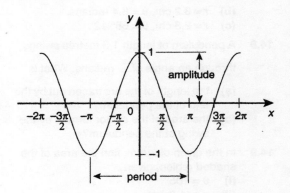

Note:

(i) The graph of $y = \cos x$ has the same periodic wave motion as the graph of $y = \sin x$. In fact, if you move the sine curve a distance $\dfrac{\pi}{2}$ to the left you obtain the cosine curve.

(ii) The **period** of $y = \cos x$ is 2π and since the graph repeats itself at intervals of width 2π, we have that

$$\cos x = \cos (x + 2\pi).$$

(iii) The **amplitude** of $y = \cos x$ is 1, since

$$\frac{1}{2}(\text{max.} - \text{min.}) = 1.$$

EXAMPLE 14.10

Draw the graph of the following functions and state the period and amplitude of each.

(a) $y = \cos\dfrac{x}{2}$, for $0 \le x \le 4\pi$.

(b) $y = 5\cos x$, for $0 \le x \le 2\pi$.

Solution

(a) $y = \cos\dfrac{x}{2}$, for $0 \le x \le 4\pi$.

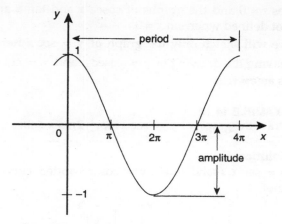

From the graph, we have

(i) period $= 4\pi$

(ii) amplitude $= 1$.

(b) The graph of $y = 5\cos x$, for $0 \le x \le 2\pi$.

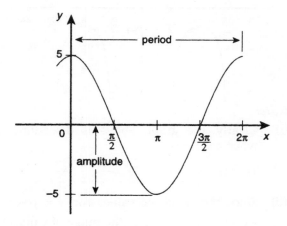

From the graph, we have

(i) period $= 2\pi$

(ii) amplitude $= 5$.

Note:

As we found with sine functions in 14.4.1, the following results are true for cosine functions.

---

1. Periodic functions of the form
   $y = a\cos x$ have *period* $= 2\pi$ and
   amplitude $= a$.

2. Periodic functions of the form
   $y = \cos bx$ have *period* $= \dfrac{2\pi}{b}$ and
   amplitude $= 1$.

You should verify these results when attempting the following exercises.

YOUR TURN

14.11 Draw sketch graphs of each of the following periodic functions and hence state   **(i)** the period, and
            (ii) the amplitude.

  (a) $y = \cos\dfrac{x}{3}$

  b) $y = 2\cos 4x$

  c) $y = \cos\left(x - \dfrac{\pi}{2}\right)$

14.4.3 The graph of the tangent function

The graph of $y = \tan x$, for $-2\pi \le x \le 2\pi$.

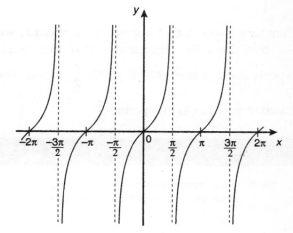

Note:

(i) Since $\tan x = \dfrac{\sin x}{\cos x}$, $\tan x$ is *not* defined whenever $\cos x = 0$ (since division by zero is not possible).

So, the graph of $y = \tan x$ is discontinuous at all values of x where $\cos x = 0$ – that is, where $x = \dfrac{-3\pi}{2}, -\dfrac{\pi}{2}, \dfrac{\pi}{2}, \dfrac{3\pi}{2}$.

(ii) The **period** of $y = \tan x$ is π and since the graph repeats itself at intervals of width π, we have that

$$\tan x = \tan(x + \pi).$$

(iii) Since the period of $y = \tan x$ is π we can deduce, from the sine and cosine functions, that the period of $y = \tan bx$ is $\dfrac{\pi}{b}$.

EXAMPLE 14.11

Sketch the graph of $y = \tan 2x$, for $0 \le x \le 2\pi$, and hence show that the period is $\dfrac{\pi}{2}$.

Solution

The graph of $y = \tan 2x$, for $0 \le x \le 2\pi$.

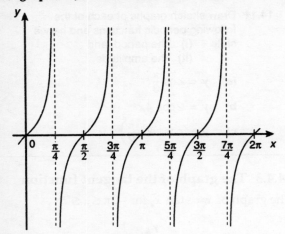

Comparing this with the graph of $y = \tan x$, we see that the only difference is that this graph repeats itself at intervals of width $\dfrac{\pi}{2}$. Hence, the function $y = \tan 2x$ has period $\dfrac{\pi}{2}$.

Y O U R   T U R N

14.12 Draw sketch graphs of the following functions and show that the period of

$y = \tan bx$ is $\dfrac{\pi}{b}$.

a) $y = \tan \dfrac{x}{2}$

b) $y = 3\tan 2x$

c) $y = 2\tan\left(x - \dfrac{\pi}{2}\right)$

14.4.4 The graphs of the reciprocal functions

By taking the reciprocal of the sine, cosine and tangent functions, we obtain

(i) $\operatorname{cosec} x = \dfrac{1}{\sin x}$

(ii) $\sec x = \dfrac{1}{\cos x}$

(iii) $\cot x = \dfrac{1}{\tan x} = \dfrac{1}{\dfrac{\sin x}{\cos x}} = \dfrac{\cos x}{\sin x}$

It follows that the graphs of $y = \sec x$, $y = \operatorname{cosec} x$ and $y = \cot x$ can be sketched quite simply using the reciprocal values of $\cos x$, $\sin x$ and $\tan x$, respectively.

Clearly, the graph of $\sec x$ is not defined when

$\cos x = 0$ and the graphs of $\operatorname{cosec} x$ and $\tan x$ are not defined when $\sin x = 0$.

We will sketch only the graph of $y = \sec x$ here, leaving the drawing of $y = \operatorname{cosec} x$ and $y = \cot x$ as an exercise.

EXAMPLE 14.12
Draw the graph of $y = \sec x$, for $-2\pi \le x \le 2\pi$.

Solution
$y = \sec x$ (solid line), $y = \cos x$ (dotted curved line).

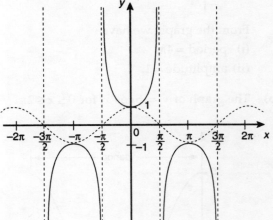

Note:

(i)   Clearly, the graph above is discontinuous when $\cos x = 0$.

That is, when $x = -\dfrac{3\pi}{2}, -\dfrac{\pi}{2}, \dfrac{\pi}{2}, \dfrac{3\pi}{2}$.

(ii) Since the range of values for $y = \cos x$ is $-1 \le y \le 1$, we see that the range of values for $y = \sec x$ is $y \le -1$ and $y \ge 1$.

(iii) As with the function $y = \cos x$, we see that the function $y = \sec x$ has period 2π.

Y O U R   T U R N

14.13 Sketch the graph of
(a) $y = \operatorname{cosec} x$
(b) $y = \cot x$
and state the period of each.

14.5 Calculus and trigonometric functions

In chapter 12 we discussed, in some detail, both differential and integral calculus. We presented rules for determining both the derivative and the integral of some simple as well as some more complex functions. We also studied several practical problems whose solutions required the use of the techniques of either differentiation or integration.

However, nowhere in chapter 12 was there any mention of trigonometric functions. Therefore, in the final section of this chapter, we will extend our knowledge of both differential and integral calculus by discussing how to differentiate and integrate trigonometric functions.

This is a very important topic as there are many practical situations involving trigonometric functions that require the determination of appropriate derivatives or integrals.

14.5.1 Derivatives of trigonometric functions

The rules for differentiating trigonometric functions, given in the box below, are stated *without proofs*. However, for those of you who are interested, proofs are readily available in most mathematics text books that cover both calculus and trigonometry.

Rules

If u is any differentiable function of x, then

1. $\dfrac{d}{dx}\sin u = \cos u \cdot \dfrac{du}{dx}$

2. $\dfrac{d}{dx}\cos u = -\sin u \cdot \dfrac{du}{dx}$

3. $\dfrac{d}{dx}\tan u = \sec^2 u \cdot \dfrac{du}{dx}$

EXAMPLE 14.13

Find $\dfrac{dy}{dx}$ for each of the following trigonometric functions:

(a) $y = \sin 4x$

(b) $y = \cos (3x - 5)$

(c) $y = \tan (2 - 3x)$

(d) $y = \cos^2 3x$

(e) $y = 5\tan (\sin 3x)$

Solution

(a) Find $\dfrac{d}{dx}\sin 4x$.

Let $u = 4x$.

Then, $\dfrac{d}{dx}\sin u = \cos u \cdot \dfrac{du}{dx}$

$= \cos 4x \cdot \dfrac{d}{dx}4x$

$= \cos 4x \cdot 4$

$= 4\cos 4x$

Hence, $\dfrac{d}{dx}\sin 4x = 4\cos 4x$.

(b) Find $\dfrac{d}{dx}\cos (3x - 5)$.

Let $u = (3x - 5)$.

Then, $\dfrac{d}{1x}\cos u = -\sin u \cdot \dfrac{du}{dx}$

$= -\sin (3x - 5) \cdot \dfrac{d}{dx}(3x - 5)$

$= -\sin (3x - 5) \cdot 3$

$= -3\sin (3x - 5)$

Hence, $\dfrac{d}{dx}\cos (3x - 5) = -3\sin (3x - 5)$.

(c) Find $\dfrac{d}{dx}\tan (2 - 3x)$.

Let $u = (2 - 3x)$.

Then, $\dfrac{d}{1x}\tan u = \sec^2 u \cdot \dfrac{du}{dx}$

$= \sec^2 (2 - 3x) \cdot \dfrac{d}{dx}(2 - 3x)$

$= \sec^2 (2 - 3x) \cdot -3$

$= -3\sec^2 (2 - 3x)$

Hence, $\dfrac{d}{dx}\tan (2 - 3x) = -3\sec^2 (2 - 3x)$.

(d) Find $\dfrac{d}{dx}\cos^2 3x = \dfrac{d}{dx}(\cos 3x)^2$

$$\text{(since } \cos^2 x = (\cos x)^2).$$

Let $u = 3x$.

Then, $\dfrac{d}{dx}(\cos u)^2 = 2(\cos u) \cdot \dfrac{d}{dx}\cos u$

$= 2\cos u \cdot -\sin u \cdot \dfrac{du}{dx}$

$= 2\cos u \cdot -\sin u \cdot \dfrac{d}{dx}3x$

$= 2\cos 3x \cdot -\sin 3x \cdot 3$

$= -6\cos 3x \cdot \sin 3x$

Hence, $\dfrac{d}{dx}\cos^2 3x = -6\cos 3x \cdot \sin 3x$.

(e) Find $\dfrac{d}{du} 5\tan(\sin 3x)$.

Let $u = \sin 3x$. Then,

$$\dfrac{d}{du} 5\tan u = 5\sec^2 u \cdot \dfrac{du}{dx}$$

$$= 5\sec^2 u \cdot \dfrac{d}{dx}(\sin 3x)$$

$$= 5\sec^2(\sin 3x) \cdot \cos 3x \cdot \dfrac{d}{dx} 3x$$

$$= 5\sec^2(\sin 3x) \cdot \cos 3x \cdot 3$$

$$= 15\sec^2 x \cdot \sin 3x \cdot \cos 3x$$

Hence,

$$\dfrac{d}{du} 5\tan(\sin 3x) = 15\sec^2 x \cdot \sin 3x \cdot \cos 3x.$$

YOUR TURN

14.14 Differentiate each of the following trigonometric functions with respect to x.

(a) $y = \cos(3x + 1)$

(b) $y = 2\sin(3x - 4)$

(c) $y = 4\tan 2x$

(d) $y = \cos^2 2x$

(e) $y = 3\cos(\sin 4x)$

14.15 The range of a shell fired from a gun having an elevation of x radians is given by $R = \dfrac{V^2}{g}\sin 2x$, where V and g are constants. For what angle of elevation will the range be a maximum?

(**Hint:** Find the value of x for which

$$\dfrac{dR}{dx} = 0 \quad \text{and} \quad \dfrac{d^2R}{dx^2} < 0 \quad \text{using the graph}$$

of $\sin 2x$.)

14.5.2 Integrals of trigonometric functions

As with the differentiation rules, the relevant rules for integrating trigonometric functions, given in the box below, are stated *without proofs*.

Once again, for those of you who are interested, these proofs are readily available in most mathematics text books that cover both calculus and trigonometry.

Rules for obtaining *indefinite integrals*

1. $\displaystyle\int \sin x \, dx = -\cos x + C$, and

$$\int \sin(ax+b)\,dx = -\dfrac{1}{a}\cos(ax+b) + C$$

2. $\displaystyle\int \cos x \, dx = \sin x + C$, and

$$\int \cos(ax+b)\,dx = \dfrac{1}{a}\sin(ax+b) + C$$

3. $\displaystyle\int \sec^2 x \, dx = \tan x + C$, and

$$\int \sec^2(ax+b)\,dx = \dfrac{1}{a}\tan(ax+b) + C$$

Note: As we learned in Chapter 12, C in all the indefinite integrals of trigonometric functions is called the *constant of integration*.

EXAMPLE 14.14
Find the indefinite integral of each of the following:

(a) $\displaystyle\int \cos 3x \, dx$

(b) $\displaystyle\int 3\sin\left(\dfrac{x}{2} + 5\right) dx$

(c) $\displaystyle\int (x - 3\sec^2 2x)\, dx$

Solution

(a) Using rule 2 with $a = 3$ and $b = 0$, we have

$$\int \cos 3x \, dx = \dfrac{1}{3}\sin x + C$$

(b) Using rule 1 with $a = \dfrac{1}{2}$ and $b = 5$, we have

$$\int 3\sin\left(\dfrac{x}{2}+5\right)dx = 3\int \sin\left(\dfrac{x}{2}+5\right)dx$$

$$= 3\left[-\dfrac{1}{\frac{1}{2}}\cos\left(\dfrac{x}{2}+5\right)+C\right]$$

$$= -6\cos\left(\dfrac{x}{2}+5\right)+C$$

(c) $\displaystyle\int (x - 3\sec^2 2x)\,dx = \int x\,dx - 3\int \sec^2 2x\,dx$

Now, we know that $\displaystyle\int x\,dx = \dfrac{x^2}{2} + C$ (from Ch 12), and using rule 3 above, with $a = 2$, we have

$$3\int \sec^2 2x\,dx = 3\left[\dfrac{1}{2}\tan 2x\right] + C$$

$$= \dfrac{3}{2}\tan 2x + C$$

Hence, $\int (x - 3\sec^2 2x)\ dx = \dfrac{x^2}{2} - \dfrac{3}{2}\tan 2x + C$

$$= \dfrac{1}{2}(x^2 - 3\tan 2x) + C$$

EXAMPLE 14.15
Evaluate each of the following definite integrals.

(a) $\displaystyle\int_0^{\pi} \cos\dfrac{\pi}{3}\ dx$

(b) $\displaystyle\int_0^{\frac{\pi}{2}} (1 + \sin 2x)\ dx$

Solution

a) Using rule 2 with $a = \dfrac{1}{3}$, we have

$$\int_0^{\pi} \cos\dfrac{x}{3}\ dx = \left[\dfrac{1}{(1/3)}\sin\dfrac{x}{3}\right]_0^{\pi}$$

$$= 3\sin\dfrac{\pi}{3} - 3\sin 0$$

$$\approx 2.598 \quad (\text{since } \sin 0 = 0)$$

b) $\displaystyle\int_0^{\frac{\pi}{2}} (1 + \sin 2x)\ dx = \int_0^{\frac{\pi}{2}} 1\ dx + \int_0^{\frac{\pi}{2}} \sin 2x\ dx$

Now, $\displaystyle\int_0^{\frac{\pi}{2}} 1\ dx = [x]_0^{\frac{\pi}{2}}$ (from Ch 12),

and using rule 1 with $a = 2$, we have

$$\int_0^{\frac{\pi}{2}} \sin 2x\ dx = \left[-\dfrac{1}{2}\cos 2x\right]_0^{\frac{\pi}{2}}$$

Therefore,

$$\int_0^{\frac{\pi}{2}} (1 + \sin 2x)\ dx = [x]_0^{\frac{\pi}{2}} + \left[-\dfrac{1}{2}\cos 2x\right]_0^{\frac{\pi}{2}}$$

$$= \left[x - \dfrac{1}{2}\cos 2x\right]_0^{\frac{\pi}{2}}$$

$$= \left(\dfrac{\pi}{2} - \dfrac{1}{2}\cos\pi\right) - \left(0 - \dfrac{1}{2}\cos 0\right)$$

$$= \left(\left[\dfrac{\pi}{2} - \dfrac{1}{2}(-1)\right] - \left(0 - \dfrac{1}{2}(1)\right)\right)$$

$$= \dfrac{\pi}{2} + \dfrac{1}{2} + \dfrac{1}{2}$$

$$= \dfrac{\pi}{2} + 1$$

Hence, $\displaystyle\int_0^{\frac{\pi}{2}} (1 + \sin 2x)\ dx = \dfrac{\pi}{2} + 1$.

YOUR TURN

14.16 Find the indefinite integrals of each of the following:

(a) $\displaystyle\int \sec^2\dfrac{x}{2}\ dx$

(b) $\displaystyle\int \cos\left(\dfrac{\pi}{2} - x\right)\ dx$

(c) $\displaystyle\int (x - \sin(2x - 3))\ dx$

14.17 Evaluate the following definite integrals:

(a) $\displaystyle\int_0^{\pi} (1 + \cos x)\ dx$

(b) $\displaystyle\int_0^{\frac{\pi}{4}} (3\cos x + 4\sin x)\ dx$

(c) $\displaystyle\int_{-\frac{\pi}{4}}^{\frac{\pi}{4}} 2(x + \sec^2 x)\ dx$

14.18 A dam holds the water supply for a certain town. Water flows into it from the catchment area and flows out to supply the needs of the town. Suppose that the water flows in at the rate of $f(t)$ megalitres per day and flows out at the rate of $g(t)$ megalitres per day.

If $\ f(t) = 20 - \dfrac{1}{2}t$, and

$g(t) = 10 + 2\sin 2\pi t$,

find the total change in the volume of water in the dam during the first 30 days.

That is, find $\displaystyle\int_0^{30} [f(t) - g(t)]\ dt$.

Summary
The main purpose of this chapter was to have a second look at **trigonometry**, but this time concentrating on a more detailed study of trigonometric functions.

Since many problems involving trigonometric functions are more easily solved using **radians** rather than **degrees**, the *second* section defined radian measure and demonstrated the associated conversions.

Conversions

1. 1 degree $= 1° = \dfrac{\pi}{180}$ radians $= \left(\dfrac{\pi}{180}\right)^c$

≈ 0.01745 radians (using the $\boxed{\text{exp}}$ button).

So, $x° = \left(\dfrac{\pi}{180}\right)^c \times x \approx 0.01745x$ radians.

2. 1 radian $= 1^c = \left(\dfrac{180}{\pi}\right)$ degrees $= \left(\dfrac{180}{\pi}\right)°$

$\approx 57.2958°$ (using the $\boxed{\text{exp}}$ button).

So, $x^c = \left(\dfrac{180}{\pi}\right)° \times x \approx 57.2958x°$.

There will be times when it is more efficient to use your calculator to perform the required calculations. The most useful keys will be $\boxed{\circ\,\prime\,\prime\prime}$, $\boxed{\text{shift}}\,\boxed{\circ\,\prime\,\prime\prime}$ and $\boxed{\text{exp}}$.

(see *Section 14.2.2* for a description of each).

Note: To evaluate any trigonometric function where the angle is given in radians, you must first put your calculator into *radian mode* by pressing the buttons $\boxed{\text{mode}}\ \boxed{5}$ (on our calculator).

Since radian measure is designed to simplify calculations in certain situations, *section 3* gives some practical applications.

1. Arc length

The length, l, of an arc AB measured in radians, that subtends an angle, θ, at the centre of a circle of radius, r, is given by

$$l = r\theta.$$

From this we may conclude that

(i) r (the radius) $= \dfrac{l \text{ (arc length)}}{\theta \text{ (central angle in radians)}}$

and

(ii) θ (radians) $= \dfrac{l \text{ (arc length)}}{r \text{ (radius of circle)}}$

Note : The above formulae are only valid when θ is measured in radians.

2. Area of a sector

Given a circle of radius, r, then the area, A, of a sector containing an angle of θ radians at the centre, is given by

$$A = \frac{1}{2}r^2\theta \quad \text{square units.}$$

Note: From the above formula it is also possible, by algebraic re-arrangement, to determine the value of

(i) r, given A and θ, and

(ii) θ, given A and r.

The *fourth* section of this chapter discussed the graphical representation of trigonometric functions. We noted firstly that all are **periodic** functions, where

1. the **period** of a function is the smallest number k, such that $f(x + k) = f(x)$.

2. the **amplitude** of a function is half the difference between the maximum and minimum values.

That is, amplitude $= \dfrac{1}{2}$ (max. $-$ min.).

To sum up

Trig. function	Period	Amplitude
(i) $y = \sin x$	2π	1
(ii) $y = a\sin bx$	$\dfrac{2\pi}{b}$	a
(iii) $y = \cos x$	2π	1
(iv) $y = a\cos bx$	$\dfrac{2\pi}{b}$	a
(v) $y = \tan x$	π	none – since the function is discontinuous and has no max. or min.
(vi) $y = a\tan bx$	$\dfrac{\pi}{b}$	none

Except for $y = \sec x$, the graphs of the reciprocal functions (namely $y = \sec x$, $y = \operatorname{cosec} x$ and $y = \cot x$) were left as an exercise. You should have found that both $y = \sec x$ and $y = \operatorname{cosec} x$ have period 2π, while $y = \cot x$ has period π.

The *final* section of this chapter discussed the very important topic of how to apply the concepts and rules of calculus when trigonometric functions are involved.

1. Rules for differentiating trigonometric functions

If u is any differentiable function of x, then

a. $\dfrac{d}{dx}\sin u = \cos u \cdot \dfrac{du}{dx}$

b. $\dfrac{d}{dx}\cos u = -\sin u \cdot \dfrac{du}{dx}$

c. $\dfrac{d}{dx}\tan u = \sec^2 u \cdot \dfrac{du}{dx}$

2. Rules for integrating trigonometric functions

a. $\displaystyle\int \sin x \, dx = -\cos x + C$

Hence,

$$\int \sin(ax+b)\,dx = -\frac{1}{a}\cos(ax+b) + C$$

b. $\int \cos x\,dx = \sin x + C$

Hence,

$$\int \cos(ax+b)\,dx = \frac{1}{a}\sin(ax+b) + C$$

c. $\int \sec^2 x\,dx = \tan x + C$

Hence,

$$\int \sec^2(ax+b)\,dx = \frac{1}{a}\tan(ax+b) + C,$$

where C is the constant of integration associated with indefinite integrals.

3. Rules for differentiating and integrating logarithmic and exponential functions.
    (*Note:* These were not dicussed in the body of the chapter, but are included here for the sake of completeness – see problems 14.6, 14.8, 14.10, 14.13, 14.15 - marked #.)

    a. Derivatives

    $$\frac{d}{dx}e^x = e^x$$

    $$\frac{d}{dx}e^{ax+b} = ae^{ax+b}$$

    $$\frac{d}{dx}\ln x = \frac{1}{x}$$

    $$\frac{d}{dx}\ln f(x) = \frac{f'(x)}{f(x)}$$

    b. Integrals

    $$\int e^x\,dx = e^x + C$$

    $$\int e^{ax+b}\,dx = \frac{1}{a}e^{ax+b} + C$$

    $$\int \frac{1}{x}\,dx = \ln x + C$$

    $$\int \frac{f'(x)}{f(x)}\,dx = \ln f(x) + C$$

14.1 Convert each of the following from degrees to radians:
  (a) $215°$ **(b)** $300°$
  (c) $10°$ **(d)** $36°$
  (e) $540°$

14.2 Convert each of the following from radians to degrees:
  (a) 4π **(b)** $\frac{7}{2}\pi$
  (c) $\frac{8}{3}\pi$ **(d)** 1.5
  (e) 8

14.3 Find the length of a circular arc subtended by an angle of $\left(\frac{\pi°}{6}\right)^c$, if the radius of the circle is 36 cm.

14.4 A circle has radius 100 cm. What angle is subtended at the centre of the circle by an arc 10 cm long?

14.5 Find the area of the sector of a circle, given

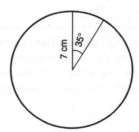

*14.6 Find dy/dx in each of the following:
  (a) $y = \sin 9x$ **(b)** $y = \operatorname{cosec} 7x$
  (c) $y = \cot x$ **(d)** $y = \sec x$
  (e)# $y = \ln(\cot 3x)$ **(f)#** $y = \cot(\ln 3x)$

14.7 If $f(x) = 3\tan x - 2\cot x$, find $f'\left(\frac{\pi}{4}\right)$

14.8# Show that $\frac{d}{dx}(\ln \sec x) = \frac{d}{dx}(-\ln \cos x)$
$$= \tan x$$

*14.9 If $y = \tan x - \cot x - 3x$, show that
$$\frac{dy}{dx} = (\tan x - \cot x)^2 + 1$$

14.10# Find $\frac{d}{dx}e^{\operatorname{cosec}^2 x}$, and hence evaluate
$$\int_{\frac{\pi}{4}}^{\frac{\pi}{2}} e^{\operatorname{cosec}^2 x} \cdot \operatorname{cosec}^2 x \cdot \cot x\,dx.$$

14.11 Find the following indefinite integrals:

(a) $\int \sin 7x \, dx$

(b) $\int \cos \dfrac{x}{6} \, dx$

(c) $\int \sec^2 5x \, dx$

14.12 Show that $\dfrac{3 - \sin^3 x}{\sin^2 x} = 3\operatorname{cosec}^2 x - \sin x$,

and hence evaluate $\displaystyle\int_{\frac{\pi}{4}}^{\frac{\pi}{3}} \dfrac{3 - \sin^3 x}{\sin^2 x} \, dx$.

14.13# Find $\dfrac{d}{dx} \ln (\tan x + 1)$, and hence find

$\displaystyle\int \dfrac{\sec^2 x}{\tan x + 1} \, dx$.

14.14 Find $\dfrac{d}{dx} \sec^4 x$, and hence find

$\displaystyle\int \sec^4 x \cdot \tan x \, dx$.

*14.15# Differentiate $e^{\cot 3x}$, and hence find the

value of $\displaystyle\int_{\frac{\pi}{15}}^{\frac{\pi}{12}} e^{\cot 3x} \cdot \operatorname{cosec}^2 3x \, dx$.

If you have any difficulties answering these questions, refer to the final section of the chapter **summary** where the rules for differentiating and integrating logarithmic and exponential functions are given.

Appendix A

Tables

Table I: Future amount of an annuity of $1 per period

$$s_{n|i} = \frac{(1+i)^n - 1}{i}$$

n \ i	$.005 = \frac{1}{2}\%$	$.01 = 1\%$	$.015 = 1\frac{1}{2}\%$	$.02 = 2\%$	$.025 = 2\frac{1}{2}\%$	$.03 = 3\%$	$.035 = 3\frac{1}{2}\%$
1	1.000000	1.000000	1.000000	1.000000	1.000000	1.000000	1.000000
2	2.005000	2.010000	2.015000	2.020000	2.025000	2.030000	2.035000
3	3.015025	3.030100	3.045225	3.060400	3.075625	3.090900	3.106225
4	4.030100	4.060401	4.090903	4.121608	4.152516	4.183627	4.214943
5	5.050250	5.101005	5.152267	5.204040	5.256328	5.309136	5.362466
6	6.075502	6.152015	6.229551	6.308121	6.387737	6.468410	6.550152
7	7.105879	7.213535	7.322994	7.434283	7.547430	7.662462	7.779408
8	8.141408	8.285671	8.432839	8.582969	8.736116	8.892336	9.051687
9	9.182115	9.368527	9.559332	9.754628	9.954519	10.159106	10.368496
10	10.228026	10.462213	10.702721	10.949721	11.203382	11.463879	11.731393
11	11.279166	11.566835	11.863262	12.168715	12.483466	12.807796	13.141992
12	12.335562	12.682503	13.041211	13.412090	13.795553	14.192029	14.601962
13	13.397239	13.809328	14.236829	14.680331	15.140442	15.617790	16.113030
14	14.464226	14.947421	15.450382	15.973938	16.518953	17.086324	17.676986
15	15.536547	16.096896	16.682138	17.293417	17.931926	18.598914	19.295681
16	16.614229	17.257864	17.932370	18.639285	19.380225	20.156881	20.971030
17	17.697300	18.430443	19.201355	20.012071	20.864730	21.761587	22.705016
18	18.785787	19.614748	20.489375	21.412312	22.386348	23.414435	24.499691
19	19.879716	20.810895	21.796716	22.840558	23.946007	25.116868	26.357180
20	20.979114	22.019004	23.123667	24.297369	25.544657	26.870374	28.279682
21	22.084010	23.239194	24.470522	25.783317	27.183274	28.676485	30.269471
22	23.194430	24.471586	25.837580	27.298983	28.862856	30.536780	32.328902
23	24.310402	25.716302	27.225143	28.844963	30.584427	32.452883	34.460414
24	25.431954	26.973465	28.633520	30.421862	32.349038	34.426470	36.666528
25	26.559114	28.243200	30.063023	32.030299	34.157763	36.459264	38.949857

Table I: Future amount of an annuity of $1 per period (cont.)

$$s_{n\rceil i} = \frac{(1+i)^n - 1}{i}$$

n	.005 = $\frac{1}{2}$%	.01 = 1%	.015 = $1\frac{1}{2}$%	.02 = 2%	.025 = $2\frac{1}{2}$%	.03 = 3%	.035 = $3\frac{1}{2}$%
26	27.691909	29.525632	31.513969	33.670905	36.011708	38.553042	41.313102
27	28.830369	30.820888	32.986678	35.344323	37.912000	40.709633	43.759060
28	29.974520	32.129097	34.481478	37.051210	39.859800	42.930922	46.290627
29	31.124393	33.450388	35.998700	38.792234	41.856295	45.218850	48.910799
30	32.280015	34.784892	37.538681	40.568079	43.902703	47.575415	51.622677
31	33.441415	36.132741	39.101761	43.379440	46.000270	50.002678	54.429471
32	34.608622	37.494068	40.688288	44.227029	48.150277	52.502758	57.334502
33	35.781665	38.869009	42.298612	46.111569	50.354034	55.077841	60.341210
34	36.960573	40.257699	43.933091	48.033801	52.612885	57.730176	63.453152
35	38.145376	41.660276	45.592087	49.994477	54.928207	60.462081	66.674013
36	39.336103	43.076878	47.275969	51.994366	57.301412	63.275943	70.007603
37	40.532783	44.507647	48.985108	54.034254	59.733947	66.174222	73.457869
38	41.735447	45.952724	50.719885	56.114939	62.227296	69.159448	77.028895
39	42.944124	47.412251	52.480683	58.237237	64.782978	72.234232	80.724906
40	44.158845	48.886373	54.267893	60.401982	67.402553	75.401259	84.550278
41	45.379639	50.375237	56.081912	62.610022	70.087616	78.663296	88.509537
42	46.606537	51.878990	57.923140	64.862222	72.839807	82.023195	92.607371
43	47.839570	53.397780	59.791987	67.159467	75.660802	85.483891	96.848629
44	49.078768	54.931757	61.688867	69.502656	78.552322	89.048408	101.23833
45	50.324162	56.481075	63.614200	71.892709	81.516130	92.719860	105.78167
46	51.575782	58.045886	65.568413	74.330563	84.554033	96.501456	110.48403
47	52.833661	59.626344	67.551940	76.817175	87.667884	100.39650	115.35097
48	54.097829	61.222608	69.565219	79.353518	90.859581	104.40839	120.38826
49	55.368318	62.834834	71.608697	81.940588	94.131071	108.54065	125.60185
50	56.645160	64.463182	73.682827	84.579400	97.484347	112.79687	130.99791

Table I: Future amount of an annuity of $1 per period (cont.)

$$s_{\overline{n}|i} = \frac{(1+i)^n - 1}{i}$$

n \ i	.04 = 4%	.045 = $4\frac{1}{2}$%	.05 = 5%	.055 = $5\frac{1}{2}$%	.06 = 6%	.065 = $6\frac{1}{2}$%	.07 = 7%
1	1.000000	1.000000	1.000000	1.000000	1.000000	1.000000	1.000000
2	2.040000	2.045000	2.050000	2.055000	2.060000	2.065000	2.070000
3	3.121600	3.137025	3.152500	3.168025	3.183600	3.199225	3.214900
4	4.246464	4.278191	4.310125	4.342266	4.374616	4.407175	4.439943
5	5.416323	5.470710	5.525631	5.581091	5.637093	5.693641	5.750739
6	6.632975	6.716892	6.801913	6.888015	6.975319	7.063728	7.153291
7	7.898294	8.019152	8.142008	8.266894	8.393838	8.522870	8.654021
8	9.214226	9.380014	9.549109	9.721573	9.897468	10.076856	10.259803
9	10.582795	10.802114	11.026564	11.256259	11.491316	11.731852	11.977989
10	12.006107	12.288209	12.577892	12.875354	13.180795	13.494423	13.816448
11	13.486351	13.841179	14.206787	14.593498	14.971643	15.371560	15.783599
12	15.025805	15.464032	15.917126	16.385591	16.869941	17.370711	17.888451
13	16.626837	17.159913	17.712983	18.286798	18.882138	19.499808	20.140643
14	18.291911	18.932109	19.598632	20.292572	21.015066	21.767295	22.550488
15	20.023587	20.784054	21.578563	22.408663	23.275970	24.182169	25.129022
16	21.824531	22.719337	23.657492	24.641140	25.672528	26.754010	27.888054
17	23.697512	24.741707	25.840366	26.996403	28.212880	29.493021	30.840217
18	25.645413	26.855084	28.132384	29.481205	30.905653	32.410067	33.999033
19	27.671229	29.063563	30.539004	32.102671	33.759992	35.516722	37.378965
20	29.778078	31.371423	33.065954	34.868318	36.785591	38.825309	40.995492
21	31.969201	33.783137	35.719252	37.786075	39.992727	42.348954	44.865177
22	34.247969	36.303378	38.505214	40.864310	43.392291	46.101636	49.005739
23	36.617888	38.937030	41.430475	44.111847	46.995828	50.098242	53.436141
24	39.082604	41.689197	44.501999	47.537998	50.815578	54.354628	58.176671
25	41.645908	44.565211	47.727098	51.152588	54.864512	58.887679	63.249038

Table I: Future amount of an annuity of $1 per period (cont.)

$$s_{n\rceil i} = \frac{(1+i)^n - 1}{i}$$

i n	.04 = 4%	.045 = $4\frac{1}{2}$%	.05 = 5%	.055 = $5\frac{1}{2}$%	.06 = 6%	.065 = $6\frac{1}{2}$%	.07 = 7%
26	44.311744	47.570645	51.113453	54.965980	59.156383	63.715378	68.676470
27	47.084214	50.711324	54.669126	58.989109	63.705766	68.856877	74.483823
28	49.967582	53.993334	58.402582	63.233510	68.528112	74.332574	80.697691
29	52.966286	57.423034	62.322711	67.711353	73.639799	80.164192	87.346529
30	56.084937	61.007070	66.438847	72.435478	79.058187	86.374864	94.460786
31	59.328335	64.752389	70.760789	77.419429	84.801678	92.989230	102.07304
32	62.701468	68.666246	75.298829	82.677498	90.889779	100.03353	110.21815
33	66.209527	72.756227	80.063770	88.224760	97.343165	107.53571	118.93343
34	69.857908	77.030257	85.066959	94.077122	104.18376	115.52553	128.25876
35	73.652224	81.496619	90.320307	100.25136	111.43478	124.03469	138.23688
36	77.598313	86.163967	95.836322	106.76519	119.12087	133.09695	148.91346
37	81.702245	91.041345	101.62814	113.63727	127.26812	142.74825	160.33740
38	85.970335	96.138206	107.70954	120.88732	135.90421	153.02688	172.56102
39	90.409149	101.46443	114.09502	128.53613	145.05846	163.97363	185.64029
40	95.025514	107.03032	120.79977	136.60561	154.76197	175.63192	199.63511
41	99.826535	112.84669	127.83976	145.11892	165.04768	188.04799	214.60957
42	104.81960	118.92479	135.23175	154.10046	175.95055	201.27111	230.63224
43	110.01238	125.27641	142.99334	163.57599	187.50758	215.35373	247.77650
44	115.41288	131.91384	151.14300	173.57267	199.75803	230.35172	266.12085
45	121.02939	138.84997	159.70015	184.11916	212.74352	246.32459	285.74931
46	126.87057	146.09822	168.68516	195.24572	226.50813	263.33568	306.75176
47	132.94539	153.67264	178.11942	206.98423	241.09861	281.45250	329.22439
48	139.26320	161.58790	188.02539	219.36837	256.56453	300.74692	353.27009
49	145.83373	169.85936	198.42666	232.43363	272.95840	321.29547	378.99900
50	152.66708	178.50303	209.34799	246.21748	290.33591	343.17967	406.52893

Table II: Present value of an annuity of $1 per period

$$a_{\overline{n}|i} = \frac{1 - (1+i)^{-n}}{i}$$

n \ i	$.005 = \frac{1}{2}\%$	$.01 = 1\%$	$.015 = 1\frac{1}{2}\%$	$.02 = 2\%$	$.025 = 2\frac{1}{2}\%$	$.03 = 3\%$	$.035 = 3\frac{1}{2}\%$
1	0.995025	0.990099	0.985222	0.980392	0.975610	0.970874	0.966184
2	1.985099	1.970395	1.955833	1.941561	1.927424	1.913470	1.899694
3	2.970248	2.940985	2.912200	2.883883	2.856023	2.828611	2.801637
4	3.950495	3.901966	3.854385	3.807729	3.761974	3.717098	3.673079
5	4.925866	4.853431	4.782645	4.713459	4.645828	4.579707	4.515052
6	5.896384	5.795477	5.697187	5.601431	5.508125	5.417191	5.328553
7	6.862074	6.728195	6.598214	6.471991	6.349391	6.230283	6.114544
8	7.822959	7.651678	7.485925	7.325481	7.170137	7.019692	6.873956
9	8.779063	8.566018	8.360517	8.162237	7.970865	7.786109	7.607687
10	9.730411	9.471305	9.222184	8.982585	8.752064	8.530203	8.316605
11	10.677026	10.367628	10.071118	9.786848	9.514207	9.252624	9.001551
12	11.618931	11.255077	10.907505	10.575341	10.257764	9.954004	9.663334
13	12.556150	12.133740	11.731532	11.348374	10.983185	10.634955	10.302738
14	13.488707	13.003703	12.543381	12.106249	11.690912	11.296073	10.920520
15	14.416624	13.865052	13.343233	12.849263	12.381378	11.937935	11.517411
16	15.339924	14.717874	14.131264	13.577709	13.055003	12.561102	12.094117
17	16.258631	15.562251	14.907649	14.291872	13.712198	13.166118	12.651321
18	17.172767	16.398269	15.672561	14.992031	14.353363	13.753513	13.189682
19	18.082355	17.226008	16.426168	15.678462	14.978891	14.323799	13.709837
20	18.987418	18.045553	17.168639	16.351433	15.589162	14.877475	14.212403
21	19.887978	18.856983	17.900137	17.011209	16.184548	15.415024	14.697974
22	20.784058	19.660379	18.620824	17.658048	16.765413	15.936917	15.167125
23	21.675680	20.455821	19.330861	18.292204	17.332110	16.443608	15.620410
24	22.562865	21.243387	20.030405	18.913925	17.884986	16.935542	16.058368
25	23.445637	22.023156	20.719611	19.523456	18.424376	17.413148	16.481515

Table II: Present value of an annuity of $1 per period (cont.)

$$a_{\overline{n}|i} = \frac{1 - (1+i)^{-n}}{i}$$

n $\quad$ i	$.005 = \frac{1}{2}\%$	$.01 = 1\%$	$.015 = 1\frac{1}{2}\%$	$.02 = 2\%$	$.025 = 2\frac{1}{2}\%$	$.03 = 3\%$	$.035 = 3\frac{1}{2}\%$
26	24.324017	22.795204	21.398632	20.121036	18.950611	17.876842	16.890352
27	25.198027	23.559608	22.067617	20.706898	19.464011	18.327031	17.285364
28	26.067688	24.316443	22.726717	21.281272	19.964889	18.764108	17.667019
29	26.933023	25.065785	23.376075	21.844384	20.453550	19.188454	18.035767
30	27.794053	25.807708	24.015838	22.396455	20.930292	19.600441	18.392045
31	28.650799	26.542285	24.646146	22.937701	21.395407	20.000428	18.736276
32	29.503282	27.269590	25.267139	23.468335	21.849178	20.388765	19.068865
33	30.351525	27.989693	25.878954	23.988563	22.291881	20.765792	19.390208
34	31.195547	28.702666	26.481728	24.498592	22.723786	21.131837	19.700684
35	32.035370	29.408580	27.075594	24.998619	23.145157	21.487220	20.000661
36	32.871015	30.107505	27.660684	25.488842	23.556251	21.832252	20.290494
37	33.702502	30.799510	28.237127	25.969453	23.957318	22.167235	20.570525
38	34.529853	31.484663	28.805051	26.440640	24.348603	22.492461	20.841087
39	35.353088	32.163033	29.364583	26.902589	24.730344	22.808215	21.102500
40	36.172226	32.835686	29.915845	27.355479	25.102775	23.114772	21.355072
41	36.987290	33.499689	30.458961	27.799489	25.466122	23.412400	21.599104
42	37.798298	34.158108	30.994050	28.234793	25.820607	23.701359	21.834883
43	38.605272	34.810008	31.521231	28.661562	26.166446	23.981902	22.062689
44	39.408231	35.455454	32.040622	29.079963	26.503849	24.254274	22.282791
45	40.207195	36.094509	32.552337	29.490160	26.833024	24.518712	22.495450
46	41.002184	36.727236	33.056490	29.892313	27.154169	24.775449	22.700918
47	41.793218	37.353699	33.553192	30.286582	27.467482	25.024708	22.899438
48	42.580316	37.973960	34.042553	30.673119	27.773154	25.266707	23.091244
49	43.363498	38.588079	34.524683	31.052078	28.071369	25.501657	23.276564
50	44.142785	39.196118	34.999688	31.423606	28.362312	25.729764	23.455618

Table II: Present value of an annuity of $1 per period (cont.)

$$a_{n\rceil i} = \frac{1 - (1 + i)^{-n}}{i}$$

i n	.04 = 4%	.045 = $4\frac{1}{2}$%	.05 = 5%	.055 = $5\frac{1}{2}$%	.06 = 6%	.065 = $6\frac{1}{2}$%	.07 = 7%
1	0.961538	0.956938	0.952381	0.947867	0.943396	0.938967	0.934579
2	1.886095	1.872668	1.859410	1.846320	1.833393	1.820626	1.808018
3	2.775091	2.748964	2.723248	2.697933	2.673012	2.648476	2.624316
4	3.629895	3.587526	3.545951	3.505150	3.465106	3.425799	3.387211
5	4.451822	4.389977	4.329477	4.270284	4.212364	4.155679	4.100197
6	5.242137	5.157873	5.075692	4.995530	4.917324	4.841014	4.766540
7	6.002055	5.892701	5.786373	5.682967	5.582381	5.484520	5.389289
8	6.732745	6.595886	6.463213	6.334566	6.209794	6.088751	5.971299
9	7.435332	7.268791	7.107822	6.952195	6.801692	6.656104	6.515232
10	8.110896	7.912718	7.721735	7.537626	7.360087	7.188830	7.023582
11	8.760477	8.528917	8.306414	8.092536	7.886875	7.689043	7.498674
12	9.385074	9.118581	8.863252	8.618518	8.383844	8.158725	7.942686
13	9.985648	9.682853	9.393573	9.117079	8.852683	8.599742	8.357651
14	10.563123	10.222825	9.898641	9.589648	9.294984	9.013842	8.745468
15	11.118387	10.739546	10.379658	10.037581	9.712249	9.402669	9.107914
16	11.652296	11.234015	10.837770	10.462162	10.105895	9.767764	9.446649
17	12.165669	11.707191	11.274066	10.864609	10.477260	10.110577	9.763223
18	12.659297	12.159992	11.689587	11.246074	10.827603	10.432466	10.059087
19	13.133939	12.593294	12.085321	11.607654	11.158117	10.734710	10.335595
20	13.590326	13.007936	12.462210	11.950382	11.469921	11.018507	10.594014
21	14.029160	13.404724	12.821153	12.275244	11.764077	11.284983	10.835527
22	14.451115	13.784425	13.163003	12.583170	12.041582	11.535196	11.061240
23	14.856842	14.147775	13.488574	12.875042	12.303379	11.770137	11.272187
24	15.246963	14.495478	13.798642	13.151699	12.550358	11.990739	11.469334
25	15.622080	14.828209	14.093945	13.413933	12.783356	12.197877	11.653583

Table II: Present value of an annuity of $1 per period (cont.)

$$a_{n]i} = \frac{1 - (1 + i)^{-n}}{i}$$

$\frac{i}{n}$	.04 = 4%	.045 = $4\frac{1}{2}$ %	.05 = 5%	.055 = $5\frac{1}{2}$ %	.06 = 6%	.065 = $6\frac{1}{2}$ %	.07 = 7%
26	15.982769	15.146611	14.375185	13.662495	13.003166	12.392373	11.825779
27	16.329586	15.451303	14.643034	13.898100	13.210534	12.574998	11.986709
28	16.663063	15.742874	14.898127	14.121422	13.406164	12.746477	12.137111
29	16.983715	16.021889	15.141074	14.333101	13.590721	12.907490	12.277674
30	17.292033	16.288889	15.372451	14.533745	13.764831	13.058676	12.409041
31	17.588493	16.544391	15.592810	14.723929	13.929086	13.200635	12.531814
32	17.873551	16.788891	15.802677	14.904198	14.084043	13.333929	12.646555
33	18.147646	17.022862	16.002549	15.075069	14.230230	13.459089	12.753790
34	18.411198	17.246758	16.192904	15.237033	14.368141	13.576609	12.854009
35	18.664613	17.461012	16.374194	15.390552	14.498246	13.686957	12.947672
36	18.908282	17.666041	16.546852	15.536068	14.620987	13.790570	13.035208
37	19.142579	17.862240	16.711287	15.673999	14.736780	13.887859	13.117017
38	19.367864	18.049990	16.867893	15.804738	14.846019	13.979210	13.193473
39	19.584485	18.229656	17.017041	15.928662	14.949075	14.064986	13.264928
40	19.792774	18.401584	17.159086	16.046125	15.146297	14.145527	13.331709
41	19.993052	18.566110	17.294368	16.157464	15.138016	14.221152	13.394120
42	20.185627	18.723550	17.423208	16.262999	15.224543	14.292161	13.452449
43	20.370795	18.874210	17.545912	16.363032	15.306173	14.358837	13.506962
44	20.548841	19.018383	17.662773	16.457851	15.383182	14.421443	13.557908
45	20.720040	19.156347	17.774070	16.547726	15.455832	14.480228	13.605522
46	20.884654	19.288371	17.880066	16.632915	15.524370	14.535426	13.650020
47	21.042936	19.414709	17.981016	16.713664	15.589028	14.587254	13.691608
48	21.195131	19.535607	18.077158	16.790203	15.650027	14.635919	13.730474
49	21.341472	19.651298	18.168722	16.862751	15.707572	14.681615	13.766799
50	21.482185	19.762008	18.255925	16.931518	15.761861	14.724521	13.800746

Table III: Number of each day of the year counting from January 1

Day of Month	Jan	Feb	Mar	Apr	May	June	July	Aug	Sept	Oct	Nov	Dec	Day of Month
1	1	32	60	91	121	152	182	213	244	274	305	335	1
2	2	33	61	92	122	153	183	214	245	275	306	336	2
3	3	34	62	93	123	154	184	215	246	276	307	337	3
4	4	35	63	94	124	155	185	216	247	277	308	338	4
5	5	36	64	95	125	156	186	217	248	278	309	339	5
6	6	37	65	96	126	157	187	218	249	279	310	340	6
7	7	38	66	97	127	158	188	219	250	280	311	341	7
8	8	39	67	98	128	159	189	220	251	281	312	342	8
9	9	40	68	99	129	160	190	221	252	282	313	343	9
10	10	41	69	100	130	161	191	222	253	283	314	344	10
11	11	42	70	101	131	162	192	223	254	284	315	345	11
12	12	43	71	102	132	163	193	224	255	285	316	346	12
13	13	44	72	103	133	164	194	225	256	286	317	347	13
14	14	45	73	104	134	165	195	226	257	287	318	348	14
15	15	46	74	105	135	166	196	227	258	288	319	349	15
16	16	47	75	106	136	167	197	228	259	289	320	350	16
17	17	48	76	107	137	168	198	229	260	290	321	351	17
18	18	49	77	108	138	169	199	230	261	291	322	352	18
19	19	50	78	109	139	170	200	231	262	292	323	353	19
20	20	51	79	110	140	171	201	232	263	293	324	354	20
21	21	52	80	111	141	172	202	233	264	294	325	355	21
22	22	53	81	112	142	173	203	234	265	295	326	356	22
23	23	54	82	113	143	174	204	235	266	296	327	357	23
24	24	55	83	114	144	175	205	236	267	297	328	358	24
25	25	56	84	115	145	176	206	237	268	298	329	359	25
26	26	57	85	116	146	177	207	238	269	299	330	360	26
27	27	58	86	117	147	178	208	239	270	300	331	361	27
28	28	59	87	118	148	179	209	240	271	301	332	362	28
29	29	..	88	119	149	180	210	241	272	302	333	363	29
30	30	..	89	120	150	181	211	242	273	303	334	364	30
31	31	..	90	..	151	..	212	243	..	304	..	365	31

Note: For leap years, the number of each day beginning with March 1 is one greater than that given here.

Appendix B

Brief solutions to 'your turn' exercises

Note: Solutions to the practice problems are in Appendix C beginning page 279.

Chapter 1

1.1	9	**1.2**	14
1.3	2	**1.4**	5
1.5	155	**1.6**	175
1.7	0	**1.8**	$5.00

1.9 8 hours

1.10 240 minutes (or 4 hours)

1.11 $\dfrac{13}{24}$

1.12 $3\dfrac{1}{3}$

1.13 5

1.14 $\dfrac{3}{8}$ is zinc.

1.15 $2\dfrac{1}{2}$ metres will remain.

1.16 89.586

1.17 21.87664

1.18 $7.19 is my change.

1.19 9

1.20 (i) $\dfrac{1}{2}$; 31   (ii) $\dfrac{9}{50}$; 45

1.21 (i) 0.15 ; 3.54   (ii) 0.25 ; 0.4

1.22 (i) $\dfrac{3}{10}$; $\dfrac{1}{5}$; 0.2   (ii) $\dfrac{2}{25}$; $\dfrac{27}{50}$; 0.54

1.23 You should receive 25c (i.e. $0.25) change.

1.24 Each tin in the carton costs $1.04. Therefore, it is cheaper to buy Spot dog food by the carton as you save 5c per tin (i.e. $1.09 – $1.04).

1.25 (i) 0.3526   (ii) 4.5%

     (iii) 78%   (iv) $\dfrac{105}{200}$; $\dfrac{21}{40}$

1.26 (i) 9100   (ii) 7800

     (iii) 1100   (iv) 60800

1.27 (i) 26.71   (ii) 0.00

     (iii) 32.82   (iv) 0.11

1.28 (i) 14   (ii) 0

     (iii) 10   (iv) 16

1.29 (i) 26.70   (ii) 0.00

     (iii) 32.82   (iv) 0.10

1.30 (i) 44.6   (ii) 38.8

     (iii) 11.3

1.31 (i) $1:7$ (ii) $4:3$

     (iii) $8:3$ (iv) $3:10$

     (v) $1:9$

1.32 (i) $5:3$ (ii) $9:21=3:7$

     (iii) $2:8=1:4$ (iv) $1000:25=40:1$

     (v) $240:36=20:3$

1.33 (i) $2:4=1:2$ (ii) $8:7$

     (iii) $7:8$ (iv) $1:3$

     (v) $10:3$

1.34 (i) $250:1000=1:4$ (ii) $40:120=1:3$

     (iii) $240:80=3:1$ (iv) $40:80=1:2$

     (v) $2500:750=10:3$

1.35 Jan receives 150 g
Kate receives 100 g.

1.36 John receives $200
Mark receives $175
Andrew receives $125.

1.37 Joy receives $87500
Pam receives $75000
Ted receives $50000
Peter receives $37500.

1.38 Sarah receives $5000
Alex receives $7500.

1.39 fruit : white sugar
    5 : 3
multiply by 0.8
    4 : 2.4
Thus, 2.4 kg of white sugar are needed.

1.40 24 kg of cement and 36 kg of sand are needed.

1.41 –2

1.42 –32

1.43   36

1.44 $2\frac{5}{8}$

1.45 $-\frac{3}{4}$

1.46   10

1.47 $2 + (-8 \div -4) = 1$
$(2 + -8) \div -4 = 0.0625$
The difference between the two expressions is
0.9375 (i.e. $1 - 0.0625$).

1.48   distance = $6 \times 8 = 48$ metres

1.49   The temperature at 3pm was 3°C.
The drop in temperature was 23°.

1.50   **(i)**   shoes   **(ii)**   sole
       (iii)   hills

Chapter 2

2.1   **(i)** $3b - 2$ **(ii)** $-8x + 6y + 5z$
       (iii) $t + 3s - r$ **(iv)** $12n - 9m$
       (v) $2xy + zx + 2yz$

2.2   **(i)** $18rst$ **(ii)** $8ab$
       (iii) $8xyz$ **(iv)** $-30mnpq$
       (v) $-2xyz$

2.3   **(i)** $2b$ **(ii)** $6b$
       (iii) $x + y$ **(iv)**   3
       (v) $3r$

2.4   **(i)**   4   **(ii)** -6
       (iii) -0.6 **(iv)** $-\frac{1}{3}$
       (v)   1

2.5   **(i)**   8.2   **(ii)**   6.27
       (iii)   74   **(iv)** $37\frac{7}{9}$°C
       (v)   27 metres per second

2.6   **(i)** $12cd + 6c$ **(ii)** $7u(2 + v)$
       (iii) $ab(1 + c)$ **(iv)** $7m - 14n$
       (v) $2a - 4 = 2(a - 2)$

2.8   **(i)**   commutative law for addition
       (ii)   identity law for multiplication
       (iii)   inverse law for multiplication
       (iv)   commutative law for multiplication
       (v)   inverse law for addition

2.9   **(i)**   associative law for multiplication
       (ii)   commutative law for multiplication
       (iii), (iv) and **(v)**   distributive law

2.10   **(i)** -6 ; $\frac{1}{6}$ **(ii)** $-\frac{1}{5}$; 5

       (iii) -2.4 ; $\frac{5}{12}$ **(iv)**   4 ; $-\frac{1}{4}$

       (v) $-2\frac{1}{4}$; $\frac{4}{9}$

Chapter 3

3.1   **(i)** a^4 **(ii)** $5m^3$
       (iii) x^2y^2 **(iv)** $(st)^5$
       (v) $4p^2$ **(vi)** $^-c^3$
       (vii) $16y^4$ **(viii)** $7a^2b^3$
       (ix) $3x^2yz^3$ **(x)** $27d^3e^3f^3$

3.2   **(i)**   4096   **(ii)**   64
       (iii) -128 **(iv)** -64
       (v)   13703.96893   **(vi)** $101\frac{17}{27}$
       (vii)   1681   **(viii)** $\frac{-125}{512}$
       (ix)   53.1441   **(x)**   0.1444

3.3   **(i)** x^3y^4 **(ii)** $\frac{q^2}{4}$
       (iii) $\frac{3yz}{4}$ **(iv)** a^3b
       (v) $125n^9$

3.4   **(i)**   1   **(ii)** a^2c^2
       (iii)   144   **(iv)** h

       (v)
$$\frac{6^{\frac{1}{3}} \times 18^{\frac{1}{2}}}{12^{\frac{3}{4}}} = \frac{(2 \times 3)^{\frac{1}{3}} \times (2 \times 3 \times 3)^{\frac{1}{2}}}{(2 \times 2 \times 3)^{\frac{3}{4}}}$$
$$= \frac{2^{\frac{1}{3}} \times 3^{\frac{1}{3}} \times 2^{\frac{1}{2}} \times 3^{\frac{1}{2}} \times 3^{\frac{1}{2}}}{2^{\frac{3}{4}} \times 2^{\frac{3}{4}} \times 3^{\frac{3}{4}}}$$
$$= 2^{-\frac{2}{3}} \times 3^{\frac{7}{12}}$$
$$= \frac{3^{\frac{7}{12}}}{2^{\frac{2}{3}}}$$

3.5   **(i)** 10^5 **(ii)** $3 \times 10^{-2} = 0.03$

3.6   **(i)** $4\sqrt{5}$ **(ii)** $15\sqrt{2}$
       (iii) $3\sqrt{6}$ **(iv)** $6\sqrt{3}$
       (v) $11\sqrt{2}$

3.7   **(i)** $5\sqrt{5}$ **(ii)** $10\sqrt{3}$
       (iii) $2\sqrt{3} + 4\sqrt{2}$ **(iv)** $\sqrt{3}$

(v) $\quad 2a^{\frac{1}{2}} + a^{\frac{3}{2}} = 2a^{\frac{1}{2}} + \left(a \cdot a^{\frac{1}{2}}\right)$

$$= (2 + a)\, a^{\frac{1}{2}}$$

$$= (2 + a)\, \sqrt{a}$$

3.8 **(i)** $\dfrac{\sqrt{5}}{4}$ **(ii)** $\dfrac{5}{6\sqrt{2}}$

(iii) $\dfrac{3\sqrt{3}}{2\sqrt{2}}$ **(iv)** $\dfrac{\sqrt{5}}{\sqrt{2}}$

(v) $\dfrac{2}{\sqrt{2}}$

3.9 **(i)** 7 **(ii)** 5

(iii) $6\sqrt{3}$ **(iv)** 18

(v) 42

3.10 **(i)** $6\sqrt{2}$ **(ii)** $7\sqrt{2}$

(iii) $30\sqrt{2}$ **(iv)** 12

(v) $2\sqrt{2}$

Chapter 4

4.1 **(a)** **(i)** multinomial **(ii)** 5
(iii) coefficient of ax is 6; of x is -9; of ay is 8; of y^2 is -12
(iv) constant term = 2 **(v)** no degree
(b) **(i)** binomial **(ii)** 2
(iii) coefficient of mn is 5, of m^2 is $-\dfrac{1}{4}$
(iv) no constant term **(v)** no degree
(c) **(i)** trinomial **(ii)** 3
(iii) coefficient of p^3 is -2; of q is 1
(iv) constant term = 6 **(v)** no degree
(d) **(i)** polynomial **(ii)** 4
(iii) coefficient of y^2 is 2; of y^4 is -1; of y is $\dfrac{1}{2}$
(iv) constant term = -5 **(v)** degree = 4
(e) **(i)** polynomial **(ii)** 5
(iii) coefficient of t^5 is 3; of t is -5; of t^3 is 1; of t^2 is 0.62
(iv) constant term = $-4\sqrt{2}$
(v) degree = 5

4.3 **(i)** $9x^2 - 1$ **(ii)** $2t^3 + 9t^2 + 4t$
(iii) $x^4 - 16y^2$ **(iv)** $1 - 10y + 25y^2$
(v) $6m^2 - mn - n^2$

4.4 **(i)** $3a^3 - 22a^2 + 7a$ polynomial of degree 3
(ii) $9a^4 - 25b^4$ binomial
(iii) $1 - 4x + 4x^2$ polynomial of degree 2

(iv) $9x^2y^2 - 6xyz + z^2$ trinomial
(v) $3x^2 - 7xy + 2y^2$ trinomial

4.5 **(i)** $27x(3x^2 - 1)$ **(ii)** $2mn(4m + 3n + 5)$
(iii) $(8 - b)(b - 10c)$ **(iv)** $(4 - c)(1 - y)$
(v) $(x - 5)(a - b)$

4.6 **(i)** $(p - 10)^2$ **(ii)** $3(x + 2)^2$
(iii) $(6t - 2s)(6t + 2s)$ **(iv)** $(1 - 5q)(1 + 5q)$
(v) $5(3xy - 4)(3xy + 4)$

4.7 **(i)** $(2x + 1)(x + 3)$ **(ii)** $(2y - 3)^2$
(iii) $(2m - 3)(3m + 1)$
(iv) $(4 - 5t)(2t + 3)$ or $(5t - 4)(3 - 2t)$
(v) $6(a + 6)(a - 1)$

4.8 **(i)** 68 **(ii)** 2
(iii) 3 **(iv)** 1635

4.9 **(i)** 2 **(ii)** $\dfrac{1}{4}$

(iii) 0 **(iv)** 2

4.10 **(i)** 2.7 **(ii)** 2.0
(iii) 13.1 **(iv)** 0.7

Chapter 5

5.3 Division by 8

5.4 Multiplication by $(y - 2)$

5.5 Division by 2 ; subtraction of $\dfrac{1}{2}t$; addition of $\dfrac{9}{2}$

5.6 $x = -\dfrac{1}{2}$

5.7 $x = -1$

5.8 $y = -2\dfrac{1}{18}$

5.9 $t = \dfrac{8}{11}$

5.10 $a = \dfrac{7}{8}$

5.11 $x = -5$

5.12 $p = 0$

5.13
$$6 - \sqrt{2y + 5} = 0$$
$$-\sqrt{2y + 5} = -6$$
$$\sqrt{2y + 5} = 6$$
$$\left(\sqrt{2y + 5}\right)^2 = 6^2$$
$$2y + 5 = 36$$
$$2y = 31$$
$$y = 15\dfrac{1}{2}$$

5.14
$$(t-3)^{3/2} = 8$$
$$[(t-3)^{3/2}]^{2/3} = (8)^{2/3}$$
$$t-3 = (2^3)^{2/3}$$
$$t-3 = 2^2$$
$$t-3 = 4$$
$$t = 7$$

5.15
$$\sqrt{a^2-9} = 9-a$$
$$\left(\sqrt{a^2-9}\right)^2 = (9-a)^2$$
$$a^2-9 = 81-18a+a^2$$
$$-90 = -18a$$
$$\frac{90}{18} = a$$
$$a = 5$$

5.16 $x = 4$, $y = 3$

5.17 $x = -\dfrac{1}{4}$, $y = -1$

5.18 $x = 1$, $y = 2$

5.19
$x + 3y = 3$	$x + 3y = 3$	—(1)
$y - 2x = 8$	$-2x + y = 8$	—(2)
(1) x 2	$2x + 6y = 6$	—(3)
(2) + (3)	$7y = 14$	
	$y = 2$	

Substitute $y = 2$ into equation **(1)**
$$x + 6 = 3$$
$$\therefore \ x = -3$$
Solution: $x = -3$, $y = 2$.

5.20
$2x - 3y = 5$		—(1)
$5x + 2y = -16$		—(2)
(1) x 2	$4x - 6y = 10$	—(3)
(2) x 3	$15x + 6y = -48$	—(4)
(3) + (4)	$19x = -38$	
	$x = -2$	

Substitute $x = -2$ into equation **(1)**
$$-4 - 3y = 5$$
$$-3y = 9$$
$$\therefore \ y = -3$$
Solution: $x = -2$, $y = -3$.

5.22 – 5.26 Solutions as for **5.16 – 5.20**

5.27 $x = \dfrac{7 \pm \sqrt{61}}{2}$. So, $x = 7.41$ or -0.41

5.28 $x = \dfrac{-9 \pm \sqrt{21}}{6}$. So, $x = -0.74$ or -2.26

5.29 $x = \dfrac{2 \pm \sqrt{52}}{8}$. So, $x = 1.15$ or -0.65

5.30 $x = \dfrac{-5 \pm \sqrt{41}}{4}$. So, $x = 0.35$ or -2.85

5.31 $x = \dfrac{-5 \pm \sqrt{169}}{12}$. So, $x = 0.67$ or -1.5

5.32 $x = 2$ or 7

5.33 $x = -2$ or -4

5.34 $x = 5$

5.35 $x = 1\frac{1}{2}$ or 2

5.36 $x = 5.37$ or -0.37

5.37 $x = 6$ or -12

5.38 $x = 6.54$ or 0.46

5.39 $x = 9$ or 0

5.40 $x = 15$ or -5

5.41 $x = 3.50$ or -2.50

5.42 $x = 0.43$ or -0.77

5.43 $x = 1.54$ or 0.26

5.44 $x = 0.47$ or -2.14

5.45 $x = 2$ or 4

5.46 $q = \dfrac{6-p}{3}$

5.47 $r = \dfrac{s-p}{pt}$

5.48 $m = \dfrac{rB(n+1)}{2l}$

5.49 $a_1 = \dfrac{2S}{n} - a_n$

5.50 $h = 10$ cm

5.51 $d = 10$ i.e. prey density is 10

5.52 **(a)** $t = 9$ seconds
(b) $t = 3$ and 6 seconds

5.53 Susan is aged 12 and Tom is aged 16.

5.54 The numbers are 36 and 72.

5.55 The number is 6.

5.56 **(a)** The sale price of a blouse is $14.99.
(b) The sale price of a pair of slacks is $34.50.

5.57 Let x be the present age of the son.
Let y be the present age of the father.
Then, $y = 3x$ —(1)
   $y - 4 = 4(x - 4)$ —(2) (4 years ago)
From **(2)** : $y - 4 = 4x - 16$
   $y = 4x - 12$ —(3)
Equate **(1)** and **(3)** : $3x = 4x - 12$
   $-x = -12$ or $x = 12$

Substitute into **(1)** : $y = 36$
Therefore, the present ages of the father and son are 36 years and 12 years, respectively.

5.58 Let x be the number of rabbits.
Let y be the number of ducks.
Then, $x + y = 40$ —**(1)** (no. of heads)
        $4x + 2y = 98$ —**(2)** (no. of feet)
(1) × 2 $2x + 2y = 80$ —**(3)**
(2) − (3) $2x = 18$
            $\therefore x = 9$
Substitute into **(1)**
        $9 + y = 40$
        $\therefore y = 31$
Therefore, there are 9 rabbits and 31 ducks.

5.59 Let x be the price of the Toyota.
Let y be the price of the Mazda.
Then, $x + y = 14500$ —**(1)** (total cost of cars)
$1.2x + 0.95y = 16200$ —**(2)** (total selling price)
 (1) × 1.2 $1.2x + 1.2y = 17400$ —**(3)**
(3) − (2) $0.25y = 1200$
            $y = 4800$
Substitute into **(1)**
        $x + 4800 = 14500$
        $\therefore x = 9700$
Therefore, the cost price of the Toyota was \$9 700 and the cost price of the Mazda was \$4 800.

5.60 Let W be the number of new wells drilled.
Then, total desired daily production
= (no. of wells)(daily output per well)
That is, $4420 = (20 + W)(200 - 5W)$
            $= 4000 - 100W + 200W - 5W^2$
        $420 = 100W - 5W^2$
    $420 - 100W + 5W^2 = 0$
    $5(W^2 - 20W + 84) = 0$
    $5(W - 14)(W - 6) = 0$
Hence, $W = 14$ or 6.
Therefore, the number of new wells that should be drilled is 6.

Chapter 6

6.1 $x < 4$

6.2 $y > 4$

6.3 $6 \le t \le 7.5$

6.4 $x \ge -\frac{7}{5}$

6.5 $-2 < x < 2$

6.6 $y \ge 9\frac{2}{3}$

6.7 $x \ge 6$

6.8 $y \le -5$

6.9 $t > -5$

6.10 $-\frac{3}{2} < x < \frac{9}{2}$

6.11 $p < 3\frac{6}{7}$

6.12 $r > 1\frac{7}{17}$

6.13 $x \ge -4$

6.14 $y \le -2$

6.15 $s > -1\frac{1}{2}$

6.16 $t \le 20$

6.17 $-1 \le x < 8$

6.18 Let x be the number of calculators to be sold.
        Cost price $= 500000 + 20x$
        Selling price $= 25x$
Therefore, to make a profit,
        $25x > 20x + 500000$
        $5x > 500000$
        $x > 100000$
So, the smallest number of calculators to be sold is 100001.

6.19 Let x be the number of kilometres.
Rental cost $= 150 \times 12 + 0.05x$
Cost involved with buying $= 1200 + 0.1x$
For renting to be the better proposition,
        $150 \times 12 + 0.05x < 1200 + 0.1x$
        $1800 + 0.05x < 1200 + 0.1x$
        $-0.05x < -600$
        $x > 12000$

Therefore, the smallest number of kilometres is 12001 (to the nearest whole kilometre).

6.20 Let S be the number of singlets sold.
Cost price $= 1.5S + 0.5S + 7500$
$\qquad\qquad = 2S + 7500$
Selling price $= 4.5S$
To make a profit
$$4.5S > 2S + 7500$$
$$2.5S > 7500$$
$$S > 3000$$
Therefore, the company must sell more than 3000 singlets to make a profit.

Chapter 7

7.1 **(i)** 4   **(ii)** 28
      (iii) 1

7.2 **(i)** $1\frac{2}{3}$ **(ii)** $3\frac{1}{2}$

      (iii) $\frac{4}{9}$

7.3 -4

7.5 **(i)** $x = 0$ **(ii)** $x = 3$
      (iii) $x = \pm 3$ **(iv)** $x > 1, x < -1$

7.6 The expression $\sqrt{4 - x^2}$ is undefined if $x > 2$.
The domain is $-2 \le x \le 2$.

7.7 Range: 0 to 10

7.8 **(i)** $x \ge 5$ **(ii)** $x \ge -4$

7.9 **(i)** $f(x) \le 0$ **(ii)** $g(x) \ge -5$

7.10

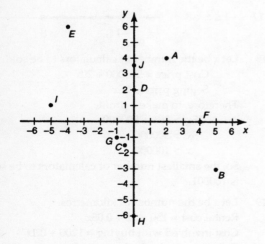

7.11 $y = 3x + 2$

x	−2	−1	0	1	2
y	−4	−1	2	5	8

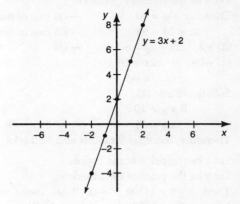

7.12 $y = 2 - x$

x	−4	−3	−2	−1	0	1	2	3	4
y	6	5	4	3	2	1	0	−1	−2

$y = 2x - 1$

x	−4	−3	−2	−1	0	1	2	3	4
y	−9	−7	−5	−3	−1	1	3	5	7

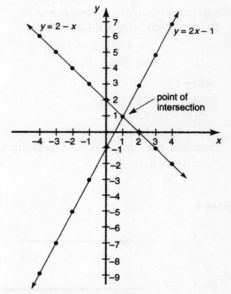

The two graphs meet at (1,1)

7.13 **(i)** $y = x^2 + x - 6$

x	-4	-3	-2	-1	0	1	2	3	4
y	6	0	-4	-6	-6	-4	0	6	14

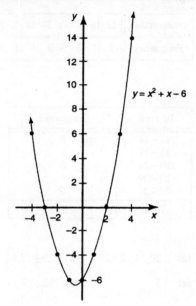

(ii) $y = x^2 + 2$

x	-4	-3	-2	-1	0	1	2	3	4
y	18	11	6	3	2	3	6	11	18

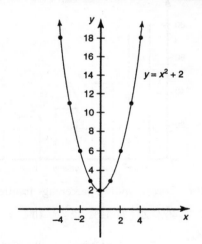

7.14 $y = \dfrac{8}{x}$

x	-8	-4	-2	-1	$-\frac{1}{2}$	$\frac{1}{2}$	1	2	4	8
y	-1	-2	-4	-8	-16	16	8	4	2	1

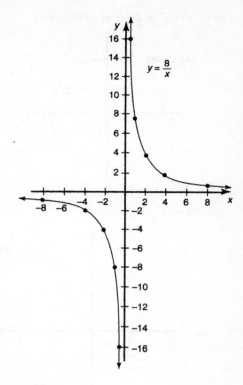

7.15 $y = \dfrac{12}{x-1}$; $x \neq 1$

x	-5	-2	-1	0	2	3	4	5	7
y	-2	-4	-6	-12	12	6	4	3	2

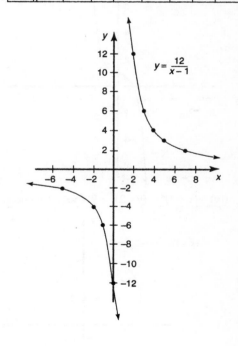

7.16 **(i)** $x^2 + y^2 = 25$ (**not** a function)

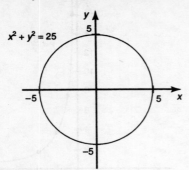

(ii) $y = 4^x - 1$ (function)

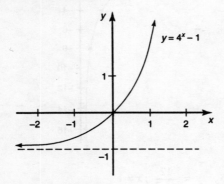

(iii) $y = x^3 + 1$ (function)

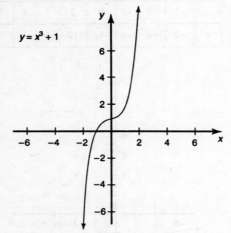

(iv) $(x + 1)^2 + (y - 2)^2 = 9$ (**not** a function)

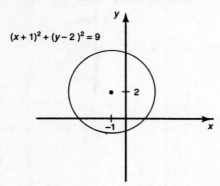

Chapter 8

8.1 **(a)**

12	12	16	16	17	17	17	21	21	21	23
23	24	25	25	26	26	26	27	27	29	31
31	31

(b)

Observation	12	16	17	21	23	24	25	26	27	29	31
Frequency	2	2	3	3	2	1	2	3	2	1	3

8.2

Interval	Frequency
10 – 14	2
15 – 19	5
20 – 24	6
25 – 29	8
30 – 34	3
Total	**24**

8.3 **(a)** 20,24 ; 25,29 **(b)** $19\frac{1}{2}$, $24\frac{1}{2}$; $24\frac{1}{2}$, $29\frac{1}{2}$

(c) 5 **(d)** 12 ; 17.

8.4 **(a)** 13 **(b)** 22

8.5 **(a)**

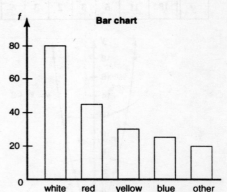

(b) Corresponding percentage frequency:

40% $22\frac{1}{2}$ % 15% $12\frac{1}{2}$ % 10%

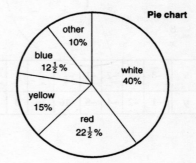

8.6

Category	Frequency	Percentage
Protestant	20	25%
No preference	20	25%
Roman Catholic	16	20%
Others	16	20%
None of your business	8	10%
Total	**80**	**100**

(i)

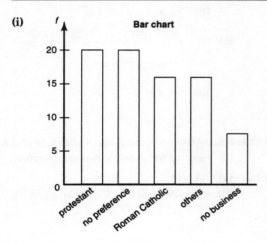

(ii)

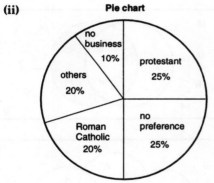

8.7 (i)

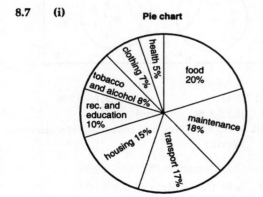

(ii)

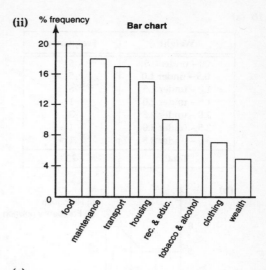

8.8 (a)

Lifetime	Frequency
0 – under 10	13
10 – under 20	10
20 – under 30	1
30 – under 40	3
40 – under 50	2
50 – under 60	3
60 – under 70	4
Total	**36**

(b)

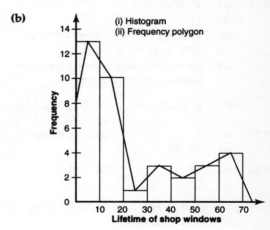

8.9

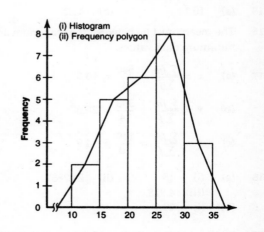

8.10 (a)

Weight	Frequency
0 – under 0.5	5
0.5 – under 1.0	5
1.0 – under 1.5	6
1.5 – under 2.0	1
2.0 – under 2.5	3
2.5 – under 3.0	1
3.0 – under 3.5	1
Total	**22**

(b)

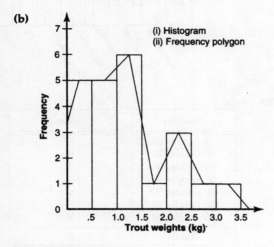

(i) Histogram
(ii) Frequency polygon

8.11 (a) 40, 40.5 ; 0.5

(b) 21

(c) lowest = 36.5°C, highest = 40.5°C

(d) 37°C to 37.5°C ($f = 6$)

(e) **no** patient had a body temperature between 39.5°C and 40°C.

(f) proportion = $\frac{7}{21}$ = 0.333 (or $33\frac{1}{3}$ %)

8.12 (a) 6   **(b)** no mode

(c) 16 , 32

8.13 (a) 30   **(b)** 10

(c) 6.5

8.14 median = middle score of ordered data. 15 is the maximum data value.

8.15 (a) 10.14   **(b)** 45.8

8.16 The mean must lie between the maximum and minimum data values.

8.17 (a) $\bar{x} = \dfrac{\sum fm}{\sum f} = \dfrac{540}{40} = 13.5$

(b) $\bar{x} = \dfrac{\sum fm}{\sum f} = \dfrac{393}{14} = 28.07$

(c) $\bar{x} = \dfrac{\sum fm}{\sum f} = \dfrac{149}{10} = 14.9$

8.18 (a) (i) 15   **(ii)** 24.2857

(iii) 4.928

(b) (i) 29.1

(ii) With $\sum x^2 = 17394.08$ and $\bar{x} = 45.8$,

$s^2 = 87.5657$

(iii) $s = 9.358$

8.19 (a) (i) $s^2 = \dfrac{\sum fm^2 - n\bar{x}^2}{n-1} = \dfrac{8500 - 40\,(13.5)^2}{39}$

$= 31.025641$

(ii) So, $s = \sqrt{s^2} = 5.57$

(b) (i) $s^2 = \dfrac{13953.5 - 14\,(28.071429)^2}{13}$

$= 224.72525$

(ii) So, $s = \sqrt{s^2} = 14.991$

8.20 $\bar{x}_{new} = 32$, $s_{new} = 9.5$

8.21 & 8.22 Answers should agree with **8.18** and **8.19**, except for possible rounding errors.

8.23 (a) $A = \varnothing$

(b) $B = \{20,21,22,23,24,25,26,27,28,29,30\}$

$C = \{21,23,25,27,29\}$

So, $C \subset B$

8.24

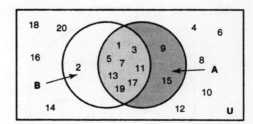

8.25

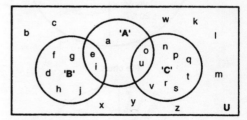

8.26 (i) $\{1,3,5,6,8\}$ **(ii)** $\varnothing$

(iii) $\{6,8\}$ **(iv)** $\{1,3,4,5,6,7,8\}$

(v) $\{1,3,5,6,8\}$ **(vi)** $\{3,5\}$

(vii) $\{1,3,5,6,8\}$ (Note that answer to **(v)** = answer to **(vii)**)

8.27 (i) $(B \cap C) = \{c\}$, in $(B \cap C)^| = \{a,b\}$

(ii) $\{a,b\}$ (**Note**: $(B \cap C)^| = B^| \cup C^|$)

(iii) $\varnothing$

(iv) Since $A \cup B = A \cup C = U$, then $U \cap U = \{a,b,c\} = U$

8.28 (i) The set of all sports people who play cricket or squash or both.

(ii) The set of all squash players who do *not* play hockey.

(iii) The set of all sports people who do *not* play

both cricket *and* hockey.

    (iv) The set of all hockey players who do *not* play cricket or squash (or both).

8.29 **(a)** U = {1,2,3,4,5,6,7,8,9,10}

    (b) **(i)** A = {6,7,8,9,10}

        (ii) B = {2,4,6,8,10}

        (iii) $B^{|}$ = {1,3,5,7,9}

        (iv) {2,4}

        (v) {1,3,5,6,7,8,9,10}

8.30 Let J = Joy, P = Pamela, S = Sandra

    (a) {JPS, JSP, PJS, PSJ, SJP, SPJ}

    (b) **(i)** {JPS,JSP}

        (ii) {JPS,PJS}

        (iii) {JPS,JSP,SJP,SPJ}

        (iv) {JPS,JSP,PJS,SJP}

8.31 **(a)** $\dfrac{4}{52} = \dfrac{1}{13}$ **(b)** $\dfrac{13}{52} = \dfrac{1}{4}$

    (c) $\dfrac{1}{52}$ **(d)** $\dfrac{13}{52} + \dfrac{13}{52} = \dfrac{26}{52} = \dfrac{1}{2}$

    (e) $1 - \dfrac{1}{52} = \dfrac{51}{52}$

8.32 **(a)** $\dfrac{9}{15} = 0.6$ **(b)** $\dfrac{7}{15} = 0.467$

    (c) $\dfrac{3}{15} = 0.2$ **(d)** $\dfrac{3}{15} = 0.2$

    (e) $\dfrac{8}{15} = 0.533$

8.33 **(a)** $\dfrac{1}{10} = 0.1$ **(b)** $\dfrac{4}{10} = 0.4$

    (c) $\dfrac{2}{10} = 0.2$ **(d)** $1 - 0.4 = 0.6$

8.34 **(a)** Yes, $\sum$ probs = 1   **(b)** No, $\sum$ probs > 1

    (c) No, $P(A) < 0$ **(d)** Yes, $\sum$ probs = 1

8.35 **(a)** $\dfrac{3}{30} = 0.1$ **(b)** $\dfrac{9}{30} = 0.3$

    (c) $1 - \dfrac{8}{30} = 0.733$ **(d)** $\dfrac{23}{30} = 0.767$

    (e) $1 - \dfrac{3+5}{30} = 0.733$

8.36 **(a)** $P(A^{|}) = 1 - 0.36 = 0.64$

    (b) $P(B^{|}) = 1 - 0.52 = 0.48$

    (c) $0.36 + 0.52 - 0.15 = 0.73$

8.37 **(a)** No, $\sum$ probs > 1

    (b) True *if* P (no change) = 0.10

    (c) No, $P(A \cap B) > P(A)$; impossible

8.38 **(a)** $P(A^{|}) = 1 - 0.3 = 0.7$

    (b) $P(A \cup B) = 0.3 + 0.1 = 0.4$

    (c) $P(B \cup C) = 0.1 + 0.5 = 0.6$

    (d) $P(A \cap C) = 0$

    (e) $P(A \cup B \cup C) = 0.3 + 0.1 + 0.5 = 0.9$

Chapter 9

9.1 **(i)** square   **(ii)** triangle

    (iii) rectangle   **(iv)** quadrilateral

9.2 **(i)** **(a)** $P = 9$ m   **(b)** $A = 5$ m^2

    (ii) **(a)** $P = 80$ mm   **(b)** $A = 400$ m^2

    (iii) **(a)** $P = 10$ m   **(b)** $A = 4.48$ m^2

    (iv) **(a)** $P = 12$ units   **(b)** $A = 6$ square units

    (v) **(a)** $P = 21.7$ m   **(b)** $A = 28$ m^2

    (vi) **(a)** $C = 12\pi \approx 37.7$ cm (correct to 1 decimal place)

        (b) $A = 36\pi \approx 113.1$ cm^2 (correct to 1 decimal place)

9.3 **(i)** $P = 33.6$ cm   **(ii)** $A = 48$ cm^2

    (iii) $A = 20$ cm^2 **(iv)** $P = 44$ cm

9.4 **(i)** Area of top and bottom : 6 cm^2 each

        Area of front and back :12 cm^2 each

        Area of 2 sides : 8 cm^2 each

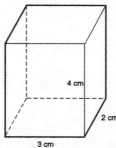

    (ii) Total surface area = 2 x 6 + 2 x 12 + 2 x 8

                         = 52 cm^2

9.5 $V = 20$ cm^3

9.6 **(a)** The fence is 396 m long.

    (b) There are 132 posts.

    (c) 99 lengths of timber were needed.

9.7 It will cost $255.

9.8 The width is 6.95 cm.

9.9 **(i)** The diameter is 164.6 mm (correct to 1 decimal place).

    (ii) The radius is 1.4 m (correct to 1 decimal place).

9.10 $V = 1800000$ mm^3

9.11 Total volume = 282743.3 mm^3 (correct to 1 decimal place).

9.12 **(i)** $V = 413400000$ mm^3 **(ii)** $V = 0.4134$ m^3

9.13 Length = 2.21 km (correct to 2 decimal places)

9.14 (difficult)

(i) Surface area = 427.7 cm² (correct to 1 decimal place)

(ii) Surface area = 213.6 cm² (correct to 1 decimal place)

(iii) Surface area = 213.63 cm² (1 bearing)

(iv) Surface area = 2564345 mm²

9.15 (a) (i) $SA = 12544 \pi \approx 39408.1$ mm² (correct to 1 decimal place).

(ii) $V = \dfrac{702464}{3} \pi \approx 735618.6$ mm³ (correct to 1 decimal place).

(b) (i) $SA = 344569 \pi \approx 1082495.4$ mm² (correct to 1 decimal place).

(ii) $V \approx 105904137.1$ mm³ (correct to 2 decimal places).

Chapter 10

10.1 (a) $\tan 40° = \dfrac{x}{75}$, $x = 62.9$ cm

(b) $\cos 38° = \dfrac{x}{65}$, $x = 51.2$ cm

(c) $\tan 27° = \dfrac{50}{x}$, $x = \dfrac{50}{\tan 27°} = 98.1$ cm

10.2 (a) $\sin \theta = \dfrac{40}{68}$, $\theta = 36°$

(b) $\tan \theta = \dfrac{30}{38}$, $\theta = 38°$

(c) $\cos \theta = \dfrac{23}{39}$, $\theta = 54°$

10.3 (a) $\cos 27° = \dfrac{46}{x}$, $x = \dfrac{46}{\cos 27°} = 51.63$ mm

(b) $\tan 37° = \dfrac{x}{14}$, $x = 10.55$ mm

(c) $\sin 49° = \dfrac{82}{x}$, $x = \dfrac{82}{\sin 49°} = 108.65$ mm

10.4 Let x be the length of side CA

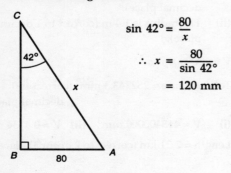

$\sin 42° = \dfrac{80}{x}$

$\therefore x = \dfrac{80}{\sin 42°}$

$= 120$ mm

10.5 (i) Let θ equal angle z

$\tan \theta = \dfrac{46.4}{54.2}$. So, $\theta = 41°$

(ii) Let x be the length of XZ

$\sin 41° = \dfrac{x}{46.4}$. So, $x = 30$ cm

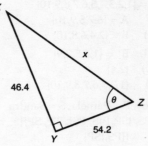

10.6 Let θ equal $\angle ABD$

$\cos \theta = \dfrac{25}{36}$. So, $\theta = 46°$

$\therefore \angle ABC = 2 \angle ABD$

$= 92°$

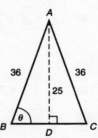

10.7 Let x be the length the ladder reaches up the wall.

$\tan 65° = \dfrac{x}{3.5}$. So, $x = 751$ cm

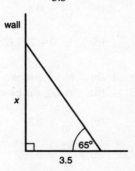

10.8 Let x be the length of the wire.

$\sin 58° = \dfrac{9}{x}$. So, $x = \sin 58° = 10.6$ m

(correct to one decimal place).

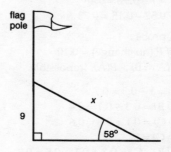

10.9 Let θ = angle of elevation of the sun.

$\tan θ = \dfrac{15}{23} = 0.652$. So, θ = 33°

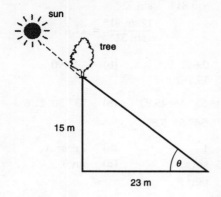

10.10 Let x = length of shadow.

$\tan 16.5° = \dfrac{15}{x}$. So, $x = 50.64$ m

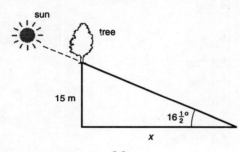

10.11 (a) $\dfrac{y}{\sin 95°} = \dfrac{3.2}{\sin 46°}$

$y = \dfrac{3.2 \sin 95°}{\sin 46°} = 44.32$ m

(b) $\dfrac{y}{\sin 76°} = \dfrac{15.2}{\sin 57°}$

$y = \dfrac{15.2 \sin 76°}{\sin 57°} = 17.59$ cm

10.12 (a) $\dfrac{\sin θ}{8} = \dfrac{\sin 50°}{9}$

$\sin θ = \dfrac{8 \sin 50°}{9}$

So, θ = 42° 55'

(b) $\dfrac{\sin θ}{52.6} = \dfrac{\sin 28°}{39.4}$

$\sin θ = \dfrac{52.6 \sin 28°}{39.4}$

So, θ = 38° 49'

10.13 (a) $d^2 = 18^2 + 16^2 - 2 \times 18 \times 16 \cos 44°$,
$d = 12.9$ m

(b) $d^2 = 290^2 + 160^2 - 2 \times 290 \times 160 \cos 32°$,
$d = 176.1$ mm

10.14 (a) $\cos θ = \dfrac{13^2 + 17.2^2 - 9.3^2}{2 \times 13 \times 17.2}$,

So, θ = 32° 13'

(b) $\cos θ = \dfrac{300^2 + 325^2 - 460^2}{2 \times 300 \times 325}$,

So, θ = 94° 42'

10.15 (a)

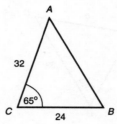

$A = \dfrac{1}{2}ab \sin C$

$= \dfrac{1}{2} \times 24 \times 32 \times \sin 65°$

$= 348.0$ sq cm

(b)

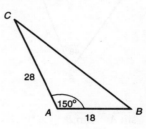

$A = \dfrac{1}{2}bc \sin A$

$= \dfrac{1}{2} \times 28 \times 18 \times \sin 150°$

$= 126$ sq m

(c)

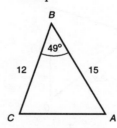

$A = \dfrac{1}{2}ac \sin B$

$= \dfrac{1}{2} \times 12 \times 15 \times \sin 49°$

$= 67.9$ mm^2
(correct to one decimal place)

10.16 (a)

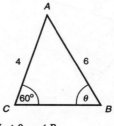

Let θ = ∠ B

$$\frac{\sin \theta}{4} = \frac{\sin 60°}{6}$$

$$\sin \theta = \frac{4 \sin 60°}{6}$$

$$\therefore \theta = 35° \, 16'$$

(b)

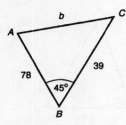

$$b^2 = a^2 + c^2 - 2ac \cos B$$
$$= 39^2 + 78^2 - 2 \times 39 \times 78 \times \cos 45°$$
$$b = 57.5 \text{ (correct to one decimal place)}$$

(c)

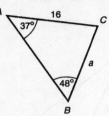

$$\frac{a}{\sin A} = \frac{b}{\sin B}$$

$$\frac{a}{\sin 37°} = \frac{16}{\sin 48°}$$

$$a = \frac{16 \sin 37°}{\sin 48°} = 13.0$$

(d)

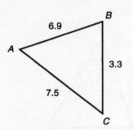

$$\cos B = \frac{a^2 + c^2 - b^2}{2ac}$$

$$= \frac{3.3^2 + 6.9^2 - 7.5^2}{2 \times 3.3 \times 6.9}$$

So, $B = 87° \, 10'$

(e)

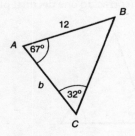

$$\frac{b}{\sin B} = \frac{c}{\sin C}$$

$$\frac{b}{\sin 81°} = \frac{12}{\sin 32°}$$

$$b = \frac{12 \sin 81°}{\sin 32°} = 22.4$$

10.18 **(a)** 0.642   **(b)** −1.190
         (c) 13.162

10.19 **(a)** 53° 16' 45.93"   **(b)** 53° 30' 21.6"
         (c) 58° 55' 5.8"

10.20 **(a)** 1   **(b)** $2 \cos^2 A$
         (c) $\sin x$ **(d)** $\sin \theta$
         (e) $\tan^2 B$

Chapter 11

11.1 $\begin{bmatrix} 1 & 2 & 3 \\ 4 & 5 & 6 \end{bmatrix}$, example only.

11.2 $\begin{bmatrix} b_{11} & b_{12} \\ b_{21} & b_{22} \\ b_{31} & b_{32} \\ b_{41} & b_{42} \end{bmatrix}$

11.3 $\begin{bmatrix} a_{11} & a_{12} & a_{13} \end{bmatrix}$

11.4 $\begin{bmatrix} b_{11} \\ b_{21} \\ b_{31} \\ b_{41} \\ b_{51} \end{bmatrix}$

11.5 $a_{12} = -2$ $a_{13} = 8$
      $a_{25} = -5$ $a_{41} = 0$
      $a_{21} = -1$ $a_{22} = \frac{1}{2}$
      $a_{44} = 10$ $a_{24} = 1$

11.6 **(a)** $A + B = \begin{bmatrix} 11 & 0 & -4 \\ 18 & 0 & -3 \\ 0 & 3 & 6 \end{bmatrix}$

      (b) $B + A = \begin{bmatrix} 11 & 0 & -4 \\ 18 & 0 & -3 \\ 0 & 3 & 6 \end{bmatrix}$. So, $A + B = B + A$.

11.7 **(a)** $A - B = \begin{bmatrix} -4 & -4 \\ -4 & -4 \end{bmatrix}$ **(b)** $B - A = \begin{bmatrix} 4 & 4 \\ 4 & 4 \end{bmatrix}$

11.8 $T - J = \begin{bmatrix} 0 & 20 & 5 \\ -15 & 25 & -5 \end{bmatrix}$

11.9 **(a)** $2A = \begin{bmatrix} 6 & 4 & 12 & 16 \\ 0 & 2 & 8 & 4 \\ 8 & 4 & 6 & 12 \end{bmatrix}$ **(b)** $0.5A = \begin{bmatrix} \frac{3}{2} & 1 & 3 & 4 \\ 0 & \frac{1}{2} & 2 & 1 \\ 2 & 1 & \frac{3}{2} & 3 \end{bmatrix}$

(c) $5A = \begin{bmatrix} 15 & 10 & 30 & 40 \\ 0 & 5 & 20 & 10 \\ 20 & 10 & 15 & 30 \end{bmatrix}$

11.10 **(a)** $A + B = \begin{bmatrix} 4 & 7 \\ 13 & 7 \\ 7 & 13 \end{bmatrix}$ **(b)** $A - B = \begin{bmatrix} 2 & -3 \\ -1 & 1 \\ -5 & -3 \end{bmatrix}$

(c) $B + A = \begin{bmatrix} 4 & 7 \\ 13 & 7 \\ 7 & 13 \end{bmatrix}$ **(d)** $B - A = \begin{bmatrix} -2 & 3 \\ 1 & -1 \\ 5 & 3 \end{bmatrix}$

(e) $2A = \begin{bmatrix} 6 & 4 \\ 12 & 8 \\ 2 & 10 \end{bmatrix}$ **(f)** $2A + B = \begin{bmatrix} 7 & 9 \\ 19 & 11 \\ 8 & 18 \end{bmatrix}$

11.11 **(a)** $\begin{bmatrix} 200 & 300 & 500 & 250 \end{bmatrix}$

(b) $\begin{bmatrix} 100 \\ 150 \\ 200 \\ 300 \end{bmatrix}$

(c) $\text{Cost} = \begin{bmatrix} 200 & 300 & 500 & 250 \end{bmatrix} \cdot \begin{bmatrix} 100 \\ 150 \\ 200 \\ 300 \end{bmatrix}$

$= \$240\,000$

11.12 **(a)** $A \cdot B = \begin{bmatrix} 1 & 21 & 8 \\ 2 & 6 & 4 \\ 13 & 33 & 24 \end{bmatrix}$

(b) $A \cdot C = \begin{bmatrix} 6 & -5 \\ 3 & -4 \\ 18 & -25 \end{bmatrix}$

(c) $B \cdot C$ not possible

(d) $B \cdot F = \begin{bmatrix} -6 \\ 0 \end{bmatrix}$

(e) $C \cdot D = \begin{bmatrix} 2 & -1 \\ 19 & 13 \end{bmatrix}$

(f) $C \cdot E = \begin{bmatrix} 4 \\ -10 \end{bmatrix}$

(g) $A \cdot E = \begin{bmatrix} 14 \\ 4 \\ 22 \end{bmatrix}$

11.13 $Q \cdot R = \begin{bmatrix} 1060 & 810 & 890 & 420 & 1680 \end{bmatrix}$

11.14 **(i)** **(a)** 1,3,2 **(b)** 1,1,1

(ii) **(a)** 0,3,−5 **(b)** 0,1,0

11.15 $A \cdot C = \begin{bmatrix} 1 & 21 & 8 \\ 2 & 6 & 4 \\ 13 & 33 & 24 \end{bmatrix}$. Size is 3 x 3

11.16 $C \cdot A = \begin{bmatrix} -3 & 1 \\ 2 & 34 \end{bmatrix}$. Size is 2 x 2

11.18 $C^T = \begin{bmatrix} -1 & 2 \\ 3 & 6 \\ 0 & 4 \end{bmatrix}$

11.19 $C^T \cdot D = \begin{bmatrix} -5 & -5 \\ 39 & 3 \\ 8 & -4 \end{bmatrix}$. Size is 3 x 2

11.20 $A^T = \begin{bmatrix} 3 & 0 & -1 \\ 2 & 1 & 6 \end{bmatrix}$

11.21 $A^T \cdot B = \begin{bmatrix} 8 & -4 & -2 \\ -26 & 27 & 11 \end{bmatrix}$. Size is 2 x 3

11.22 $|D| = -15$

11.23 **(i)** $|B| = b_{11}B_{11} + b_{12}B_{12} + b_{13}B_{13} = 24$

(ii) $|B| = b_{11}B_{11} + b_{21}B_{21} + b_{31}B_{31} = 24$

11.24 $|C| = 0$. Hence, C is singular.

11.25 $A \cdot A^{-1} = \begin{bmatrix} 1 & 0 & 0 \\ 0 & 1 & 0 \\ 0 & 0 & 1 \end{bmatrix} = I$. Thus, $A^{-1} = \begin{bmatrix} 1 & 1 & 4 \\ 3 & 2 & 4 \\ 1 & 1 & 6 \end{bmatrix}$

11.26 **(a)** $|A| = 12$. So, A is non-singular.

(b) $A^{-1} = \frac{1}{|A|}(A^c)^T = \frac{1}{12}\begin{bmatrix} 2 & 0 \\ -3 & 6 \end{bmatrix}$

(c) $A \cdot A^{-1} = \begin{bmatrix} 6 & 0 \\ 3 & 2 \end{bmatrix} \cdot \frac{1}{12}\begin{bmatrix} 2 & 0 \\ -3 & 6 \end{bmatrix}$

$= \begin{bmatrix} 6 & 0 \\ 3 & 2 \end{bmatrix} \cdot \begin{bmatrix} \frac{1}{6} & 0 \\ -\frac{1}{4} & \frac{1}{2} \end{bmatrix} = \begin{bmatrix} 1 & 0 \\ 0 & 1 \end{bmatrix}$

11.27 $|A| = -2$. So, $A^{-1} = \dfrac{1}{|A|}\begin{bmatrix} 8 & -6 \\ -7 & 5 \end{bmatrix} = -\dfrac{1}{2}\begin{bmatrix} 8 & -6 \\ -7 & 5 \end{bmatrix}$

$A \cdot A^{-1} = \begin{bmatrix} 5 & 6 \\ 7 & 8 \end{bmatrix} \cdot \begin{bmatrix} -4 & 3 \\ 7 & -\frac{5}{2} \\ \frac{7}{2} & -\frac{5}{2} \end{bmatrix} = \begin{bmatrix} 1 & 0 \\ 0 & 1 \end{bmatrix}$

11.28 **(a)** $\begin{bmatrix} 3 & 3 & 6 \\ 0 & 2 & 7 \\ 2 & 2 & 4 \end{bmatrix} \cdot \begin{bmatrix} x \\ y \\ z \end{bmatrix} = \begin{bmatrix} 3 \\ 4 \\ 4 \end{bmatrix}$

(b) $\begin{bmatrix} 1 & -2 & 3 \\ 2 & 3 & -1 \\ 0 & 0 & 2 \end{bmatrix} \cdot \begin{bmatrix} x \\ y \\ z \end{bmatrix} = \begin{bmatrix} 5 \\ 6 \\ 4 \end{bmatrix}$

11.29 **(a)** Let x = no. of tonnes of ore shipped through Afriport.
Let y = no. of tonnes of ore shipped through Bulgaport.
Let z = no. of tonnes of ore shipped through Calciport.
Then,
$x + y + z = 1000$ (tonnes of ore leaving mine)
$2x + 3y + 5z = 3100$ (cost of rail transport)
$50x + 200y + 1000z = 340000$ (freight capacities)
Note: Dividing the third equation by 10 gives $5x + 20y + 100z = 34000$

(b) $\begin{bmatrix} 1 & 1 & 1 \\ 2 & 3 & 5 \\ 5 & 20 & 100 \end{bmatrix} \cdot \begin{bmatrix} x \\ y \\ z \end{bmatrix} = \begin{bmatrix} 1000 \\ 3100 \\ 34000 \end{bmatrix}$

11.30 **(i)** $\begin{bmatrix} 2 & 5 \\ 3 & -1 \end{bmatrix} \cdot \begin{bmatrix} x \\ y \end{bmatrix} = \begin{bmatrix} 17 \\ 17 \end{bmatrix}$; $A^{-1} = \dfrac{1}{-17}\begin{bmatrix} -1 & -5 \\ -3 & 2 \end{bmatrix}$

So, $X = A^{-1} \cdot B = \begin{bmatrix} \frac{1}{17} & \frac{5}{17} \\ \frac{3}{17} & \frac{-2}{17} \end{bmatrix} \cdot \begin{bmatrix} 17 \\ 17 \end{bmatrix} = \begin{bmatrix} 6 \\ 1 \end{bmatrix}$

(ii) $x = \dfrac{\begin{vmatrix} 17 & 5 \\ 17 & -1 \end{vmatrix}}{-17} = 6$, $y = \dfrac{\begin{vmatrix} 2 & 17 \\ 3 & 17 \end{vmatrix}}{-17} = 1$

11.31 **(i)** $\begin{bmatrix} 6 & 0 \\ 3 & 2 \end{bmatrix} \cdot \begin{bmatrix} x \\ y \end{bmatrix} = \begin{bmatrix} 6 \\ 3 \end{bmatrix}$; $A^{-1} = \dfrac{1}{12}\begin{bmatrix} 2 & 0 \\ -3 & 6 \end{bmatrix}$

So, $X = A^{-1} \cdot B = \begin{bmatrix} \frac{1}{6} & 0 \\ -\frac{1}{4} & \frac{1}{2} \end{bmatrix} \cdot \begin{bmatrix} 6 \\ 3 \end{bmatrix} = \begin{bmatrix} 1 \\ 0 \end{bmatrix}$

(ii) $x = \dfrac{\begin{vmatrix} 6 & 0 \\ 3 & 2 \end{vmatrix}}{12} = 1$, $y = \dfrac{\begin{vmatrix} 6 & 6 \\ 3 & 3 \end{vmatrix}}{12} = 0$

11.32 $X = A^{-1} \cdot B = \begin{bmatrix} -2 & \frac{5}{2} & \frac{3}{2} \\ 4 & -5 & -2 \\ -1 & \frac{3}{2} & \frac{1}{2} \end{bmatrix} \cdot \begin{bmatrix} 1 \\ 1 \\ 1 \end{bmatrix} = \begin{bmatrix} 2 \\ -3 \\ 1 \end{bmatrix}$

11.33 $X = A^{-1} \cdot B = \begin{bmatrix} \frac{9}{2} & -\frac{3}{2} & -4 \\ -\frac{7}{2} & \frac{3}{2} & 3 \\ -3 & 1 & 3 \end{bmatrix} \cdot \begin{bmatrix} 2 \\ 2 \\ 1 \end{bmatrix} = \begin{bmatrix} 2 \\ -1 \\ -1 \end{bmatrix}$

Chapter 12

12.1 **(a)** $t = 200 + 80w$

12.2 **(a)** $\dfrac{dy}{dx} = 5$ **(b)** $\dfrac{dy}{dx} = -50$

(c) $\dfrac{dy}{dx} = \dfrac{3}{4}$ **(d)** $\dfrac{dy}{dx} = 0$

(e) $\dfrac{dy}{dx} = 24x^3$ **(f)** $\dfrac{dy}{dx} = -8x^{-9}$ or $\dfrac{-8}{x^9}$

12.3 **(a)** $\dfrac{dm}{dn} = 300n^2 + 2n + 7$

(b) $\dfrac{dm}{dn} = -13 + 15n^2$

(c) $\dfrac{dy}{dx} = 28x^3 - 10$

(d) $\dfrac{dy}{dx} = -2x^{-3} - 6x$

(e) $\dfrac{dy}{dx} = \dfrac{1}{2}x^{-\frac{1}{2}} - 2x$ or $\dfrac{1}{2\sqrt{x}} - 2x$

12.4 $\dfrac{dy}{dx} = 3x^2 - 10x$. When $x = 3$, $\dfrac{dy}{dx} = -3$.

12.5 $\dfrac{ds}{dt} = 3 + 24t$. When $t = 5$, $\dfrac{ds}{dt} = 123$

12.6 **(a)** $g(1) = 12\frac{1}{2}$ **(b)** $g(10) = 5120$

12.7 **(a)** $f(-3) = 44$ **(b)** $f(0) = 2$

(c) $f(3) - f(2) = 10$

12.8 **(a)** $f'(x) = 1$ **(b)** $f'(x) = 10x - 3$

(c) $f'(x) = \dfrac{1}{12}$

12.9 **(a)** $g'(Q) = 2Q^3 + 12$ **(b)** $g'(3) = 66$

12.10 $h'(t) = 4t^3 - 8t + 2$. So, $h'(2) = 18$

12.11 **(a)** $\dfrac{dy}{dx} = 16x^3 + 3x^2$, $\dfrac{d^2y}{dx^2} = 48x^2 + 6x$

(b) $\dfrac{dy}{dx} = -12x + 1$, $\dfrac{d^2y}{dx^2} = -12$

(c) $\dfrac{dy}{dx} = \dfrac{d^2y}{dx^2} = 0$

(d) $\dfrac{dy}{dx} = -4x^{-5} + 4x^3$, $\dfrac{d^2y}{dx^2} = 20x^{-6} + 12x^2$

$$\text{or } \dfrac{20}{x^6} + 12x^2$$

12.12 (a) $f'(t) = 10 - 32t$, $f''(t) = -32$
(b) $f'(0) = 10$

12.13 (a) $g'(y) = 4y^3 - 9y^2 + 2$, $g'(0) = 2$
(b) $g''(y) = 12y^2 - 18y$
(c) $g''(3) = 54$
(d) $g'''(y) = 24y - 18$, $g'''(4) = 78$.

12.14 $y = x^3 - 48x + 10$. So, $\dfrac{dy}{dx} = 3x^2 - 48$

$$= 0, \text{ when } x = 4 \text{ or } -4.$$

$\dfrac{d^2y}{dx^2} = 6x > 0$, when $x = 4$.

Therefore, y is a **minimised** at $x = 4$.
Minimum value of function is $f(4) = -118$.

$\dfrac{d^2y}{dx^2} = 6x < 0$, when $x = -4$.

Therefore, y is **maximised** at $x = -4$
Maximum value of function is $f(-4) = 138$.

12.15 (a) $\dfrac{dy}{dx} = 2x - 6 = 0$, when $x = 3$.

$\dfrac{d^2y}{dx^2} = 2 > 0$ for all x.

So, y has a **minimum** at $x = 3$ and
$f(3) = -2$.

(b) $\dfrac{dy}{dx} = 8x + 40 = 0$, when $x = 5$.

$\dfrac{d^2y}{dx^2} = 8 > 0$ for all x.

So, y has a **minimum** at $x = 5$ and
$f(5) = 300$.

(c) $\dfrac{dy}{dx} = -3 - 2x = 0$, when $x = -\dfrac{3}{2}$.

$\dfrac{d^2y}{dx^2} = -2 < 0$ for all x.

So, y has a **maximum** at $x = -\dfrac{3}{2}$ and

$$f\left(-\dfrac{3}{2}\right) = 6.25.$$

12.16 $A = 100x - x^2$; $\dfrac{dA}{dx} = 100 - 2x = 0$, when $x = 50$.

$\dfrac{d^2A}{dx^2} = -2 < 0$ for all x. So, A is **maximised** when

$x = 50$. Max. area $= f(50) = 2500$ sq.m.

12.17 $Pr = TR - TC = 24q - 4q^2 - q^2 - 4q = 20q - 5q^2$

$\dfrac{dPr}{dq} = 20 - 10q = 0$, when $q = 2$.

$\dfrac{d^2Pr}{dq^2} = -10 < 0$. So, profit is a **maximised** when

$q = 2$. Max. profit $= f(2) = \$20$.

12.18 (a) $s = 10 + 5t + 12t^2 - t^3$. So,

$\dfrac{ds}{dt} = 5 + 24t - 3t^2$.

At $t = 2$, $\dfrac{ds}{dt} = 41$. So, velocity $= 41$ m/sec.

(b) $\dfrac{d^2s}{dt^2} = 24 - 6t$.

At $t = 3$, $\dfrac{d^2s}{dt^2} = 6$ m/sec/sec.

(c) $\dfrac{d^2s}{dt^2} = 24 - 6t = 0$, when $t = 4$.

Hence, after 4 secs. the acceleration is zero.

12.19 (a) $s = 10 + 6t + 13t^2 - t^3$

$\dfrac{ds}{dt} = 6 + 26t - 3t^2$. At $t = 3$, $\dfrac{ds}{dt} = 57$.

So, velocity after 3 secs. is 57 m/sec.

(b) $\dfrac{d^2s}{dt^2} = 26 - 6t$. At $t = 2$, $\dfrac{d^2s}{dt^2} = 14$

So, acceleration after 2 secs. is 14m/sec/sec.

(c) $\dfrac{d^2s}{dt^2} = 26 - 6t = 0$, when $t = 4\dfrac{1}{3}$.

So, after $4\dfrac{1}{3}$ secs, acceleration is zero.

12.20 $s = 2t^3 + 3t^2 + 4t + 5$

$\dfrac{ds}{dt} = 6t^2 + 6t + 4$. At $t = 2$, $\dfrac{ds}{dt} = 40$

So, velocity after 2 secs is 40 m/sec.

$\dfrac{d^2s}{dt^2} = 12t + 6$. At $t = 2$, $\dfrac{d^2s}{dt^2} = 30$.

So, acceleration after 2 secs is 30 m/sec/sec.

At $t = 3$, $\dfrac{ds}{dt} = 76$.

So, velocity $= 76$ m/sec.

At $t = 3$, $\dfrac{d^2s}{dt^2} = 42$.

So, acceleration = 42 m/sec/sec.

12.21 $TR = 60q$. $MR = \dfrac{dTR}{dq} = 60$.

So, marginal revenue = $60

12.22 **(a)** $MC = \dfrac{dTC}{dq} = 0.0006q^2 - 0.06q + 10$

At $q = 1000$, $MC = 550$
So, marginal cost if producing 1000 items is $550.

(b) At $q = 500$, $MC = 130$
So, marginal cost if producing 500 items is $130.

12.23 **(a)** $MC = \dfrac{dTC}{dq} = 6q + 21$

(b) $MC = \dfrac{dTC}{dq} = 6q^2 - 8$

(c) $MC = \dfrac{dTC}{dq} = 1.5q^2 + 25$

12.24 **(a)** $TR = 200q - 2q^3$; $MR = \dfrac{dTR}{dq} = 200 - 6q^2$

At $q = 5$, $MR = 50$.
So, marginal revenue if 5 items are sold is $50.

(b) $TR = 300q - 75q^3$;

$MR = \dfrac{dTR}{dq} = 300 - 225q^2$.

At $q = \sqrt{3}$, $MR = -375$.
So, marginal revenue is –$375.

12.25 **(a)** $MR = -20q + 100$
(b) $MC = 10q - 200$
(c) $Pr = TR - TC = -15q^2 + 300q - 1125$
(d) $MPr = -30q + 300$
(e) **(i)** $\dfrac{dPr}{dq} = 0$ when $q = 10$. So, profit is maximised when $q = 10$.
(ii) $MC = MR$ when $q = 10$. Again, profits are maximized at $q = 10$.

12.26 **(a)** $\dfrac{dy}{dx} = x^4$. So, $y = \dfrac{x^5}{5} + C$

(b) $p' = 6q^2 - 5$. So, $p = 2q^3 - 5q + C$

(c) $y' = 7x^6 - 4x^3$. So, $y = x^7 - x^4 + C$

12.27 **(a)** $6x + C$ **(b)** $\dfrac{x^{24}}{8} + C$

(c) $-2x^{-4} + C$ **(d)** $\dfrac{2x^5}{5} + \dfrac{1}{8}x^4 + C$

(e) $x^4 + 3x^{-1} + C$ **(f)** $\dfrac{x^4}{4} - 2x^3 + 5x + C$

(g) $\dfrac{(5x+7)^4}{20} + C$

12.28 **(a)** $y = 2x^2 - 3x + C$. If $y(-1) = 8$, $C = 3$.
So, $y = 2x^2 - 3x + 3$.

(b) $y = \dfrac{1}{2}x^4 - \dfrac{1}{4}x^2 + C$. If $y(2) = 5$, $C = -2$.

So, $y = \dfrac{1}{2}x^4 - \dfrac{1}{4}x^2 - 2$.

(c) $y = -2x^{1/2} + x^2 + C$. If $y(4) = 10$, $C = 10$.

So, $y = -2\sqrt{x} + x^2 + 10$.

12.29 **(a)** $\left[\dfrac{7x^2}{2}\right]_1^2 = 10\dfrac{1}{2}$

(b) $\left[4y - \dfrac{9y^2}{2}\right]_1^1 = 8$

(c) $-4\left[\dfrac{t^{-3}}{-3}\right]_1^2 = -1\dfrac{1}{6}$

(d) $\left[\dfrac{x^3}{3} + \dfrac{x^2}{2} + x\right]_{\frac{1}{2}}^{\frac{3}{2}} = 3\dfrac{1}{12}$

(e) $\left[\dfrac{(x+3)^4}{4}\right]_1^3 = 260$

(f) $[-2t^{-1}]_{\frac{1}{2}}^3 = 3\dfrac{1}{3}$

(g) $\dfrac{1}{3}\left[(2x+4)^{\frac{3}{2}}\right]_0^6 = 18\dfrac{2}{3}$

(h) $\left[\dfrac{x^4}{4} - \dfrac{x^3}{3}\right]_1^3 = 11\dfrac{1}{3}$

(i) $\left[(x-2)^3\right]_{-1}^2 = 27$

(j) $\left[\dfrac{-1}{8}(3-2t)^4\right]_{1/2}^1 = \dfrac{15}{8}$

12.30 **(a)** The car is slowing down.

(b) $\dfrac{dV}{dt} = -6$ **(c)** $V = -6t + C$

(d) $V(0) = 20$. So, $C = 20$.

(e) $V = -6t + 20$. At $V = 0$, $t = 6\dfrac{2}{3}$.

Car stops after $6\dfrac{2}{3}$ secs.

(f) $\dfrac{ds}{dt} = -6t + 20$ **(g)** $s = -3t^2 + 20t + C$

(h) If $s(0) = 0$, then $C = 0$. So, $s = -3t^2 + 20t$.
At $s = 5$, $t \approx 6.9$. So, after 6.9 secs, distance travelled = 5 m.
Time elapsed = 6.9 – 6.67 = 0.23 secs.
Distance travelled, $s(.23) = 4.4413$ m.

12.31 TC (of producing 9 items)

$= .001q^3 - .2q^2 + 40q + C$

Since fixed production costs $= \$5\,000, C = \$5\,000$

So, $TC = .001q^3 - .2q^2 + 40q + 5000$

Total cost of 100 items $= TC(q = 100)$
$= 8000$

Therefore, average cost per item $= \dfrac{TC}{100} = \$80$.

12.32 $G = \dfrac{-P^2}{60} + 2.5P + C$

If $G(15) = 40$, then

$C = 40 + \dfrac{15^2}{60} - 2.5\,(15) = 6.25$

Therefore, $G = \dfrac{-P^2}{60} + 2.5P + 6.25$

12.33 $TC = 10q - 200\,(q+10)^{1/2} + C$

If average cost per item for 90 items is

$\dfrac{TC\,(90)}{90} = \$100$, then

$\dfrac{1}{90}\left[10 \times 90 - 200\,(90+10)^{\frac{1}{2}} + C\right] = 100$

So, $C = 9000 - 900 + 2000 = 10100$

That is, manufacturer's fixed costs $= \$10\,100$

12.34 $f(70) - f(60) = \displaystyle\int_{60}^{70} (0.2q + 3)\,dq$

$= [0.1q^2 + 3q]_{60}^{70}$

$= 160$

Therefore, the cost of increasing production from 60 to 70 items is $160.

12.35 **(i)** Given $f(t) = 8t + 10$

$f(12) - f(0) = \displaystyle\int_0^{12} (8t + 10)\,dt$

$= [4t^2 + 10t]_0^{12}$

$= 696$

Therefore, the number of crimes expected to be committed next year is 696.

(ii) $f(12) - f(6) = \displaystyle\int_6^{12} (8t + 10)\,dt$

$= [4t^2 + 10t]_6^{12}$

$= 492$

Therefore, the number of crimes in the last 6 months of next year is 492.

12.36 $f(900) - f(400) = \displaystyle\int_{400}^{900} \dfrac{1000}{\sqrt{100q}}\,dq = 100\int_{400}^{900} q^{-\frac{1}{2}}\,dq$

$= 100\left[2q^{\frac{1}{2}}\right]_{400}^{900} = 2000$

Hence, the total revenue is increased by $2\,000.

12.37 $A = \displaystyle\int_{-2}^{1} (x^2 + 2x + 2)\,dx$

$= \left[\dfrac{x^3}{3} + x^2 + 2x\right]_{-2}^{1} = 6$ square units

Chapter 13

13.1 **(a)** $I = Prt = \$200$ **(b)** $I = \$114.83$

(c) $I = \$900$

13.2 **(a)** $t = \dfrac{I}{Pr} = 6$ months

(b) $t = 2$ years **(c)** $t = 4.5$ years

13.3 **(a)** $r = \dfrac{I}{Pt} = 4\tfrac{1}{2}\%$p.a.

(b) $r = 9\%$p.a. **(c)** $r = 10.5\%$p.a.

13.4 **(a)** $S = P(1 + rt) = \$11\,600$
(b) $S = \$11\,725$
(c) $S = \$10\,393.75$

13.5 **(a)** $r = \dfrac{I}{Pt} = 6\%$p.a.

(b) $r = 8\tfrac{1}{2}\%$p.a.

(c) $r = 12\tfrac{1}{2}\%$p.a.

13.6 $S = P(1 + rt) = \$2\,313$ (maturity value)

13.7 $I = \$180$. So, $t = \dfrac{I}{Pr} = 2$ years.

13.8 $P = \dfrac{S}{(1 + rt)} = \$984.49$

13.9 $P = \dfrac{I}{rt} = \$4\,166.67$

13.10 **(a)** $P = \dfrac{S}{(1 + rt)} = \$12\,727.27$

(b) $P = \$10\,687.02$
(c) $P = \$12\,121.21$

13.11 **(a)** $P = \dfrac{I}{rt} = \$3\,333.33$

(b) $P = \dfrac{S}{(1 + rt)} = \$2\,608.70$

13.12 **(i)** $P_1 =$ present value of $500 in 4 months
$= \$485.44$

(ii) $P_2 =$ present value of $750 in 6 months
$= \$717.70$

So, $P =$ present value of debt
$= P_1 + P_2 = \$1\,203.14$.

13.13 **(a)** Days $= 146$ **(b)** Days $= 288$
(c) Days $= 554$

13.14 $I = Prt = \$50.19$

13.15 Maturity value, $S = P(1 + rt) = \$12\,693.29$

13.16 Loan $= \dfrac{S}{(1 + rt)} = \$469.25$

13.17 (a) (i) $i = 0.0175$ (ii) $n = 8$
    (b) (i) $i = 0.0025$ (ii) $n = 26$
    (c) (i) $i = 0.0125$ (ii) $n = 64$
    (d) (i) $i = 0.03$ (ii) $n = 15$

13.18 (a) $S = P(1 + i)^n = 700(1 + .06)^7 = \$1\,052.54$
    (b) $S = 700(1 + .03)^{14} = \$1\,058.81$
    (c) $S = 700(1 + .015)^{28} = \$1\,062.06$
    (d) $S = 700(1 + .005)^{84} = \$1\,064.26$

13.19 (i) $S = 500(1.025)^{24} = \$904.36$
    (ii) $S = 500(1.05)^{12} = \$897.93$
    (iii) $S = 500(1.1)^6 = \$885.78$

13.20 (a) $S = 2500(1.04)^{10} = \$3\,700.61$
    (b) $S = 2500(1.04)^{18} = \$5\,064.54$
    (c) $S = 2500(1.04)^{30} = \$8\,108.49$.
         So, $S \approx \$5\,000$ after 9 years

13.21 $S = 1500(1.055)^{10} = \$25\,622.17$
    So, interest $= 25622.17 - 15000 = \$10\,622.17$.

13.22 (a) $P = 60000(1.06)^{-10} = \$33\,503.69$
    (b) $P = 60000(1.015)^{-40} = \$33\,075.74$
    (c) $P = 60000(1.03)^{-20} = \$33\,220.55$

13.23 (a) $P = 5000(1.01)^{-18} = \$4\,180.09$
    (b) $P = 5000(1.0225)^{-6} = \$4\,375.12$

13.24 $P = 10000(1.025)^{-42} = \$3\,544.85$

13.25 $P = 8200(1.015)^{-24} = \$5\,736.26$
    So, choose (b) since present value is less than in (a).

13.26 (a) $i = \left(\dfrac{3000}{2000}\right)^{\frac{1}{15}} - 1 = 0.0274$. So,
         $r = 0.0274 \times 4 \times 100 = 10.96\%$p.a. quarterly

    (b) $i = \left(\dfrac{150}{100}\right)^{\frac{1}{55}} - 1 = 0.0074$. So,
         $r = 0.0074 \times 12 \times 100 = 8.88\%$p.a. monthly

    (c) $i = \left(\dfrac{1581}{1000}\right)^{\frac{1}{7}} - 1 = 0.0676$. So,
         $r = 0.0676 \times 2 \times 100 = 13.52\%$p.a. semi-annually

13.27 $i = \left(\dfrac{6000}{4000}\right)^{\frac{1}{36}} - 1 = 0.01133$. So,
    $r = 13.6\%$p.a. monthly

13.28 $i = \left(\dfrac{32500}{2500}\right)^{\frac{1}{8}} - 1 = 0.03334$. So,
    $r = 6.67\%$p.a. semi-annually

13.29 (a) 10%p.a. flat $\approx$ 18%p.a. compound.
         Borrow at 10%p.a. monthly.
    (b) 13%p.a. flat $\approx$ 24%p.a. compound.
         Borrow at 13%p.a. flat.
    (c) 6.5%p.a. flat $\approx$ 12%p.a. compound.
         Borrow at 9%p.a. monthly.

13.30 (a) 18%p.a. flat $\approx$ 34%p.a. compound.
         Borrow at 18%p.a. monthly.
    (b) 7%p.a. flat $\approx$ 12%p.a. compound.
         Borrow at 7%p.a. flat.

13.31 $S = Rs_{\overline{n}|i} = 400\,s_{\overline{5}|0.06} = \$2\,254.84$

13.32 $S = R[s_{\overline{n+1}|i} - 1] = 450[s_{\overline{25}|.015} - 1] = \$13\,078.36$

13.33 $S = R[s_{\overline{n+1}|i} - 1] = 50[s_{\overline{61}|.005} - 1] = \$3\,505.94$

13.34 $S = Rs_{\overline{n}|i} = 250\,s_{\overline{160}|.02} = \$284\,623.84$

13.35 (a) $S = Rs_{\overline{n}|i} = 6000\,s_{\overline{10}|.08} = \$86\,919.37$
    (b) $S = R[s_{\overline{n+1}|i} - 1] = 6000[s_{\overline{11}|.08} - 1]$
             $= \$93\,872.92$

13.36 $R = \dfrac{12000}{s_{\overline{10}|0.03}} = \$1\,046.77$

13.37 $R = \dfrac{8000}{(s_{\overline{11}|0.05} - 1)} = \$605.75$

13.38 (a) $R = \dfrac{5000}{(s_{\overline{19}|0.015} - 1)} = \$240.42$

    (b) $R = \dfrac{5000}{s_{\overline{18}|0.015}} = \$244.03$

13.39 $A = Ra_{\overline{n}|i} = 30a_{\overline{36}|.005} = \$986.13$

13.40 $A = R[1 + a_{\overline{n-1}|i}] = 200[1 + a_{\overline{9}|.045}] = \$1\,653.76$

13.41 $A = R[1 + a_{\overline{n-1}|i}] = 1000[1 + a_{\overline{31}|.025}] = \$22\,395.41$

13.42 (a) $A = 75[1 + a_{\overline{207}|.0025}] = \$12\,183.19$
    (b) $A = 75a_{\overline{208}|.0025} = \$12\,152.81$

13.43 $R = \dfrac{9000}{a_{\overline{12}|0.03}} = \$904.16$

13.44 $R = \dfrac{340000}{(1 + a_{\overline{11}|0.08})} = \$41774.36$

13.45 (a) $R = \dfrac{8000}{(1 + a_{\overline{259}|0.005})} = \$54.78$

    (b) $R = \dfrac{8000}{a_{\overline{260}|0.005}} = \$55.05$

Chapter 14

14.1 **(a)** $120° = 120° \times \dfrac{\pi}{180} = \dfrac{2\pi^c}{3}$

(b) $18° \times \dfrac{\pi}{180} = \dfrac{\pi^c}{10}$

(c) $300 \times \dfrac{\pi}{180} = \dfrac{5}{3}\pi^c$

(d) $22\dfrac{1}{2} \times \dfrac{\pi}{180} = \dfrac{\pi^c}{8}$

14.2 **(a)** $3\pi^c = 3\pi \times \dfrac{180}{\pi} = 540°$

(b) $\dfrac{7\pi}{18} \times \dfrac{180}{\pi} = 70°$

(c) $\dfrac{\pi}{12} \times \dfrac{180}{\pi} = 15°$

(d) $\dfrac{5}{3}\pi \times \dfrac{180}{\pi} = 300°$

14.3 **(a)** $61° \, 38' \times \dfrac{\pi}{180} = 1.0757^c$

(b) $266° \, 13' \, 58" \times \dfrac{\pi}{180} = 4.6466^c$

(c) $0° \, 01' \times \dfrac{\pi}{180} = 0.00029$

14.4 **(a)** $1.38492^c \times \dfrac{180}{\pi} = 79° \, 21' \, 0.26"$

(b) $\dfrac{2\pi}{13} \times \dfrac{180}{\pi} = 27° \, 41' \, 32.31"$

(c) $3.10682 \times \dfrac{180}{\pi} = 178° \, 0' \, 27.63"$

14.5 **(a)** $\cos 5.8^c = 0.8855$

(b) $\tan \dfrac{7\pi}{5} = 3.0777$

(c) $\sin\left(\dfrac{-3\pi}{2}\right) = 1$

(d) $\cot 3 = \dfrac{1}{\tan 3} = -7.0153$

(e) $\tan (5\pi - 1) = -1.5574$

14.6 **(a)** $l = r\theta = 10.5 \times \dfrac{4\pi}{3} = 43.98$ cm

(b) $l = 6.5 \times 1.8 = 11.7$ cm

(c) $l = 9.2 \times 84 \times \dfrac{\pi}{180} = 13.49$

(d) $\theta = \dfrac{l}{r} = \dfrac{25.2}{18.4} = 1.37^c$

(e) $\theta = \dfrac{4.7}{7.9} = 0.5949^c$

(f) $r = \dfrac{l}{\theta} = \dfrac{8.3}{(3\pi)/4} = 3.5$ cm

(g) $r = \dfrac{1.8}{2.4} = 0.75$ cm

(h) $r = \dfrac{3.7}{64° \, 32' \, 12" \times \dfrac{\pi}{180}} = 3.3$ cm

14.7 **(a)** $A = \dfrac{1}{2}r^2\theta = \dfrac{1}{2} \times 103^2 \times \dfrac{5\pi}{6}$

$\qquad\qquad = 13887.1$ square mm

(b) $A = \dfrac{1}{2} \times 8.2^2 \times 3.4 = 114.308$ square cm

(c) $A = \dfrac{1}{2} \times 2.3^2 \times 56° \, 12' \times \dfrac{\pi}{180}$

$\qquad\qquad = 2.594$ square cm

14.8 **(a)** $l = 1.5 \times \dfrac{\pi}{6} = 0.79$ m

(b) $A = \dfrac{1}{2} \times 1.5^2 \times \dfrac{\pi}{6} = 0.59$ square m

14.9 Area (shaded region) $= \dfrac{1}{2}r^2\theta - \dfrac{1}{2}ab\sin C$

$= \dfrac{1}{2} \times 11.8^2 \times 105 \times \dfrac{\pi}{180} - \dfrac{1}{2} \times 11.8^2 \times \sin 105°$

$= 60.34$ sq. cm

14.10 **(a)** $y = 3\sin x$

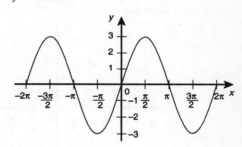

(i) period $= 2\pi$

(ii) amplitude $= 3$

(b) $y = \sin 4x$

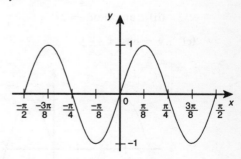

(i) period $= \dfrac{\pi}{2}$

(ii) amplitude $= 1$

(c) $y = 2\sin\left(\dfrac{\pi}{2} + x\right)$

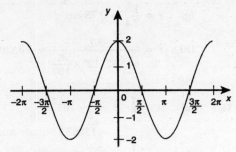

 (i) period = 2π
 (ii) amplitude = 2

14.11 **(a)** $y = \cos \dfrac{x}{3}$

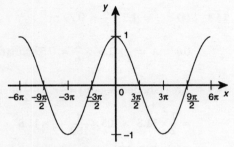

 (i) period = 6π
 (ii amplitude = 1

(b) $y = 2\cos 4x$

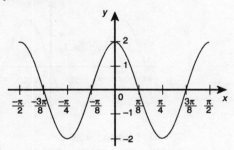

 (i) period = $\dfrac{\pi}{2}$

 (ii) amplitude = 2

(c) $y = \cos\left(x - \dfrac{\pi}{2}\right)$

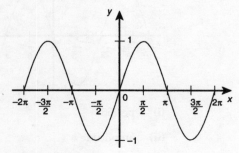

 (i) period = 2π
 (ii) amplitude = 1

14.12 **(a)** $y = \tan \dfrac{x}{2}$

 period = 2π

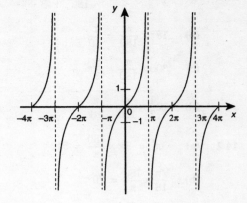

(b) $y = 3\tan 2x$

 period = $\dfrac{\pi}{2}$

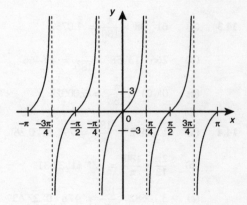

(c) $y = 2\tan\left(x - \dfrac{\pi}{2}\right)$

 period = $\dfrac{\pi}{2}$

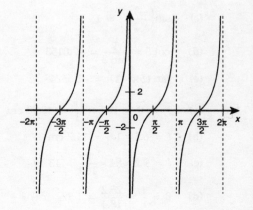

14.13 (a) $y = \operatorname{cosec} x$

period $= 2\pi$

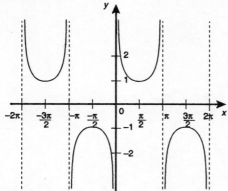

(b) $y = \cot x$

period $= \pi$

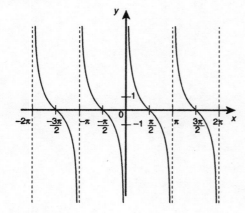

14.14 (a) $\dfrac{dy}{dx} = -3\sin(3x+1)$

(b) $\dfrac{dy}{dx} = 6\cos(3x-4)$

(c) $\dfrac{dy}{dx} = 8\sec^2 2x$

(d) $\dfrac{dy}{dx} = -4\sin 2x \cos 2x$

(e) $\dfrac{dy}{dx} = -12\sin(\sin 4x)\cos 4x$

14.15 $R = \dfrac{V^2}{g}\sin 2x$

$\dfrac{dR}{dx} = \dfrac{V^2}{g}2\cos 2x$

$\quad = 0,\ \text{when}\left(x = \dfrac{\pi}{4}\right)$

$\dfrac{d^2R}{dx^2} = \dfrac{-4V^2}{g}\sin 2x.$

When $x = \dfrac{\pi}{4}$,

$\dfrac{d^2R}{dx^2} = \dfrac{-4V^2}{g}\sin 2 \cdot \dfrac{\pi}{4} < 0$

Therefore, maximum range occurs when the

angle of elevation is $\dfrac{\pi^c}{4}$ or $45°$.

14.16 (a) $\displaystyle\int \sec^2\dfrac{x}{2}\,dx = 2\tan\dfrac{x}{2} + C$

(b) $\displaystyle\int \cos\left(\dfrac{\pi}{2} - x\right)dx = -\sin\left(\dfrac{\pi}{2} - x\right) + C$

(c) $\displaystyle\int (x - \sin(2x-3))\,dx$

$= \displaystyle\int x\,dx - \int \sin(2x-3)\,dx$

$= \dfrac{x^2}{2} + \dfrac{\cos(2x-3)}{2} + C$

14.17 (a) $\displaystyle\int_0^\pi (1 + \cos x)\,dx = [x + \sin x]_0^\pi = \pi$

(b) $\displaystyle\int_0^{\frac{\pi}{4}} (3\cos x + 4\sin x)\,dx$

$= [3\sin x - 4\cos x]_0^{\frac{\pi}{4}} = 4 - \dfrac{1}{\sqrt{2}}$

(c) $\displaystyle\int_{-\frac{\pi}{4}}^{\frac{\pi}{4}} 2(x + \sec^2 x)\,dx = 2\int_{-\frac{\pi}{4}}^{\frac{\pi}{4}} (x + \sec^2 x)\,dx$

$= 2\left[\dfrac{x^2}{2} + \tan x\right]_{-\frac{\pi}{4}}^{\frac{\pi}{4}} = 4$

14.18 $\displaystyle\int_0^{30}\left[20 - \dfrac{1}{2}t - (10 + 2\sin 2\pi t)\right]dt$

$= \displaystyle\int_0^{30}\left(20 - \dfrac{t}{2} - 10 - 2\sin 2\pi t\right)dt$

$= \left[10t - \dfrac{t^2}{4} + \dfrac{\cos 2\pi t}{\pi}\right]_0^{30}$

$= \left[300 - \dfrac{900}{4} + \dfrac{\cos 60\pi}{\pi} - \left(0 - 0 - \dfrac{1}{\pi}\right)\right]$

$= 75.64$ megalitres

Appendix C

Brief solutions to practice problems

Chapter 1

1.1 $-21\frac{1}{2}$

1.2 $-13\frac{2}{3}$

1.3   18

1.4 -30

1.5   18

1.6   0

1.7 $\frac{5}{6}$

1.8 $1\frac{7}{20}$

1.9   6.5

1.10   **(a)**   8 x \$24.50 = \$196
        (b)   38 x \$24.50 = \$931

1.11   No. of baskets = 98
       ∴ Sam's earnings = 98 x \$5.80 = \$568.40.

1.12   Wage = (36 x \$10.35) + (14.2 x 1.5 x \$10.35)
           = \$593.06.

1.13   Commission = 23% of \$625 = \$143.75.

1.14   Shop leased at \$10 556 per year (or \$203/week) is cheaper by \$2/week.

1.15   Actual production = 26212 units.
       Difference = prod. capacity – actual prod.
              = 30000 – 26212 = 3788 units.

1.16   Profit = selling price – cost price
            = (175 x \$4.55) – (175 x \$2.534)
            = \$352.80.

1.17   3 teaspoons : 1 pie.
       108 teaspoons : 36 pies.
       ∴ 108 teaspoons of honey makes 36 pies.

1.18   **(a)**   3.5 cups sugar + 3 tablespoons malt serves 5.
            ∴ 7 cups sugar + 6 tablespoons malt serves 10.

       (b) $\frac{3.5}{5}$ x 12 cups sugar + $\frac{3}{5}$ x 12 tablespoons malt serves 12.
            Hence, 8.4 cups sugar + 7.2 tablespoons malt serves 12.

1.19 $\frac{26.5}{100}$ people earn < \$10 100/year.

       ∴ expect $\frac{26.5}{100} \times 554 = 146.81$ of a sample of 554 people to earn < \$10 100/year. (i.e. 147 people).

1.20   Total thickness = 355 x 0.35 = 124.25 cm.

1.21   Interest, $I = \frac{PRT}{100}$.

       ∴ $P = \frac{100I}{RT} = \frac{100 \times 825}{8.5 \times 1} = \$9\,705.88$

       ∴ the amount invested is \$9 705.88.

1.22   Percentage = $\frac{78}{125} \times 100 = 62\%$ (nearest whole percentage).

1.23   60 ml : 12% solution
       1440 ml : 0.5% solution
       ∴ 1440 mls of a 0.5% solution can be prepared.

1.24   Money spent = 45% of \$10 500 = \$4 725.

1.25   **(a)** 4 tablets **(b)** 2 tablets **(c)** 2 tablets

Chapter 2

2.1   **(a)** $y + 5$ **(b)** $x - 3$
       (c) $3b$ **(d)** $c + 10$
       (e) $5z - 4$

2.2   **(a)** $6a$ **(b)** $4x$
       (c) $9ac$ **(d)** $3b$
       (e) $10x$ **(f)** $6b$

2.3   **(a)** $5x + 10$ **(b)** $4 - 12b$
       (c) $3b + bx$ **(d)** $-4x + 20$
       (e) $-2x + xc$

2.4   **(a)** $b - c = 3 - (-2) = 5$
       (b) $ac = 7 \times -2 = -14$
       (c) $a + c - d = 7 - 2 + 1 = 6$
       (d) $b + bc = 3 + 3 \times -2 = -3$
       (e) $abcd = 7 \times 3 \times -2 \times 1 = -42$

2.5 $V = 5 \times 2 \times 4 = 40$

2.6 $M = \dfrac{13}{13-3} = 1.3$

2.7 $y = \dfrac{12-3x}{4}$

x	-2	0	5
y	4.5	3	$\dfrac{-3}{4}$

2.8 $3a + (2 \times 3) = 12$. So, $a = 2$.

2.9 $-3 = 6 + p$. So, $p = -9$.

2.10 **(a)** $c = 100e$ **(b)** $s = 60n$

      (c) $m = \dfrac{y}{100}$

2.11 $3 \times -2 \times 4 - (-2) = -22$

2.12 $S = \dfrac{60}{2}(2+4) = 180$

2.13 $\dfrac{x}{8} = \dfrac{65}{32}$. So, $x = 16.25$.

2.14 $-4x - 23y$

2.15 $A = 3(c-d)$ or $60 = 3(c-4)$

      $\therefore c = 24$

Chapter 3

3.1 **(a)** 36   **(b)** 7

      (c) 25

3.2 **(a)** $12x^3$ **(b)** a^8

      (c) $5x^3$ **(d)** $24y^4$

      (e) $8x^6$

3.3 9.3×10^7

3.4   3.9

3.5 $x^2 - 3y = 6^2 - 3 \times -2 = 42$

3.6 $xy^2 = -2 \times 4^2 = -32$

3.7   493.039

3.8 a^2b

3.9 **(a)** $\dfrac{\sqrt{14}}{\sqrt{2}} = \dfrac{\sqrt{7} \times \sqrt{2}}{\sqrt{2}} = \sqrt{7}$

      (b) $\dfrac{10\sqrt{3}}{\sqrt{5}} = \dfrac{2 \times 5 \times \sqrt{3} \times \sqrt{5}}{\sqrt{5} \times \sqrt{5}} = 2\sqrt{15}$

      (c) $\dfrac{10}{\sqrt{2}} = \dfrac{5 \times 2 \times \sqrt{2}}{\sqrt{2} \times \sqrt{2}} = 5\sqrt{2}$

      (d) $\dfrac{6ab\sqrt{cd}}{3\sqrt{ad}} = \dfrac{6ab\sqrt{c}\sqrt{d}\sqrt{a}}{3\sqrt{a}\sqrt{d}\sqrt{a}} = 2b\sqrt{ac}$

(e) $\dfrac{4\sqrt{5}}{\sqrt{2}} = \dfrac{2 \times 2 \times \sqrt{5} \times \sqrt{2}}{\sqrt{2} \times \sqrt{2}} = 2\sqrt{10}$

3.10 **(a)** $10\sqrt{15}$ **(b)** $2\sqrt{16} = 8$

      (c) $8\sqrt{150} = 8\sqrt{25}\sqrt{6} = 40\sqrt{6}$

      (d) $6 \times 7\sqrt{7} = 42\sqrt{7}$

3.11 **(a)** $\log 20 = \log 2 \times 10 = \log 2 + \log 10$

           $\log 20 = \log (4 \times 5) = \log 4 + \log 5$

      (b) $\log 20 - \log 5 = \log 4 + \log 5 - \log 5 = \log 4$

3.12 **(a)** $\log_{10}\dfrac{\sqrt{b}}{a^2} = -1.5$

      (b) $\dfrac{\log_{10}b}{\log_{10}a} = \dfrac{1.8}{1.2} = 1.5$

      (c) $\log_{10}a^3b^3 = 9$

3.13 **(a)** $\ln 9 = \ln 3 + \ln 3 = 2y$

      (b) $\ln 3\sqrt{3} = \ln 3 + \dfrac{1}{2}\ln 3 = \dfrac{3}{2}y$

      (c) $\ln 12 = \ln 2 + \ln 2 + \ln 3 = 2x + y$

3.14 -0.005

3.15 **(a)** $\log_3 5 + \log_3 3 - \log_3 30 = \log_3\dfrac{1}{2}$

                 $= \dfrac{\ln\dfrac{1}{2}}{\ln 3} = -0.631$

      (b) $\log_3 10 - \log_3 45 + 2\log_3 3 = \log_3\left(\dfrac{10 \times 9}{45}\right)$

                 $= \log_3 2$

                 $= \dfrac{\ln 2}{\ln 3}$

                 $= 0.631$

      (c) $\log_2\sqrt{3} - \log_3\sqrt{6} = \dfrac{1}{2}\log_2 3 - \dfrac{1}{2}\log_3 6$

                 $= \dfrac{1}{2}\left(\dfrac{\ln 3}{\ln 2} - \dfrac{\ln 6}{\ln 3}\right) = -0.023$

      (d) $(2\log_{10}20 - \log_4 6 + 2\log_5 2 + \log_4 3)$

                 $= \log_{10}20^2 + \log_4\dfrac{3}{6} + \log_5 2^2$

                 $= \dfrac{\ln 400}{\ln 10} + \dfrac{\ln\dfrac{1}{2}}{\ln 4} + \dfrac{\ln 4}{\ln 5}$

                 $= 2.963$

Chapter 4

4.1 **(a)** $3 + \sqrt{15} + \sqrt{6} + \sqrt{10}$

      (b) $6\sqrt{5} + 4\sqrt{15} - 9 - 6\sqrt{3}$

(c) $\quad 9 + 2\sqrt{10}$ 　　　**(d)** $\quad 1$

(e) $\quad 10$

4.2 **(a)** $\quad a^2 - b^2$

(b) $\quad 6a^2 + 5a - 6$

(c) $\quad -2x^2 + 15x - 25$

(d) $\quad 4x^4 + 8x^3 + 3x^2 + 8x + 4$

(e) $\quad -6x^4 - 7x^3 + 19x^2 + 2x - 8$

4.3 **(a)** $\quad xy(x-y)$

(b) $\quad x(1+3yz)$

(c) $\quad x(12x^2 + 13x + 24)$

(d) $\quad 3x^2(x^2+5)$

(e) $\quad (5-4x)(5+4x)$

(f) $\quad (0.3a - 0.5)(0.3a + 0.5)$

(g) $\quad (a^n - 1)(a^n + 1)$

(h) $\quad (x+y-1)(x+y+1)$

4.4 **(a)** $\quad (x+4)(x-1)$

(b) $\quad (m+8)(m-3)$

(c) $\quad (ax-3)(ax+2)$

(d) $\quad (t+n+7)(t+n-2)$

(e) $\quad (3x-1)(x+2)$

4.5 **(a)** $\quad \dfrac{x^2 + 7x + 12}{x+3} = \dfrac{(x+4)(x+3)}{x+3} = x+4$

(b) $\quad \dfrac{10a^2 + 7a - 12}{2a+3} = \dfrac{(5a-4)(2a+3)}{2a+3}$
$$= 5a - 4$$

(c) $\quad \dfrac{x^2 - 16}{4+x} = \dfrac{(x-4)(x+4)}{x+4} = x-4$

(d) $\quad \dfrac{12a^2 + 6a - 18}{2a+3} = \dfrac{(6a-6)(2a+3)}{2a+3}$
$$= 6(a-1)$$

4.6 **(a)** $\quad \dfrac{1}{\sqrt{3}} \times \dfrac{\sqrt{3}}{\sqrt{3}} = \dfrac{\sqrt{3}}{3}$

(b) $\quad \dfrac{2}{3\sqrt{7}} \times \dfrac{\sqrt{7}}{\sqrt{7}} = \dfrac{2\sqrt{7}}{21}$

(c) $\quad \dfrac{3}{\sqrt{7}+2} \times \dfrac{\sqrt{7}-2}{\sqrt{7}-2} = \dfrac{3(\sqrt{7}-2)}{3} = \sqrt{7}-2$

(d) $\quad \dfrac{\sqrt{6}}{\sqrt{5}+\sqrt{2}} \times \dfrac{\sqrt{5}-\sqrt{2}}{\sqrt{5}-\sqrt{2}} = \dfrac{\sqrt{30}-\sqrt{12}}{3}$

(e) $\quad \dfrac{\sqrt{3}+\sqrt{2}}{\sqrt{3}-\sqrt{2}} \times \dfrac{\sqrt{3}+\sqrt{2}}{\sqrt{3}+\sqrt{2}} = 5+2\sqrt{6}$

(f) $\quad \dfrac{7-3\sqrt{5}}{3\sqrt{5}-2\sqrt{2}} \times \dfrac{3\sqrt{5}+2\sqrt{2}}{3\sqrt{5}+2\sqrt{2}}$
$$= \dfrac{21\sqrt{5} + 14\sqrt{2} - 45 - 6\sqrt{10}}{37}$$

4.7 $\quad \dfrac{a^2\sqrt{b}}{\sqrt[3]{c}} = \dfrac{10^2\sqrt{100}}{\sqrt[3]{1000}} = 100$

4.8 **(a)** $\quad 2a + 3b - c = 8 - 9 - 5 = -6$

(b) $\quad 6a - 3b - 2d = 24 + 9 + 12 = 45$

(c) $\quad 5a - \dfrac{cd}{2} + \dfrac{abc}{3} = 20 - \left(\dfrac{-30}{2}\right) + \left(\dfrac{-60}{3}\right) = 15$

4.9 $\quad x^2 - 7x + 12 = 0$, when $x = 3$.

$x^2 - 7x + 12 = 0$, when $x = 4$.

$x^2 - 7x + 12 = 30$, when $x = -2$.

4.10 $\quad 4(x-y) + 5(x+y) = 84$, when $x = 9$ and $y = 3$.

$7(x+y) + 2x - 6y = 84$, when $x = 9$ and $y = 3$.

$\therefore\ 4(x-y) + 5(x+y) = 7(x+y) + 2x - 6y$,
$$\text{when } x = 9 \text{ and } y = 3.$$

4.11 $\quad a^3 + 19a = \begin{cases} 20, \text{ when } a = 1. \\ 84, \text{ when } a = 3. \\ 140, \text{ when } a = 4. \end{cases}$

$4(2a^2 + 3) = \begin{cases} 20, \text{ when } a = 1. \\ 84, \text{ when } a = 3. \\ 140, \text{ when } a = 4. \end{cases}$

$\therefore\ a^3 + 19a = 4(2a^2 + 3)$, when $a = 1, 3$ or 4.

When $a = 2$, $\begin{cases} a^3 + 19a = 46 \\ 4(2a^2 + 3) = 44 \end{cases}$

$\therefore\ a^3 + 19a > 4(2a^2 + 3)$, when $a = 2$.

4.12 $\quad 16 - x + x^2 = \begin{cases} 18, \text{ when } x = -1. \\ 28, \text{ when } x = 4. \\ 72, \text{ when } x = -7. \\ 106, \text{ when } x = 10. \end{cases}$

Chapter 5

5.1 **(a)** $\quad x = 9$ 　　　**(b)** $\quad y = 5$

(c) $\quad a = 8$ 　　　**(d)** $\quad a = 4$

(e) $\quad x = 6$ 　　　**(f)** $\quad y = 4$

(g) $\quad x = 10\dfrac{2}{3}$ 　　**(h)** $\quad x = 4\dfrac{1}{2}$

(i) $\quad x = 7$ 　　　**(j)** $\quad x = 10$

(k) $\quad x = 15$ 　　　**(l)** $\quad x = 6$

(m) $\quad x = 5$ 　　　**(n)** $\quad x = 18$

5.2 **(a)** $\quad (x = 2, y = 3)$

(b) $(x = 2, y = 1)$

(c) $(a = 1, b = 3)$

(d) $(x = 2, y = 6)$

(e) $(m = 3, n = -1)$

5.3 **(a)** $x = 6$ or $x = 2$

(b) $x = 4$ or $x = -7$

(c) $x = 5$ or $x = 2$

(d) $x = 0$ or $x = -5$

(e) $x = 0$ or $x = \dfrac{1}{2}$

(f) $x = 3$ or $x = 4$

(g) $x = -5$ or $x = 3$

(h) $x = 0$ or $x = -5$

(i) $2(x + 8)(x - 2) = 0$
So, $x = -8$ or $x = 2$.

(j) $x = \dfrac{3}{2}$ or $x = -\dfrac{3}{2}$

(k) $(x - 11)(x + 1) = 0$
So, $x = 11$ or $x = -1$.

(l) $(x + 4)(x + 1) = 0$
So, $x = -4$ or $x = -1$.

5.4 **(a)** $x = \dfrac{-6 \pm \sqrt{64}}{2}$. So, $x = -7, 1$

(b) $x = \dfrac{-6 \pm \sqrt{20}}{2} = -3 \pm \sqrt{5}$

(c) $x = \dfrac{2 \pm \sqrt{84}}{10} = \dfrac{1 \pm \sqrt{21}}{5}$

(d) $x = \dfrac{7 \pm \sqrt{37}}{2}$

(e) $x = \dfrac{2 \pm \sqrt{28}}{6} = \dfrac{1 \pm \sqrt{7}}{3}$

(f) $x = \dfrac{1 \pm \sqrt{13}}{6}$

(g) $x = \dfrac{6 \pm \sqrt{36}}{2}$. So, $x = 0, 6$

(h) $x = \dfrac{6 \pm \sqrt{44}}{2} = 3 \pm \sqrt{11}$

(i) $x = \dfrac{-3 \pm \sqrt{49}}{4}$. So, $x = -\dfrac{10}{4}, 1$

5.5 **(a)** $x = 4$ **(b)** $x = 6$

(c) $x = 4$ **(d)** $x = 2$

(e) $x = -2$ **(f)** $x = 9$

(g) $x = 5\dfrac{1}{2}$ **(h)** $x = 2\dfrac{2}{3}$

(i) $x = -5$ **(j)** $x = 1$

5.6 **(a)** $x = -3$ or $x = 3$

(b) $x = -2$ or $x = 3$

(c) $x = -8$ or $x = 8$

(d) $x = -5$ or $x = 3$

(e) $x = -\dfrac{1}{9}$ or $x = 1$

5.7 **(a)** Let $u = x^2$. So, $u^2 - 5u + 4 = 0$,
or $(u - 4)(u - 1) = 0$.
Hence, $x = \pm 2$ or $x = \pm 1$.

(b) Let $u = x^3$. So, $u^2 - 9u + 8 = 0$,
or $(u - 8)(u - 1) = 0$.
Hence, $x = 2$ or $x = 1$.

(c) Let $u = (x + 1)^2$. So, $u^2 - 6u + 8 = 0$,
or $(u - 4)(u - 2) = 0$.
$\therefore (x + 1)^2 - 4 = 0$ or $(x + 1)^2 - 2 = 0$
So, $x = \dfrac{-2 \pm \sqrt{16}}{2}$ or $x = \dfrac{-2 \pm \sqrt{8}}{2}$.
$\therefore x = -3, 1$ or $x = -1 \pm \sqrt{2}$.

(d) Let $u = x + \dfrac{1}{x}$. So, $u^2 - 5u + 6 = 0$,
or $(u - 3)(u - 2) = 0$.
$\therefore \left(x + \dfrac{1}{x}\right) - 3 = 0$ or $\left(x + \dfrac{1}{x}\right) - 2 = 0$.
That is, $x^2 - 3x + 1 = 0$ or $x^2 - 2x + 1 = 0$.
$\therefore x = \dfrac{3 \pm \sqrt{5}}{2}$ or $x = \dfrac{2 \pm 0}{2} = 1$.

(e) Let $u = 5^x$. So, $u^2 - 26u + 25 = 0$
or $(u - 25)(u - 1) = 0$.
$\therefore 5^x - 25 = 0$ or $5^x - 1 = 0$.
That is, $x^x = 5^2$ or $5^x = 5^0$.
$\therefore x = 2$ or $x = 0$.

(f) Let $u = (2x + 1)^2$. So, $u^2 - u - 12 = 0$,
or $(u - 4)(u + 3) = 0$.
$\therefore (2x + 1)^2 - 4 = 0$ or $(2x + 1)^2 + 3 = 0$.
That is, $x = \dfrac{-4 \pm \sqrt{64}}{8}$ or $x = \dfrac{-4 \pm \sqrt{-48}}{8}$.
Since $x = \dfrac{-4 \pm \sqrt{-48}}{8}$ has no solution, we
find that $x = -\dfrac{12}{8}, \dfrac{1}{2}$ or $x = -\dfrac{3}{2}, \dfrac{1}{2}$.

Chapter 6

6.1 $5x \leq -40$. So, $x \leq -8$.

6.2 $7x < 21$. So, $x < 3$.

6.3 $6x < 27$. So, $x < 4.5$.

6.4 $4x + 4 \geq 2x$. So, $x \geq -2$.

6.5 $6x - 5 \le 9x + 11$. So, $3x \ge -16$ or $x \ge -5\frac{1}{3}$.

6.6 $x + 6 \le 4x$. So, $x \ge 2$.

6.7 $5x + 1 > 3x + 12$. So, $x > 5.5$.

6.8 $2x + 25 < 15$. So, $x < -5$.

6.9 $x + 6 < 15$. So, $x < 9$.

6.10 $3x \le 2x + 4$. So, $x \le 4$.

6.11 $x < 4$, $x \ne 0$.

6.12 $6 < 5x - 5$. So, $x > 2.2$.

6.13 $12 > 12x + 8$. So, $x < \frac{1}{3}$.

6.14 $-5 < (2x - 1) < 5$. So, $-2 < x < 3$.

6.15 $-5 < \frac{x}{2} < 5$. So, $-10 < x < 10$.

6.16 $3x - 1 > 2$ or $3x - 1 < -2$. So, $x > 1$ or $x < -\frac{1}{3}$.

6.17 Let x = units sold this month.
$\quad\quad\quad$ y = units sold next month.
$\quad$ Given: $x + y = 2500 \quad\quad$ —(1)
$\quad\quad\quad$ $4.5x + 5y \ge 12000 \quad\quad$ —(2)
$\quad$ Solving (1) and (2) simultaneously gives
$\quad$ $x \le 1000$, $y \ge 1500$.
$\quad$ ∴ maximum units sold this month is 1000.

6.18 Let x = amount of money invested at 5.5%p.a.
$\quad\quad\quad$ y = amount of money invested at 4%p.a.
$\quad$ Given: $x + y = 50000 \quad\quad\quad$ —(1)
$\quad$ $0.055x + 0.04y \ge 0.05\,(50000) = 2500$ —(2)
$\quad$ Solving (1) and (2) simultaneously gives
$\quad$ $x \ge 33333.33$ and $y \le 16666.67$.
$\quad$ ∴ at least \$33 333.33 must be invested at 5.5%p.a.

6.19 Let x = no. of kms driven.
$\quad$ Hiring costs: $135 \times 12 + 0.05x$
$\quad$ Buying costs: $1000 + 0.10x$
$\quad$ Find x so that renting costs $\le$ buying costs.
$\quad$ That is, $1620 + 0.05x \le 1000 + 0.1x$, or $x \ge 12400$.
$\quad$ ∴ if at least 12400 kms are driven, then rental costs are $\le$ buying costs.

6.20 Let x = yearly sales level.
$\quad$ Salary: Method 1: $25000 + 0.02x$
$\quad\quad\quad\quad$ Method 2: $0.10x$.
$\quad$ Find x so that method 1 $\ge$ method 2.
$\quad$ That is, $25000 + 0.02x > 0.1x$, or $x < 312500$
$\quad$ ∴ method 1 is preferable whenever the sales level is less than \$312 500.

Chapter 7

7.1 $f(2) = f(-2) = 5$;
$\quad$ $9 - x^2 = 0$, when $x = \pm 3$.

7.2 $H\left(\frac{1}{2}\right) = \frac{1}{2} - \frac{1}{\frac{1}{2}} = \frac{1}{2} - 2 = -1\frac{1}{2}$;

$\quad$ $H(-2) = -2 - \frac{1}{-2} = -2 + \frac{1}{2} = -1\frac{1}{2} = H\left(\frac{1}{2}\right)$.

7.3 $f(1) = 2^1 + 2^{-1} = 2 + \frac{1}{2} = 2\frac{1}{2}$

$\quad$ $f(2) = 2^2 + 2^{-2} = 4 + \frac{1}{4} = 4\frac{1}{4}$

$\quad$ $f(-2) = 2^{-2} + 2^2 = 4\frac{1}{4}$

7.4 $g(3) = |3| = 3$; $g(-2) = |-2| = 2$

7.5 $f\left(\frac{1}{2}\right) = |4| = 4$; $f(2) = |10| = 10$;
$\quad$ $f(-3) = |-10| = 10$

7.6 D: all real x; R: all real $f(x)$

7.7 D: $-4 \le x \le 4$; R: $0 \le g(x) \le 4$

7.8 D: all real x, except $x = 1$
$\quad$ R: all real $h(x)$, except $h(x) = 0$

7.9 D: all $x \ge 4$; R: $g(x) \ge 0$

7.10 D: all real x; R: all real $f(x) > 0$

7.11 (a) $H(4) = 6$ $\quad$ (b) $H(0) = 0$
$\quad\quad$ (c) $H(-2) = -4$ $\quad$ (d) $H(3) = 6$

7.12 (a) No $\quad\quad\quad\quad$ (b) 1 (when $x = 0$)
$\quad\quad$ (c) No. $(x + 3)^2 = 0$, when $x = -3$.
$\quad\quad$ (d) When $x = 0$, $y = 8$. ∴ graph cuts y-axis at $(0,8)$.

7.13 (a) hyperbola $\quad\quad$ (b) linear (staight line)
$\quad\quad$ (c) parabola $\quad\quad\quad$ (d) logarithmic
$\quad\quad$ (e) exponential

7.14 (a) $y = 2x + 1$, linear

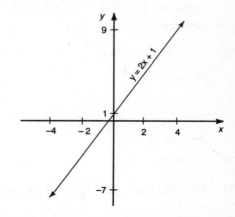

(b) $y = \dfrac{-2}{x}$

hyperbola

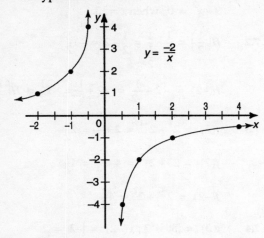

(c) $y = |x|$

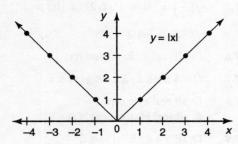

(d) $y = x^2 - 1$

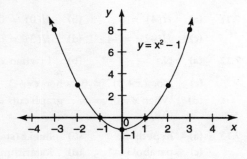

7.15 $f(-a) = (-a)^3 + 3(-a) = -a^3 - 3a;$

$-f(a) = -[a^3 + 3a] = -a^3 - 3a = f(-a),$

(true for all a).

7.16 $f(a) = a^4 = a^2 \times a^2$

$f(-a) = (-a)^4$

$= (-a)^2 \times (-a)^2$

$= a^2 \times a^2 = f(a)$

(true for all a).

7.17 Graph (i)

7.18 **(a)** 240 km **(b)** $\dfrac{240}{4} = 60$ km/hr

(c) approx. $\dfrac{1}{2}$ hr

(d) Between 1 and 2 hours into the journey when 120 km was covered in 1 hour.

7.19 The first 100 km used 15 litres of petrol.
The second and third 100 km stretches each used 10 litres of petrol.
Hence, average fuel consumption is approximately 8.57 km/litre.

7.20 **(a)** 10 centimetres **(b)** 34 cm; 24 cm
(c) 6 kg **(d)** 3 kg

Chapter 8

8.1 **(a)**

Interval	Freq.
40-49	1
50-59	5
60-69	10
70-79	9
80-89	3
90-99	2
Total	30

(b)

(c) 40, 49; 60, 69
(d) 49.5, 59.5; 69.5, 79.5
(e) 54.5

8.2 **(a)** 7.5 to 9.5 **(b)** 15%
(c) 20%
(d) 20% of 100 = 20 students

8.3 Percentage frequencies

Factory A	27%	60%	5%	8%
Factory B	50%	25%	10%	15%

(i)

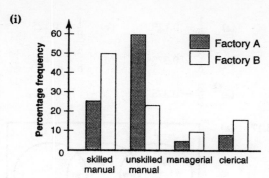

(i)

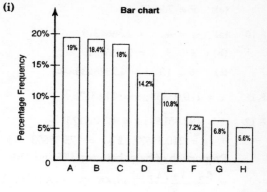

(ii)

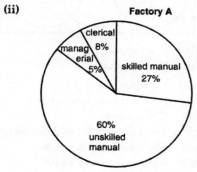

Factory A

Factory B

(ii)

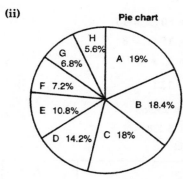

8.6 Ordered data:
96 95 94 92 91 89 89 88 88 85
83 82 79 77 76 69 66 58 50 18

(a) Mode: Bimodal (89, 88)

Median: $\dfrac{85 + 83}{2} = 84$

Mean $= \dfrac{\sum x}{n} = \dfrac{1565}{20} = 78.25$

(c) Mode: 89, 88 (bimodal)

Median $= \dfrac{88 + 85}{2} = 86.5$

Mean $= \dfrac{\sum x}{n} = \dfrac{1497}{18} = 83.17$

The mean is most affected by the low scores.

8.4

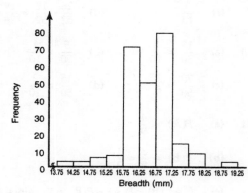

8.7 **(b)** range (1) = 10
range (2) = 10

(c) $s(1) = 4.58$

$s(2) = 2.39$

(d) Standard deviation is preferable as it measures internal variability.

8.8 **(a)** $\bar{x} = 12.143$

(b) $s = 6.311$

8.9 **(a)** mode = 63

(b) $\bar{x} = 62.7$

(c) range = 79 − 44 = 35

(d) $s^2 = 62.642$

8.5

A	Food	19%
B	Housing experiment & operation	18.4%
C	Transportation	18.0%
D	Housing	14.2%
E	Recreation – education	10.8%
F	Tobacco & alcohol	7.2%
G	Clothing	6.8%
H	Health & personal care	5.6%

(e) $s = 7.9147$

8.10   **(a)** $\bar{x} = 4$, $s = 2.3664319$

       (b) $\bar{x} = 9$, $s = 2.3664319$

       (c) $\bar{x} = 13$, $s = 4.7328638$

8.11 $\bar{x} = 110.03333$; $s = 11.65178$

8.12 $\bar{x} = 192.85165$; $s = 17.465123$

     $\bar{x} - s = 210.31677$; $\bar{x} + s = 175.38653$

     Proportion of bales between 175 and 210

     $\approx \dfrac{5 + 13 + 24 + 20}{91} \approx 0.68132$ or 68.132%

8.13   **(a)** $\bar{x} = 80$, $s = 16$

       (b) $\bar{x} = 15$, $s = 8$

       (c) $\bar{x} = 4$, $s = 0.8$

8.14   **(a)** $\bar{x} = 12.1$

          $s^2 = 29.568$

          $s = 5.438$

     (b) and (c)

Interval	x	f
4–7	5.5	4
8–11	9.5	5
12–15	13.5	7
16–19	17.5	2
20–23	21.5	1
24–27	25.5	1

     $\bar{x} = 12.3$

     $s^2 = 28.8$

     $s = 5.367$

     (d)

	Raw data	G.F.D.	|Error|
$\bar{x}$	12.1	12.3	0.2
s^2	29.568	28.8	0.768
s	5.438	5.367	0.071

     Statistics calculated from raw data are more accurate as actual data values are lost using G.F.D.

8.15   **(a)**   {copper, sodium, zinc, oxygen, uranium}

       (b)   {sodium}

       (c)   {zinc, sodium, nitrogen, potassium}

8.16

Sequence	HHH	HHT	HTH	HTT	THT	TTH	THH	TTT
No. of Heads	3	2	2	1	1	1	2	0

(a) $\dfrac{1}{8} = 0.125$ **(b)** $\dfrac{1}{8} = 0.125$

(c) $\dfrac{3}{8} = 0.375$ **(d)** $\dfrac{3}{8} + \dfrac{1}{8} = 0.5$

(e) $\dfrac{3}{8} = 0.375$

8.17   **(a)**

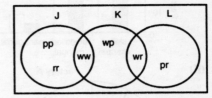

     (b)   **(i)** $J^| = \{wp, wr, pr\}$

          (ii) $J \cap K = \{ww\}$

          (iii) $K^| = \{pp, rr, pr\}$

          (iv) $J \cup L = \{pp, rr, ww, wr, pr\}$

          (v) $K \cap L = \{wr\}$

     (c)   **(i)**   The 2 flowers are of different colours.

          (ii)   The 2 flowers are white.

          (iii)   Neither flower is white.

          (iv)   The 2 flowers are *not* white and pink.

          (v)   The 2 flowers *are* white *and* red.

8.18   **(a)** $\dfrac{20}{80} = 0.25$ **(b)** $1 - \left(\dfrac{5}{80} + \dfrac{15}{80}\right) = 0.75$

       (c) $\dfrac{60}{80} + \dfrac{5}{80} = 0.8125$ **(d)** $\dfrac{15}{80} = 0.1875$

       (e) $\dfrac{15}{80} = 0.1875$

8.19   **(a)** $P(A) = \dfrac{1}{2}$ **(b)** $\dfrac{24}{52} = \dfrac{6}{13}$

       (c) $\dfrac{3}{13}$ **(d)** $\dfrac{26}{52} + \dfrac{12}{52} = \dfrac{19}{26}$

8.20   **(a)** $\dfrac{15}{50} = \dfrac{3}{10}$ **(b)** $\dfrac{5}{50} = \dfrac{1}{10}$

       (c) $\dfrac{30}{50} = \dfrac{3}{5}$ **(d)** $\dfrac{15}{50} = \dfrac{3}{10}$

8.21   **(a)** $P(A) = \dfrac{4}{9}$

       (b) $P(B) = \dfrac{3}{9} = \dfrac{1}{3}$

       (c) $P(A \cup B) = P(A) + P(B) = \dfrac{7}{9}$ (since A and B are mutually exclusive).

       (d) $P(A \cap B) = 0$

8.22   **(a)** $P(A) = P(SS) = 0.81$

       (b) $P(B) = 1 - P(SS) = 0.19$

       (c) $P(C) = P(SD) + P(DS) = 0.18$

8.23   Let the letters a, b, ..., h represent each staff member in order as listed. (i.e. a = Mr A-Hassan and h = Ms Ng).

(a) $S = \{a, b, c, d, e, f, g, h\}$

(b) **(i)** $A = \{a, d, f\}$

   (ii) $B = \{a, b, d, e\}$

   (iii) $C = \{a, c, d, e, h\}$

(c)

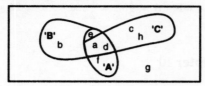

(d) $P(A) = \dfrac{3}{8}$, $P(B) = \dfrac{4}{8} = \dfrac{1}{2}$, $P(C) = \dfrac{5}{8}$

(e) **(i)** $\{g, b, e, c, h\}$

   (ii) $\{a, d\}$

   (iii) $\{a, b, c, d, e, h\}$

   (iv) $\{a, b, c, d, e, f, g, h\}$

   (v) $\{a, d\}$

(f) **(i)** $P(A^{|}) = \dfrac{5}{8}$

   (ii) $P(A \cap C) = \dfrac{2}{8} = \dfrac{1}{4}$

   (iii) $P(B \cup C) = \dfrac{6}{8} = \dfrac{3}{4}$

   (iv) $P(A \cup A^{|}) = 1$

   (v) $P(A \cap B) = \dfrac{2}{8} = \dfrac{1}{4}$

8.24 **(a)** Required prob. $= 0.4 + 0.3 - 0.2$
$= 0.5$ (the *additive* rule)

(b) Required prob. $= (0.4 - 0.2) + (0.3 - 0.2)$
$= 0.3$

Chapter 9

9.1 **(a)** $P = 4 \times 10 = 40$ cm.

(b) $P = (2 \times 32) + (2 \times 22) = 108$ cm.

(c) $P = 5 \times 10 = 50$ cm.

(d) $C = 2 \times \pi \times 5 = 31.42$ cm.

(e) $P = 3 \times 3 = 9$ cm.

(f) $P = 6 + 6 + (2 \times 3) + 10 = 28$ cm.

(g) $P = 12.6$ cm.

(h) $P = \dfrac{1}{4} \times 2\pi r + 16 = 28.57$ cm.

(i) $P = 9.8$ cm.

9.2 **(a)** $A = lb = 48$ sq. m.

(b) $A = l^2 = 17.64$ sq. m.

(c) $A = bh = 96$ sq. m.

(d) $A = \dfrac{1}{2}bh = 198$ sq. mm.

(e) $A = \dfrac{1}{2}h(a+b) = 1410$ sq. mm.

(f) $A = \pi r^2 = 38.48$ sq. cm.

(g) $A = \dfrac{1}{2}h(a+b) = 1085$ sq. cm.

(h)

$A_1 = 35 \times 18 = 630$

$A_2 = 15 \times 12 = 180$

$A_3 = 10 \times 25 = 250$

$\therefore$ total area $= A_1 + A_2 + A_3 = 1060$ sq. mm.

9.3 **(a)** **(i)** $V = lbh = 1440$ cubic cm.

   (ii) $SA = (2 \times 12 \times 12) + (2 \times 12 \times 10)$
$+ (2 \times 12 \times 10) = 768$ sq. cm.

(b) **(i)** $V = \pi r^2 h = 339.292$ cubic cm.

   (ii) $SA = 2 \times \pi r^2 + 2\pi rh = 282.743$ sq. cm.

(c) **(i)** $V = \dfrac{1}{3}\pi r^3 h = 37.699$ cubic cm.

   (ii) $SA = \pi ra + \pi r^2 = 75.398$ sq. cm.

(d) **(i)** $V = \dfrac{1}{3}abh = 2240$ cubic mm.

(e) **(i)** $V = \pi r^2 h = 314.159$ cubic cm.

   (ii) $SA = 2\pi r^2 + 2\pi rh = 634.602$ sq. cm.

(f) V (rect. prism) $= lbh = 4410$ cubic m.

V (rect. pyramid) $= \dfrac{1}{3}abh = 2940$ cubic m.

$\therefore$ total volume $= 7350$ cubic metres.

9.4 **(a)** $SA = 6 \times 5^2 = 150$ sq. cm.

(b) $A = 2\pi rh = 980.177$ sq. cm.

No. of tins needed $= \dfrac{980.177}{15} = 66$

$\therefore$ total cost $= 20 \times 66 = \$1\ 320$.

9.5 **(a)** **(i)** $d = \sqrt{(x_2 - x_1)^2 + (y_2 - y_1)^2} = \sqrt{25} = 5$

   (ii) midpoint $= \dfrac{x_1 + x_2}{2}, \dfrac{y_1 + y_2}{2} = (5.5, 4)$

   (iii) gradient, $m = \dfrac{y_2 - y_1}{x_2 - x_1} = \dfrac{4}{3}$ or $1\dfrac{1}{3}$

(b) **(i)** $d = \sqrt{(-1-2)^2 + (3-0)^2} = \sqrt{18}$

(ii) midpoint $= \dfrac{2-1}{2}, \dfrac{0+3}{2} = \left(\dfrac{1}{2}, 1\dfrac{1}{2}\right)$

or $(0.5, 1.5)$

(iii) gradient, $m = \dfrac{3-0}{-1-2} = -1$

(c) **(i)** $d = \sqrt{(3+5)^2 + (-2+1)^2} = \sqrt{65}$

(ii) midpoint

$= \dfrac{-5+3}{2}, \dfrac{-1-2}{2} = (-1, -1.5)$

(iii) gradient, $m = \dfrac{-2+1}{3+5} = -\dfrac{1}{8}$

9.6 $AB = \sqrt{(\sqrt{3}-0)^2 + (1-0)^2} = 2$

$BC = \sqrt{(0-\sqrt{3})^2 + (2-1)^2} = 2$

$AC = \sqrt{(0-0)^2 + (2-0)^2} = 2$

$\therefore \triangle ABC$ is equilateral.

9.7 Since $\dfrac{x_1 + x_2}{2} = 3$ and $\dfrac{y_1 + y_2}{2} = -1$,

with $x_1 = -1$, $y_1 = -2$, we have

$\dfrac{-1 + x_2}{2} = 3$ and $\dfrac{-2 + y_2}{2} = -1$.

$\therefore x_2 = 7$ and $y_2 = 0$.

That is, the coordinates are $(7,0)$.

9.8 gradient of $AB = \dfrac{8-2}{4-2} = 3$

gradient of $BC = \dfrac{4-8}{5-4} = -4$

gradient of $CA = \dfrac{4-2}{5-2} = \dfrac{2}{3}$

9.9 midpoint $= \dfrac{3+5}{2}, \dfrac{-2+2}{2} = (4,0)$

length, $d = \sqrt{(4-1)^2 + (0-4)^2} = 5$

9.10 **(a)** gradient of $AB = \dfrac{4-0}{6-0} = \dfrac{2}{3}$

(b) gradient of $BC = \dfrac{-2-4}{10-6} = -\dfrac{3}{2}$

(c) length of $AB = \sqrt{(6-0)^2 + (4-0)^2}$

$= \sqrt{52}$

length of $BC = \sqrt{(10-6)^2 + (-2-4)^2}$

$= \sqrt{52}$

length of $AC = \sqrt{(10-0)^2 + (-2-0)^2}$

$= \sqrt{104}$

(d) $m_1 = \dfrac{2}{3}$, $m_2 = -\dfrac{3}{2}$ $\therefore m_1 m_2 = -1$

Hence, AB is perpendicular to BC.

(e) $A = \dfrac{1}{2}bh = \dfrac{1}{2}\sqrt{52} \cdot \sqrt{52} = 26$ square units

Chapter 10

10.1 **(a)** $\cos \alpha = \dfrac{4}{5}$ **(b)** $\tan \alpha = \dfrac{3}{4}$

(c) $\sec \alpha = \dfrac{5}{4}$ **(d)** $\operatorname{cosec} \alpha = \dfrac{5}{3}$

(e) $\cot \alpha = \dfrac{4}{3}$

10.2 **(a)** $\sin 35° = \dfrac{x}{10}$. $\therefore x = 10 \sin 35° = 5.7$

(b) $\cos 45° = \dfrac{x}{8}$. $\therefore x = 8 \cos 45° = 5.7$

(c) $\tan 65° = \dfrac{x}{17}$. $\therefore x = 17 \tan 65° = 36.5$

(d) $\tan 36° = \dfrac{5}{x}$

$\therefore x = \dfrac{5}{\tan 36°} = 6.9$

(e) $\sin 85° = \dfrac{10}{x}$

$\therefore x = \dfrac{10}{\sin 85°} = 10.0$

(f) $\cos 36° = \dfrac{18}{x}$

$\therefore x = \dfrac{18}{\cos 36°} = 22.2$

10.3 **(a)** $\tan \alpha = \dfrac{1}{1}$; $\alpha = 45°$

(b) $\cos \alpha = \dfrac{1}{2}$; $\alpha = 60°$

(c) $\sin \alpha = \dfrac{5}{7}$; $\alpha = 46°$

(d) $\sin \alpha = \dfrac{3}{5}$; $\alpha = 37°$

(e) $\tan \alpha = \dfrac{4}{3}$; $\alpha = 53°$

10.4 **(a)** $\dfrac{\sin B}{4} = \dfrac{\sin 60°}{6}$

$\therefore \sin B = \dfrac{4 \sin 60°}{6}$

$\therefore B = 35° \, 16'$

(b) $\dfrac{\sin A}{5} = \dfrac{\sin 120°}{6}$

$\therefore \sin A = \dfrac{5 \sin 120°}{6}$

$$\therefore A = 46°\ 12'$$

(c) $\dfrac{a}{\sin 37°} = \dfrac{16}{\sin 48°}$

$$\therefore a = \dfrac{16\sin 37°}{\sin 48°} = 13.0$$

(d) $c^2 = 10^2 + 5^2 - 2 \times 10 \times 5\cos 47°$

$$= 56.80016399$$

$$\therefore c = 7.5$$

(e) $a^2 = 5.4^2 + 4.5^2 - 2 \times 5.4 \times 4.5\cos 138°$

$$= 85.52683852$$

$$\therefore a = 9.2$$

(f) $\cos C = \dfrac{5^2 + 4^2 - 3.4^2}{2 \times 5 \times 4} = 0.736$

$$\therefore C = 42°\ 36'$$

(g) **(i)** $\angle B = 40°$

(ii) $\dfrac{5}{\sin 30°} = \dfrac{AC}{\sin 40°}$

$$\therefore AC = \dfrac{5\sin 40°}{\sin 30°} = 6.4$$

(iii) $\dfrac{AD}{\sin 70°} = \dfrac{6.4278761}{\sin 65°}$

$$\therefore AD = 6.7$$

10.5 **(a)** $A = \dfrac{1}{2}ab\sin C = 23.8$ sq. units

(b) $A = \dfrac{1}{2}ac\sin B = 54.4$ sq. units

(c) $A = \dfrac{1}{2}ac\sin B = 46.7$ sq. units

(d) $A = \dfrac{1}{2}bc\sin A = 162.5$ sq. units

(e) $A = \dfrac{1}{2}ab\sin C = 15.1$ sq. units

10.6

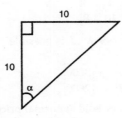

$$\tan\alpha = \dfrac{10}{10} = 1$$

$$\therefore \alpha = 45°$$

10.7 $33°\ 41'$

10.8

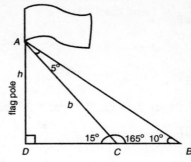

$$\dfrac{50}{\sin 5°} = \dfrac{b}{\sin 10°}$$

$$\therefore b = \dfrac{50\sin 10°}{\sin 5°} = 99.6\ \text{m}.$$

If h = height of flagpole, then $\sin 15° = \dfrac{h}{b}$

$$\therefore h = 99.61946981\sin 15° = 25.8\ \text{m}.$$

10.9

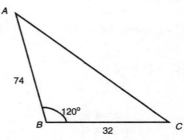

$$b^2 = a^2 + c^2 - 2ac\cos B = 8868$$

$$\therefore b = 94.2\ \text{cm}.$$

Perimeter $= a + b + c = 200.2$ m.

10.10

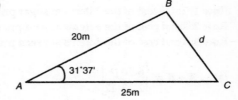

Let d = distance

Then, $d^2 = 20^2 + 25^2 - 2 \times 20 \times 25\cos 31°\ 37'$

$$= 173.4255232$$

$$\therefore d = 13.2\ \text{m}.$$

10.11 **(a)** $\cos^2\alpha - \sin^2\alpha = \cos^2\alpha - (1 - \cos^2\alpha)$

$$= 2\cos^2\alpha - 1$$

or $\cos^2\alpha - \sin^2\alpha = (1 - \sin^2\alpha) - \sin^2\alpha$

$$= 1 - 2\sin^2\alpha$$

(b) $2\cos 4\theta$

(c) $2\cos^2 B$

(d) $2\sin B \cos B$

10.12 If $\sec^2 B = 1$, $\sec B = 1$

$\therefore \dfrac{1}{\cos B} = 1$ and $\cos B = 1$

$\therefore B = 0°$

Chapter 11

11.1 $a_{33} = 8 \,; a_{21} = 7 \,; b_{23} = 9 \,; a_{32} = 9 \,; b_{22} = 0$

11.2 $[4 \times 2]$

11.3 $[3 \times 3]$

11.4 $[4 \times 1]$

11.5 $[1 \times 5]$

11.6   **(a)** $A + B = \begin{bmatrix} 5 & 1 & 5 \\ 5 & 2 & 11 \end{bmatrix}$

      (b) $A - B = \begin{bmatrix} -1 & 5 & -3 \\ 3 & 2 & -1 \end{bmatrix}$

      (c) $C + D$ cannot be found.

      (d) $C^T + D = \begin{bmatrix} 5 & 6 \\ 4 & 5 \\ 3 & 3 \end{bmatrix}$

      (e) $B - D^T = \begin{bmatrix} 2 & -4 & 2 \\ -2 & -4 & 5 \end{bmatrix}$

11.7 $B \cdot A = \begin{bmatrix} 3.9 & 1.95 & 4.8 \\ 23 & 11.5 & 26 \\ 6.8 & 3.4 & 7.6 \end{bmatrix}$

Row 1: Total cost of the 3 items at **sugar** prices.
Row 2: Total cost of the 3 items at **beef** prices.
Row 3: Total cost of the 3 items at **bread** prices.

11.8   **(a)** $A + B = \begin{bmatrix} 6 & 10 & 8 \\ 2 & 13 & 12 \end{bmatrix}$

      (b) $A - B = \begin{bmatrix} -4 & -4 & 4 \\ 2 & 1 & -4 \end{bmatrix}$

      (c) $E - G = \begin{bmatrix} -5 & 1 & 0 \\ 2 & -15 & -2 \\ -11 & -9 & 0 \end{bmatrix}$

      (d) $G + F - E = \begin{bmatrix} 17 & 1 & 0 \\ 1 & 22 & 7 \\ 20 & 20 & 4 \end{bmatrix}$

      (e) $A + B + C + D = \begin{bmatrix} 28 & 16 & 23 \\ 15 & 21 & 17 \end{bmatrix}$

      (f) $7E = \begin{bmatrix} 7 & 21 & 56 \\ 49 & 28 & 35 \\ 14 & 42 & 63 \end{bmatrix}$

(g) $\dfrac{1}{6}D = \begin{bmatrix} \dfrac{2}{3} & \dfrac{1}{6} & 1 \\ \dfrac{1}{3} & \dfrac{5}{6} & \dfrac{1}{2} \end{bmatrix}$

(h)

$5C - B - 2D = \begin{bmatrix} 90 & 25 & 45 \\ 55 & 15 & 10 \end{bmatrix} + \begin{bmatrix} 5 & 7 & 2 \\ 0 & 6 & 8 \end{bmatrix} - \begin{bmatrix} 8 & 2 & 12 \\ 4 & 10 & 6 \end{bmatrix}$

$\qquad\qquad = \begin{bmatrix} 87 & 30 & 35 \\ 51 & 11 & 12 \end{bmatrix}$

(i) $F - 3E = \begin{bmatrix} 9 & -7 & -24 \\ -18 & -5 & -10 \\ 3 & -7 & -23 \end{bmatrix}$

So, $\dfrac{1}{4}(F - 3E) = \begin{bmatrix} \dfrac{9}{4} & \dfrac{-7}{4} & -6 \\ \dfrac{-9}{2} & \dfrac{-5}{4} & \dfrac{-5}{2} \\ \dfrac{3}{4} & \dfrac{-7}{4} & \dfrac{-23}{4} \end{bmatrix}$

11.9   **(a)** $A \cdot B = \begin{bmatrix} 51 & 42 \\ 49 & 51 \\ 78 & 63 \\ 25 & 22 \end{bmatrix} \,; \; B.A$ is not possible.

      (b) $C \cdot D = \begin{bmatrix} 4 & 5 & 6 \\ 8 & 10 & 12 \\ 12 & 15 & 18 \end{bmatrix} \,; \; D \cdot C = [32]$

      (c) $E \cdot F = \begin{bmatrix} 0 & 12 \\ 1 & 5 \end{bmatrix} \,; \; F \cdot E = \begin{bmatrix} 3 & -3 \\ -6 & 2 \end{bmatrix}$

      (d) $G \cdot H$ is not possible;

$H \cdot G = \begin{bmatrix} 1 & 4 & 4 & 2 & 3 \\ 3 & 8 & 10 & 4 & 9 \end{bmatrix}$

      (e)   Neither $J \cdot K$ nor $K \cdot J$ is possible.

11.10   **(a)** $\begin{bmatrix} 400 & 100 \\ 300 & 100 \end{bmatrix} \cdot \begin{bmatrix} 6 \\ 8 \end{bmatrix} = \begin{bmatrix} 3200 \\ 2600 \end{bmatrix}$

Hence, if tins are sold at normal price, gross amount realised is $3 200 for white plus $2 600 for green or $5 800 altogether.

      (b) $\begin{bmatrix} 400 & 100 \\ 300 & 100 \end{bmatrix} \cdot \begin{bmatrix} 5 \\ 7 \end{bmatrix} = \begin{bmatrix} 2700 \\ 2200 \end{bmatrix}$

Hence, if tins are sold at sale price, gross amount realised is $2 700 for white plus $2 200 for green or $4 900 altogether.

11.11   **(a)** $1.5C = \begin{bmatrix} 1.05 & 4.50 & 2.33 & 1.35 \\ 3.00 & 6.30 & 0.30 & 4.65 \\ 3.15 & 12.00 & 3.00 & 1.95 \end{bmatrix}$

(b) $C \cdot P = \begin{bmatrix} 0.7 & 3.0 & 1.55 & 0.9 \\ 2.0 & 4.2 & 0.2 & 3.1 \\ 2.1 & 8.0 & 2.0 & 1.3 \end{bmatrix} \cdot \begin{bmatrix} 1000 \\ 500 \\ 800 \\ 50000 \end{bmatrix}$

$= \begin{bmatrix} 48440 \\ 159260 \\ 72700 \end{bmatrix}$

Interpretation:

(i) Total labour costs for production
= \$48 440.

(ii) Total material costs for production
= \$159 260.

(iii) Total overheads for production
= \$72 700.

11.12 A and C are inverses of each other since
$A \cdot C = I$

11.13 (a) (i) $|A| = -2 - 15 = -17$

(ii) $A^{-1} = \begin{bmatrix} \dfrac{1}{17} & \dfrac{5}{17} \\[2mm] \dfrac{3}{17} & \dfrac{-2}{17} \end{bmatrix}$

(b) (i) $|B| = -9 - 8 = -17$

(ii) $B^{-1} = \begin{bmatrix} \dfrac{3}{17} & \dfrac{4}{17} \\[2mm] \dfrac{2}{17} & \dfrac{-3}{17} \end{bmatrix}$

(c) (i) $|C| = 3$

(ii) $C^{-1} = \begin{bmatrix} \dfrac{-2}{3} & \dfrac{-1}{3} & \dfrac{-1}{3} \\[2mm] \dfrac{10}{3} & \dfrac{5}{3} & \dfrac{-2}{3} \\[2mm] 5 & 3 & -1 \end{bmatrix}$

(d) (i) $|D| = 2$

(ii) $D^{-1} = \begin{bmatrix} -2 & \dfrac{5}{2} & \dfrac{3}{2} \\[2mm] 4 & -5 & -2 \\[2mm] -1 & \dfrac{3}{2} & \dfrac{1}{2} \end{bmatrix}$

11.14 (a) $\begin{bmatrix} 3 & 7 \\ 2 & 4 \end{bmatrix} \cdot \begin{bmatrix} x \\ y \end{bmatrix} = \begin{bmatrix} 5 \\ 4 \end{bmatrix}$

So, $\begin{bmatrix} x \\ y \end{bmatrix} = A^{-1} \cdot \begin{bmatrix} 5 \\ 4 \end{bmatrix} = \begin{bmatrix} -2 & 3\frac{1}{2} \\ 1 & -1\frac{1}{2} \end{bmatrix} \cdot \begin{bmatrix} 5 \\ 4 \end{bmatrix} = \begin{bmatrix} 4 \\ 1 \end{bmatrix}$

Hence, $x = 4$ and $y = -1$.

(b) $\begin{bmatrix} 3 & 7 \\ 2 & 4 \end{bmatrix} \cdot \begin{bmatrix} x \\ y \end{bmatrix} = \begin{bmatrix} 27 \\ 16 \end{bmatrix}$ (interchange **(1)** and **(2)**)

So, $\begin{bmatrix} x \\ y \end{bmatrix} = \begin{bmatrix} -2 & \frac{7}{2} \\ 1 & -\frac{3}{2} \end{bmatrix} \cdot \begin{bmatrix} 27 \\ 16 \end{bmatrix} = \begin{bmatrix} 2 \\ 3 \end{bmatrix}$

Hence, $x = 2$ and $y = 3$.

(c) $\begin{bmatrix} 3 & 7 \\ 2 & 4 \end{bmatrix} \cdot \begin{bmatrix} x \\ y \end{bmatrix} = \begin{bmatrix} 15 \\ 8 \end{bmatrix}$ (multiply **(2)** by 2)

So, $\begin{bmatrix} x \\ y \end{bmatrix} = \begin{bmatrix} -2 & \frac{7}{2} \\ 1 & -\frac{3}{2} \end{bmatrix} \cdot \begin{bmatrix} 15 \\ 8 \end{bmatrix} = \begin{bmatrix} -2 \\ 3 \end{bmatrix}$

Hence, $x = -2$ and $y = 3$.

11.15 (a) (i) $\begin{bmatrix} 2 & 3 & 1 \\ 2 & 2 & 2 \\ 0 & 2 & -1 \end{bmatrix} \cdot \begin{bmatrix} x \\ y \\ z \end{bmatrix} = \begin{bmatrix} 5 \\ 12 \\ 8 \end{bmatrix}$

(ii) $|A| = 2(-6) - 3(-2) + 1(4) = -2$

So, $A^{-1} = \begin{bmatrix} 3 & \frac{-5}{2} & -2 \\ -1 & 1 & 1 \\ -2 & 2 & 1 \end{bmatrix}$

Hence, $\begin{bmatrix} x \\ y \\ z \end{bmatrix} = \begin{bmatrix} 3 & \frac{-5}{2} & -2 \\ -1 & 1 & 1 \\ -2 & 2 & 1 \end{bmatrix} \cdot \begin{bmatrix} 5 \\ 12 \\ 8 \end{bmatrix} = \begin{bmatrix} -31 \\ 15 \\ 22 \end{bmatrix}$

Thus, $x = -31$, $y = 15$ and $z = 22$.

(b) (i) $\begin{bmatrix} 2 & 3 & 4 \\ 4 & 3 & 1 \\ 1 & 2 & 4 \end{bmatrix} \cdot \begin{bmatrix} x \\ y \\ z \end{bmatrix} = \begin{bmatrix} 5 \\ 5 \\ 5 \end{bmatrix}$

(ii) $|A| = 2(10) - 3(15) + 4(5) = -5$

So, $A^{-1} = \begin{bmatrix} -2 & \frac{4}{5} & \frac{-9}{5} \\ 3 & \frac{-4}{5} & \frac{-14}{5} \\ -1 & \frac{1}{5} & \frac{6}{5} \end{bmatrix}$

Hence, $\begin{bmatrix} x \\ y \\ z \end{bmatrix} = \begin{bmatrix} -2 & \frac{4}{5} & \frac{-9}{5} \\ 3 & \frac{-4}{5} & \frac{-14}{5} \\ -1 & \frac{1}{5} & \frac{6}{5} \end{bmatrix} \cdot \begin{bmatrix} 5 \\ 5 \\ 5 \end{bmatrix} = \begin{bmatrix} 3 \\ -3 \\ 2 \end{bmatrix}$

Thus, $x = 3$, $y = -3$ and $z = 2$.

11.16 Total expenditure $= M + A = \begin{bmatrix} 80 & 98 & 180 \\ 225 & 142 & 125 \\ 85 & 115 & 103 \\ 235 & 105 & 80 \end{bmatrix}$

11.17 **(a)** $B \cdot A = \begin{bmatrix} 16825 \\ 12920 \\ 5200 \\ 5225 \end{bmatrix} \begin{matrix} \text{week 1} \\ \text{week 2} \\ \text{week 3} \\ \text{week 4} \end{matrix}$

(b) Total amount $= 16825 + 12920 + 5200 + 5225$
$= \$40\,170$

11.18 Let $x =$ no. of nuts
$y =$ no. of bolts
Then, $15x + 20y = 480$
$10x + 15y = 480$
In matrix form, $A \cdot X = B$

$\begin{bmatrix} 15 & 20 \\ 10 & 15 \end{bmatrix} \cdot \begin{bmatrix} x \\ y \end{bmatrix} = \begin{bmatrix} 480 \\ 480 \end{bmatrix}$

$|A| = 225 - 200 = 25$

$A^{-1} = \begin{bmatrix} \frac{15}{25} & \frac{-20}{25} \\ \frac{-10}{25} & \frac{15}{25} \end{bmatrix} = \begin{bmatrix} \frac{3}{5} & \frac{-4}{5} \\ \frac{-2}{5} & \frac{3}{5} \end{bmatrix}$

So, $\begin{bmatrix} x \\ y \end{bmatrix} = \begin{bmatrix} \frac{3}{5} & \frac{-4}{5} \\ \frac{-2}{5} & \frac{3}{5} \end{bmatrix} \cdot \begin{bmatrix} 480 \\ 480 \end{bmatrix} = \begin{bmatrix} -96 \\ 96 \end{bmatrix}$

Check:
$15\,(-96) + 20\,(96) = 480$
$10\,(-96) + 15\,(96) = 480$

11.19 **(a)**

	Orange	G'fruit	Apple	Banana	Lemon
Larry, L	12	5	20	6	3
Todd, T	20	3	10	4	0
Amy, A	10	10	0	12	0

(b) Prices, $P = \begin{bmatrix} 10 \\ 20 \\ 8 \\ 6 \\ 5 \end{bmatrix} \begin{matrix} \text{Or} \\ \text{Gr} \\ \text{Ap} \\ \text{Ba} \\ \text{Le} \end{matrix}$

(c) Larry: $(L \cdot P) = \begin{bmatrix} 12 & 5 & 20 & 6 & 3 \end{bmatrix} \cdot \begin{bmatrix} 10 \\ 20 \\ 8 \\ 6 \\ 5 \end{bmatrix}$

$= \begin{bmatrix} 431 \end{bmatrix}$ or $\$4.31$

Todd: $T \cdot P = \begin{bmatrix} 364 \end{bmatrix} = \$3.64$

Amy: $A \cdot P = \begin{bmatrix} 372 \end{bmatrix} = \$3.72$

(d) $L + T + A = \begin{bmatrix} 42 & 18 & 30 & 22 & 3 \end{bmatrix}$

Total spent $= \begin{bmatrix} 42 & 18 & 30 & 22 & 3 \end{bmatrix} \cdot \begin{bmatrix} 10 \\ 20 \\ 8 \\ 6 \\ 5 \end{bmatrix}$

$= 1167$
So, total spent $= \$11.67$.

11.20 Let $x =$ units of I
$y =$ units of II
$z =$ units of III
Then, $4x + y = 10$
$3x + 2y + z = 12$
$4y + 5z = 20$
or, in matrix form

$\begin{bmatrix} 4 & 1 & 0 \\ 3 & 2 & 1 \\ 0 & 4 & 5 \end{bmatrix} \cdot \begin{bmatrix} x \\ y \\ z \end{bmatrix} = \begin{bmatrix} 10 \\ 12 \\ 20 \end{bmatrix}$

Hence, $\begin{bmatrix} x \\ y \\ z \end{bmatrix} = A^{-1} \cdot B$

$|A| = 4\,(6) - 1\,(15) + 0 = 9$

$\therefore A^{-1} = \begin{bmatrix} \frac{6}{9} & \frac{-5}{9} & \frac{1}{9} \\ \frac{-15}{9} & \frac{20}{9} & \frac{-4}{9} \\ \frac{12}{9} & \frac{-16}{9} & \frac{5}{9} \end{bmatrix}$

So, $\begin{bmatrix} x \\ y \\ z \end{bmatrix} = \begin{bmatrix} \frac{6}{9} & \frac{-5}{9} & \frac{1}{9} \\ \frac{-15}{9} & \frac{20}{9} & \frac{-4}{9} \\ \frac{12}{9} & \frac{-16}{9} & \frac{5}{9} \end{bmatrix} \cdot \begin{bmatrix} 10 \\ 12 \\ 20 \end{bmatrix} = \begin{bmatrix} \frac{20}{9} \\ \frac{10}{9} \\ \frac{28}{9} \end{bmatrix}$

Hence, $x = \frac{20}{9}$, $y = \frac{10}{9}$ and $z = \frac{28}{9}$.

Chapter 12

12.1 **(a)** $\dfrac{dy}{dx} = 3$ **(b)** $\dfrac{dy}{dx} = -6$

(c) $\dfrac{dy}{dx} = 6x$ **(d)** $\dfrac{dy}{dx} = 15x^2 - 3$

(e) $\dfrac{dy}{dx} = 3x^2 - 2$

(f) $\dfrac{dy}{dx} = 4x^3 - 6x^2 + 14x - 3$

12.2 **(a)** $f'(x) = 20x$
$f'(6) = 20 \times 6 = 120$

(b) $f'(t) = 3t^2 + 7$
$f'(6) = 115$

(c) $f'(y) = 100$
$f'(6) = 100$

12.3 **(a)** $g(10) = 1105$

(b) $g(0) = 0$

(c) $g'(x) = \dfrac{1}{2} + 2x + 3x^2$

12.4 **(a)** $y' = 18x^5 + 10x$
$\therefore y'' = 90x^4 + 10$
and $y''' = 360x^3$

(b) $p' = 2q + 10$
$\therefore p'' = 2$
and $p''' = 0$

(c) $f'(x) = -3x^{-4} + 3x^2$
$\therefore f''(x) = 12x^{-5} + 6x$,
and $f'''(x) = -60x^{-6} + 6$

(d) $p' = 16q$
$\therefore p'' = 16$,
and $p''' = 0$

12.5 **(a)** $\dfrac{dy}{dx} = 64 - 8x^3$
$= 0$, when $x = 2$
$\dfrac{d^2y}{dx^2} = -24x^2 < 0$, when $x = 2$.
$\therefore y$ is maximum when $x = 2$,
and $y_{max} = 136$.

(b) $p' = 6q^2 - 150$
$= 0$, when $q = \pm 5$
$p'' = 12q > 0$, when $q = 5$
< 0, when $q = -5$
$\therefore p$ is **minimised** when $q = 5$,
with $p_{min} = -500$, and p is **maximised** when
$q = -5$, with $p_{max} = 500$.

12.6 $N' = 1.04r - 5.2$
$= 0$, when $r = 5$
$N'' = 1.04 > 0$
$\therefore N$ is minimised when $r = 5$, with $N_{min} = 19$.

12.7 **(a)** $Pr' = 9 - 9x^2$
$= 0$, when $x = 1$
Since $p'' = -18x < 0$ when $x = 1$, profit
is maximised when $x = 1$ million dollars.

(b) Profit (max.) $= 9(1) - 3(1)^3$
$= 6$ million dollars

12.8 **(a)** $s' = 2t + 6$ **(b)** $s'' = 2$

(c) $s'(5) = 16$ **(d)** $s''(10) = 2$

12.9 $s' = 9.8t = 19.6$ metres/sec, when $t = 2$.

12.10 **(a)** $\dfrac{dp}{dt} = 0.01 + 0.02t$

(b) When $t = 4$, $\dfrac{dp}{dt} = 0.09$

(c) $\dfrac{d^2p}{dt^2} = 0.02$, constant for all t.

12.11 **(a)** $MR = 500 - 10q$

(b) $MC = 3q^2 - 20q + 40$

(c) $Pr = TR - TC = 460q + 5q^2 - 50 - q^3$

(d) $MPr = 460 + 10q - 3q^2$

(e) When $q = 10$, $MR = 400$
$MC = 140$
$Pr = 4050$
$MPr = 260$

12.12 **(a)** $Pr = 20q - 5q^2$

(b) **(i)** $\dfrac{dPr}{dq} = 20 - 10q$
$= 0$, when $q = 2$

(ii) $MR = 24 - 8q$
$MC = 2q + 4$
When $MR = MC$, $q = 2$
$\therefore$ profit is maximised when $q = 2$.

12.13 $R'(x) = -2x + 8$
$= 0$, when $x = 4$
$R''(x) = -2 < 0$.
$\therefore$ to maximise revenue, ticket price = $4.00.

12.14 $80 - 16x = 0$, when $x = 5$.
$\dfrac{d^2y}{dx^2} < 0$, when $x = 5$.
Therefore, a maximum occurs after 5 hours.

12.15 **(a)** $\displaystyle\int \sqrt{x+3}\, dx = \dfrac{2}{3}(x+3)^{\frac{3}{2}} + C$

(b) $\dfrac{1}{3}(x^2 - 1)^{\frac{3}{2}} + C$

(c) $\frac{1}{6}(1-x^4)^{\frac{3}{2}} + C$

12.16 (a) $y = \int x^2\sqrt{x^3+1}\,dx = \frac{2}{9}(x^3+1)^{\frac{3}{2}} + C$

When $x = 2$, $y = 8$, giving

$8 = \frac{2}{9}(2^3+1)^{\frac{3}{2}} + C$.

Hence, $C = 2$.

Therefore, $y = \frac{2}{9}(x^3+1)^{\frac{3}{2}} + 2$

(b) $s = \int(t^2+t^{-2})\,dt = \frac{t^3}{3} - t^{-1} + C$

When $s = 9$, $t = 3$, giving

$9 = \frac{3^3}{3} - \frac{1}{3} + C$

Hence, $C = \frac{1}{3}$.

Therefore, $s = \frac{t^3}{3} - t^{-1} + \frac{1}{3}$

12.17 (a) $\left[\frac{x^3}{3} - \frac{3x^2}{2} + 5x\right]_2^4 = 10\frac{2}{3}$

(b) $\int_1^{16} x^{3/2}\,dx = \left[\frac{2}{5}x^{5/2}\right]_1^{16} = 409\frac{1}{5}$

(c) $\left[\frac{p^4}{2} + p^5\right]_0^1 = \frac{3}{2}$

12.18 $Q(t) = \int(100 + 10t - t^2)\,dt$

$= 100t + 5t^2 - \frac{t^3}{3} + C$

When $t = 0$, $Q = 0$. Hence, $C = 0$.

Therefore, $Q(t) = 100t + 5t^2 - \frac{t^3}{3}$

(a) When $t = 10$, $Q(t) = 1166\frac{2}{3}$ thousand barrels.

(b) $\int_{10}^{15}(100 + 10t - t^2)\,dt$

$= \left[100t + 5t^2 - \frac{t^3}{3}\right]_{10}^{15}$

$= 333\frac{1}{3}$ thousand barrels

12.19 $M(t) = \int_0^{10}(4 + 6t + t^2)\,dt$

$= \left[4t + 3t^2 + \frac{t^3}{3}\right]_0^{10}$

$= 673\frac{1}{3}$

∴ the total amount of service costs is \$673.33.

12.20 (a) $C = \int\left(10 + 2x^{-\frac{1}{2}}\right)dx$

$= 10x + 4x^{\frac{1}{2}} + C_1$

When $x = 900$, $C = \$11\,000$, giving

$11000 = 10(900) + 4(900)^{\frac{1}{2}} + C_1$

Hence, $C_1 = 1880$

Therefore, $C = 10x + 4\sqrt{x} + 1880$

(b) $C(400) = 10(400) + 4\sqrt{400} + 1880$

$= \$5\,960$

Chapter 13

13.1 $I = 15400 \times 0.12 \times \frac{5}{12} = \$770$

13.2 $t = \frac{I}{Pr} = \frac{315}{14000 \times 0.18}$

$= 0.125$ year or 1.5 months

13.3 $r = \frac{I}{Pt} = \frac{520}{5900 \times \frac{18}{12}} = 0.0588$ or 5.88%p.a.

13.4 Interest $= 2500 \times 0.06 \times \frac{4}{12} = \$50$

So, maturity value $= 2500 + 50 = \$2\,550$

13.5 $P = \frac{S}{(1+rt)} = \frac{1000}{\left(1 + 0.05 \times \frac{8}{12}\right)} = \$967.74$

13.6 Days $= 329 - 32 = 297$

So, $P = \frac{1820}{\left(1 + 0.115 \times \frac{297}{365}\right)} = \$1\,664.74$

13.7 Days $= 365 + (324 - 156) = 533$

So, $S = P(1+rt)$

$= 1500\left(1 + 0.10 \times \frac{533}{365}\right)$

$= \$1\,719.04$

13.8 (a) F **(b)** T
(c) T **(d)** F

13.9 $S = 850(1.04)^{12} = \$1\,360.88$

13.10 Balance $= 2000 - (2000 \times 0.2) = \$1\,600$

So, $S = 1600\,(1.01)^{24} = \$2\,031.58$

13.11 $P = S\,(1+i)^{-n} = 3500\left(1 + \dfrac{0.14}{4}\right)^{-16} = \$2\,018.47$

13.12 $n = \dfrac{\log\left(\dfrac{S}{P}\right)}{\log\,(1+i)} = \dfrac{\log\,(1.5)}{\log\,(1.025)} = 16.4205$

So, $t = \dfrac{16.4205}{4} = 4.105$ years .

13.13 Balance $= 8000 - 2000 = \$6\,000$.

$\therefore\ i = \left(\dfrac{S}{P}\right)^{\frac{1}{n}} - 1 = \left(\dfrac{6800}{6000}\right)^{\frac{1}{8}} - 1 = 0.01577$, and

$r = 0.01577 \times 4 \times 100 = 6.31\%$ p.a. quarterly.

13.14 $S = P(1+i)^{n} = 2000\left(1 + \dfrac{0.12}{4}\right)^{12} = \$2\,851.52$

13.15 $P = S(1+i)^{-n} = 3000\left(1 + \dfrac{0.13}{2}\right)^{-9} = \$1\,702.06$

13.16 **(a)** $n = \dfrac{\log\,(2)}{\log\,(1.015)} = 46.56$ months

So, $t = \dfrac{46.56}{12} = 3.88$ years .

(b) $n = \dfrac{\log\,(2)}{\log\,(1.045)} = 15.75$ 6-month periods

So, $t = 7.874$ years .

13.17 $R = \dfrac{S}{s_{n}\rceil_i} = \dfrac{5000}{s_{24}\rceil_{0.01}} = \dfrac{5000}{26.973465} = \$185.37$

13.18 **(a)** $S = R\,[\,s_{n+1}\rceil_i - 1\,] = 564\,[\,s_{17}\rceil_{0.035} - 1\,]$

$= 564\,(21.705016) = \$12\,241.63$

(b) $S = Rs_{n}\rceil_i = 564 s_{16}\rceil_{0.035}$

$= 564\,(20.971030) = \$11\,827.66$

13.19 **(a)** $R = \dfrac{S}{s_{n}\rceil_i} = \dfrac{8000}{s_{104}\rceil_{0.0025}}$

$= \dfrac{8000}{118.603742} = \$67.45$

(b) $R = \dfrac{S}{[\,s_{n+1}\rceil_i - 1\,]} = \dfrac{8000}{[\,s_{105}\rceil_{0.0025} - 1\,]}$

$= \dfrac{8000}{118.900251} = \$67.28$

13.20 **(a)** $A = R\,[1 + a_{n-1}\rceil_i]$

$= 1000\,[1 + a_{19}\rceil_{0.05}]$

$= 1000 \times 13.085321 = \$1\,308$

(b) $A = Ra_{n}\rceil_i = 1000 a_{20}\rceil_{0.05}$

$= 1000 \times 12.462210 = \$12\,462.21$

13.21 Balance $= 8500 - 1500 = \$7\,000$

(a) $R = \dfrac{A}{a_{n}\rceil_i} = \dfrac{7000}{a_{36}\rceil_{0.015}} = \dfrac{7000}{27.660684} = \$253.07$

(b) $R = \dfrac{A}{[1 + a_{n-1}\rceil_i]}$

$= \dfrac{7000}{[1 + a_{35}\rceil_{0.015}]} = \dfrac{7000}{28.075594} = \$249.33$

13.22 $s_{n}\rceil_i = \dfrac{S}{R}$. So, $s_{40}\rceil_i = 67.402556$,

giving $i \approx 0.025$.

Thus, $r \approx 0.025 \times 4 = 0.1$ or 10% p.a. quarterly.

13.23 $S = Rs_{n}\rceil_i = 500 s_{16}\rceil_{0.02375} = 500 \times 19.191709$

$= \$9\,595.85$

13.24 **(a)** $s_{n}\rceil_i = \dfrac{S}{R}$. So, $s_{n}\rceil_{0.06} = \dfrac{8000}{2000} = 4.0$,

giving $n \approx 4$.

Thus, $t \approx \dfrac{4}{2} = 2$ years .

(b) $s_{n+1}\rceil_i = \left(\dfrac{S}{R}\right) + 1$. So, $s_{n+1}\rceil_{0.06} = 5.0$,

giving $n + 1 \approx 4\frac{1}{2}$ or $n \approx 3\frac{1}{2}$.

Thus, $t = \dfrac{3.5}{2} = 1.7$ years or 21 months.

13.25 $a_{n-1}\rceil_i = \left(\dfrac{A}{R}\right) - 1$.

So, $a_{29}\rceil_i = \left(\dfrac{6000}{230}\right) - 1 = 25.086957$, giving

$i \approx 0.01$.

Thus, $r \approx 0.01 \times 12 = 12\%$ p.a. monthly.

Chapter 14

14.1 **(a)** $\dfrac{43\pi}{36}$ **(b)** $\dfrac{5\pi}{3}$

(c) $\dfrac{\pi}{18}$ **(d)** $\dfrac{\pi}{5}$

(e) 3π

14.2 **(a)** $720°$ **(b)** $630°$

(c) $480°$ **(d)** $85°\,56'$

(e) $8 \times \dfrac{180}{\pi} = 458°\,22'$

14.3 $l = r\theta = 36 \times \dfrac{\pi}{6} = 18.8$ cm

14.4 $\theta = \dfrac{l}{r} = \dfrac{10}{100} = \left(\dfrac{1}{10}\right)^{c}$

14.5 $A = \dfrac{1}{2}r^{2}\theta = 14.97$ sq. cm

14.6 (a) $\dfrac{dy}{dx} = 9\cos 9x$

(b) $\dfrac{dy}{dx} = -7\operatorname{cosec} 7x \cdot \cot 7x$

(c) $\dfrac{dy}{dx} = -\operatorname{cosec}^2 x$

(d) $\dfrac{dy}{dx} = \sec x \cdot \tan x$

(e) $\dfrac{dy}{dx} = \dfrac{-3\operatorname{cosec}^2 3x}{\cot 3x}$

(f) $\dfrac{dy}{dx} = \dfrac{-3\operatorname{cosec}^2 (\ln 3x)}{3x}$

$\qquad = \dfrac{-\operatorname{cosec}^2 (\ln 3x)}{x}$

14.7 $f'(x) = 3\sec^2 x + 2\operatorname{cosec}^2 x$

$\therefore f''\left(\dfrac{\pi}{4}\right) = 3\left(\sqrt{2}\right)^2 + 2\left(\sqrt{2}\right)^2 = 10$

14.8 $\dfrac{d}{dx}(\ln \sec x) = \dfrac{1}{\sec x} \cdot \sec x \cdot \tan x = \tan x$

$\dfrac{d}{dx}(-\ln \cos x) = -\dfrac{1}{\cos x} \cdot -\sin x = \tan x$

14.9 $\dfrac{dy}{dx} = \sec^2 x + \operatorname{cosec}^2 x - 3$

$\qquad = (1 + \tan^2 x) + (1 + \cot^2 x) - 3$

$\qquad = \tan^2 x - 2\tan x \cot x + \cot^2 x + 2 - 1$

$\qquad = (\tan x - \cot x)^2 + 1$

14.10 $\dfrac{d}{dx}e^{\operatorname{cosec}^2 x} = e^{\operatorname{cosec}^2 x} \cdot 2\operatorname{cosec} x \cdot -\operatorname{cosec} x \cdot \cot x$

$\qquad = -2e^{\operatorname{cosec}^2 x} \cdot \operatorname{cosec}^2 x \cdot \cot x$

$\therefore \displaystyle\int_{\frac{\pi}{4}}^{\frac{\pi}{2}} e^{\operatorname{cosec}^2 x} \cdot \operatorname{cosec}^2 x \cdot \cot x \, dx$

$\qquad = \left[\dfrac{e^{\operatorname{cosec}^2 x}}{-2}\right]_{\frac{\pi}{4}}^{\frac{\pi}{2}}$

$\qquad = \dfrac{-1}{2}\left[e^{\operatorname{cosec}^2\left(\frac{\pi}{2}\right)} - e^{\operatorname{cosec}^2\left(\frac{\pi}{4}\right)}\right]$

$\qquad = -\dfrac{1}{2}[e^1 - e^2]$

14.11 (a) $-\dfrac{\cos 7x}{7} + C$ \qquad (b) $6\sin \dfrac{x}{6} + C$

(c) $\dfrac{\sec 9x}{9} + C$ \qquad (d) $\dfrac{1}{5}\tan 5x + C$

(e) $-2\operatorname{cosec} \dfrac{x}{2} + C$

14.12 $\dfrac{3 - \sin^3 x}{\sin^2 x} = \dfrac{3}{\sin^2 x} - \dfrac{\sin^3 x}{\sin^2 x} = 3\operatorname{cosec}^2 x - \sin x$

$\therefore \displaystyle\int_0^{\frac{\pi}{4}} \dfrac{3 - \sin^3 x}{\sin^2 x} \, dx = \int_0^{\frac{\pi}{4}} (3\operatorname{cosec}^2 x - \sin x)\, dx$

$\qquad = [-3\cot x + \cos x]_0^{\frac{\pi}{4}},$

$\qquad$ undefined at $x = 0$.

14.13 $\dfrac{d}{dx}\ln(\tan x + 1) = \dfrac{\sec^2 x}{\tan x + 1}$

$\therefore \displaystyle\int \dfrac{\sec^2 x}{\tan x + 1}\, dx = \ln(\tan x + 1) + C$

14.14 $\dfrac{d}{dx}\sec^4 x = 4\sec^3 x \cdot (\sec x \cdot \tan x)$

$\qquad = 4\sec^4 x \cdot \tan x$

$\therefore \displaystyle\int \sec^4 x \cdot \tan x \, dx = \dfrac{1}{4}\sec^4 x + C$

14.15 $\dfrac{d}{dx}e^{\cot 3x} = (-3)\operatorname{cosec}^2 3x \cdot e^{\cot 3x}$

$\therefore \displaystyle\int_0^{\frac{\pi}{12}} e^{\cot 3x} \cdot \operatorname{cosec}^2 3x \, dx = \left[-\dfrac{1}{3}e^{\cot 3x}\right]_0^{\frac{\pi}{12}},$

$\qquad$ undefined at $x = 0$.

Appendix D

Mathematical notation

A. Miscellaneous symbols

=	is equal to
≠	is not equal to
≈	is approximately equal to
>	is greater than
<	is less than
≥	is greater than or equal to
≤	is less than or equal to
%	percent
∠	angle
Δ	triangle
mm	millimetre
cm	centimetre
dm	decimetre
m	metre
km	kilometre
g	gram
kg	kilogram
ml	millilitre
L	litre
kL	kilolitre

B. Arithmetic operatioµns

$a + b$	a plus b
$a - b$	a minus b
$a \times b$, ab, $a \cdot b$	a multiplied by b
$a \div b$, $\frac{a}{b}$, a/b	a divided by b
$\sum\limits_{i=1}^{n} a_i$	$a_1 + a_2 + a_3 + ... + a_n$

C. Functions

$f(x)$	the value of the function f at x
$\dfrac{dy}{dx}$	the derivative of y with respect to x
$\dfrac{d^n y}{dx^n}$	the nth derivative of y with respect to x
$f'(x)$, $f''(x)$, ..., $f^{(n)}(x)$	the first, second, ..., n^{th} derivatives of $f(x)$
$\int y \, dx$	the indefinite integral of y with respect to x
$\int_a^b y \, dx$	the definite integral, with limits a and b
$[F(x)]_a^b$	$F(b) - F(a)$
e^x or $\exp x$	the exponential function
$\log_a x$	logarithm of x to the base a
$\ln x$	$\log_e x$

D. Units of measurement

(i) Length

$1 \text{ cm} = 10 \text{ mm}$
$1 \text{ dm} = 10 \text{ cm}$
$\phantom{1 \text{ dm}} = 100 \text{ mm}$
$1 \text{ m} = 10 \text{ dm}$
$\phantom{1 \text{ m}} = 100 \text{ cm}$
$\phantom{1 \text{ m}} = 1000 \text{ mm}$
$1 \text{ km} = 1000 \text{ m}$

(ii) Area

$1 \text{ cm}^2 = 100 \text{ mm}^2$
$1 \text{ m}^2 = 10000 \text{ cm}^2$
$1 \text{ km}^2 = 1000000 \text{ m}^2$

(iii) Volume

$1 \text{ cm}^3 = 1000 \text{ mm}^3$
$1 \text{ dm}^3 = 1000 \text{ cm}^3$
$\phantom{1 \text{ dm}^3} = 1000000 \text{ mm}^3$
$1 \text{ m}^3 = 1000 \text{ dm}^3$
$\phantom{1 \text{ m}^3} = 1000000 \text{ cm}^3$

(iv) Mass

1000 mg = 1 g

1000 g = 1 kg

1000 kg = 1 tonne

(v) Capacity

1 L = 1000 ml

1 kL = 1000 L

A 1 cm cube has a capacity of 1 ml

A 10 cm cube has a capacity of 1 L

A 1 m cube has a capacity of 1 kL

(vi) Time

60 seconds = 1 minute

60 minutes = 1 hour

24 hours = 1 day

7 days = 1 week

365 days = 1 year

366 days = 1 leap year

12 months = 1 year

52 weeks = 1 year

10 years = 1 decade

100 years = 1 century

E. Metric prefixes

Prefix	Abbreviation	Factor
mega	M	10^6
kilo	k	10^3
hecto	h	10^2
deca	da	10^1
deci	d	10^{-1}
centi	c	10^{-2}
milli	m	10^{-3}
micro	μ	10^{-6}
nano	n	10^{-9}
pico	p	10^{-12}

F. Approximate conversions (to 3 significant figures)

Imperial	Metric
1 inch (in)	2.54 cm
1 foot (ft)	0.305 m
1 yard (yd)	0.914 m
1 mile	1.61 km
1 acre	0.405 hectare (ha)
1 pound	0.454 kg
1 ton	1.016 tonnes
1 pint	0.568 litre
1 gallon	4.55 litres

Metric	Imperial
1 cm	0.394 in (inch)
1 m	3.28 ft (feet)
1 m	1.09 yd (yards)
1 km	0.621 mile
1 ha	2.47 acres
1 kg	2.20 lbs (pounds)
1 tonne	0.984 ton
1 litre	1.76 pints
1 litre	0.220 gallon

G. Useful constants

π (pi) = 3.14159

e = 2.71828

H. The Greek alphabet

Name	Upper case	Lower case	Name	Upper case	Lower case
Alpha	A	α	Nu	N	ν
Beta	B	β	Xi	Ξ	ξ
Gamma	Γ	γ	Omicron	O	o
Delta	Δ	δ	Pi	Π	π
Epsilon	E	ε	Rho	P	ρ
Zeta	Z	ζ	Sigma	Σ	σ
Eta	H	η	Tau	T	τ
Theta	Θ	θ	Upsilon	Y	υ
Iota	I	ι	Phi	Φ	φ
Kappa	K	κ	Chi	X	χ
Lambda	Λ	λ	Psi	Ψ	ψ
Mu	M	μ	Omega	Ω	ω

Appendix E

Can you use your calculator?

In order to work successfully through many of the chapters in this book, you must be able to use a calculator efficiently and with confidence.

Although most readers will already own and be able to work with the basic operating keys of a calculator, there will be some of you who lack both confidence and expertise. For this reason we will spend some time here discussing all the '**general keys**' and some of the '**special keys**' on a typical calculator.

Notes:

1. In all chapters, when calculators are utilised to solve given examples more easily and efficiently, the calculator keys associated with each function, operation or numeral are shown. However, as calculators vary considerably both in the location and in the accessing of a number of functions and operations, you may find that, on your calculator, a particular function, designated [fn] say, may require more than one key press to access. Hence it is very important to keep your instruction booklet handy and, especially for those of you with little calculator experience, to read this section on calculator usage very carefully.

2. Not all the calculator keys will be discussed here as some are not applicable to the material covered in the text. Furthermore, in some of the more specialised cases – *for example*, statistical calculations – the use of the calculator to solve specific problems will be discussed in detail in the particular chapter itself.

As you will see as you work through the chapters, some keys will be used frequently, others infrequently and some, as we have already stated, not at all.

Important: Here and throughout this text we will use the **Casio** *fx-82super* scientific calculator. Although most calculators perform the same basic operations, the location of some keys and the accessing of some functions may vary from brand to brand and from model to model. Therefore, if your calculator differs from the **Casio** discussed here, it is **very important** that you keep your Instruction booklet close by so that you can check quickly and easily the keys on your calculator that correspond to those referred to here or in the relevant chapters, and (more importantly) how to access them.

The following diagram shows the face of a **Casio** *fx-82super* scientific calculator. The table on the next page displays some of the more commonly used keys (numbered (i) to (viii) on the diagram) and their associated purpose.

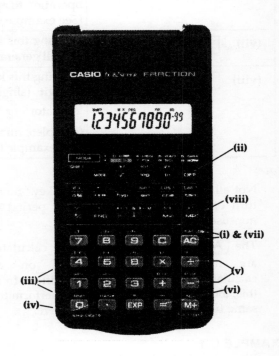

Table 1: General keys – used frequently

	Key	Purpose
(i)	[SAC] ON \n AC	Pressing this key turns the calculator on.
(ii)	OFF	Pressing this key turns the calculator off.
(iii)	7 8 9 \n 4 5 6 \n 1 2 3 \n 0	numerals: 0 to 9 \n Hence, pressing 6 2 9 \n shows 629. on the display.
(iv)	.	Decimal point key. \n Hence, pressing 2 . 0 6 \n shows 2.06 on the display.
(v)	+ − \n × ÷	These are the 4 basic arithmetic operator keys. \n Pressing + indicates addition. \n Pressing − indicates subtraction. \n Pressing × indicates multiplication. \n Pressing ÷ indicates division. \n (See example C.1 below.)
(vi)	=	Pressing the 'equals' key = , allows the *result* of an arithmetic operation to appear on the display. \n (See example C.1 below.)
(vii)	AC	Pressing this key clears the current calculation and sets the display and all general computation memories to zero.
(viii)	C	Pressing this key clears the *effect* of the last key pressed. If you input a digit (single number) incorrectly but have not pressed the 'operator' or 'equal' key, then pressing C clears the last value (complete number). Now input the correct value. \n (See example C.2 below.)

Note:

a. Not all calculators have an OFF key. Some turn off automatically after a set period of non-use.

b. The AC and C keys on **our** calculator are not necessarily standard on all calculators. If your calculator differs from this, locate the keys on yours that perform the same function as those indicated here.

EXAMPLE C.1
Use the numerals, basic operator and equals keys to calculate the value of the given expressions.
The following table shows the expression to be simplified in column 1, the appropriate key presses in column 2 and the final result on the display in column 3.
Remember: All key presses indicated here are specific to our **Casio** *fx-82super*. Your own calculator may or may not be the same, so check carefully to avoid confusion.
Note: In general , both here and in the body of the text, all required operator or function keys will be shown in a special box, ☐ , whereas all numerals to be entered will stand alone (unboxed).

Expression	Key presses	Display output
(a) 45 + 6.4	45 `+` 6.4 `=`	`51.4`
(b) 3.2 - 1.7	3.2 `−` 1.7 `=`	`1.5`
(c) 36 + 0.5 - 3.9	36 `+` .5 `−` 3.9 `=`	`32.6`
(d) 15.1 ÷ 20.3	15.1 `÷` 20.3 `=`	`0.743842364`
(e) 3 x 68 ÷ 4.2 + 1	3 `x` 68 `÷` 4.2 `+` 1 `=`	`49.57142857`

EXAMPLE C.2

In each of the following cases an input error is made and then corrected.
The table below shows the expression to be simplified in column 1, the appropriate key presses in column 2 and both interim and final results on this display in column 3.

Expression	Key presses	Display output
(a) 6.2 x 4	6.3 (Whoops!) `C` 6.2 `x` 4 `=`	`0.` `24.8`
(b) 24 ÷ 0.3	24 `÷` .6 (Whoops!) `C` .3 `=`	`0.` `80.`

(c) 56 − 42	56 `−` 45 (Whoops!) `C`
	2 `=` (No!)
	42 `=` (Yes!)
		`0.`
		`54.`
		`14.`

(d) 8.3 + 6	8.3 `x` (Whoops!) `+` 6 `=`	`8.3` `14.3`

Note:

In (c) above, an error was made during the input of 42. The final digit was keyed in as 5 instead of 2. In this case, pressing `C` clears the *whole* number and not just the incorrect digit.

In (d) above, an error was made during the input of the arithmetic operator. In this case the `C` key is unnecessary (although not incorrect). All that is required is to follow the *incorrect* operator key by pressing the *correct* one. This overrides the error made.

The diagram below and the table following display those keys that are used less frequently and often have a more specialised purpose.

(x) —
(ix) —
(xiii) —
(xi) —

(xii)
(xiv)
(xv)
(xvi)

Table 2: Special keys – used less frequently

Key	Purpose
(ix) $\boxed{\text{shift}}$	Pressing this key allows access to the function written (in brown) *above* a key. (See example C.3 below.)
(x) $\boxed{\text{mode}}$	Pressing this key followed by the appropriate number key allows you to select the mode relevant to your calculation requirements. e.g. $\boxed{\text{mode}}$ $\boxed{\cdot}$: SD appears on the display. This mode allows statistical calculations to be made.
(xi) $\boxed{+/-}$	Pressing this key *changes the sign* of the number on the display. e.g. 62 display $\boxed{62.}$ $\boxed{+/-}$ display $\boxed{-62.}$
(xii) $\boxed{\sqrt{\ }}$	Pressing this key takes the *square root* of the number on the display. e.g. display $\boxed{6400.}$ $\boxed{\sqrt{\ }}$ display $\boxed{80.}$ (See *note* below.)
(xiii) $\boxed{a^b_c}$	This key allows you to work with *fractions*, e.g. $\frac{1}{2}$ $\boxed{+}$ $\frac{3}{4}$. The answer on the display will be in fraction form. Pressing $\boxed{a^b_c}$ *after* the $\boxed{=}$ key converts the fraction on the display into its **decimal** form. (See *note* and example C.4 below.)
(xiv) $\boxed{\circ\,'\,''}$	This key is useful when measuring angles. Pressing this key will convert an angle measured in degrees, minutes and seconds (called **sexagesimal** notation) into decimal notation. This key will be useful in the chapters on geometry and trigonometry.
(xv) $\boxed{[(---}$ $\boxed{---)]}$	These keys indicate where *parentheses* (brackets) are required. Any arithmetic operations enclosed within parentheses are *always* performed first.

Key	Purpose
(xvi) Accessing x^2, x^y, $\frac{1}{x}$. 1. x^2 : [shift] [$\sqrt{\ }$] 2. x^y : [shift] [x] 3. $\frac{1}{x}$: [shift] [Min]	On **our** calculator, x^2, x^y and $\frac{1}{x}$ are all 'brown' functions located *above* a key, and are accessed by pressing [shift] followed by the appropriate key. (See *note* and example C.3 below.)

Note:

Remember that the keys shown in the above table are specific to our **Casio** *fx-82super* shown in the diagram. If your calculator is a different make or model then you must refer to your instruction booklet to find the location and name of the key on **your** calculator that performs the same function as any particular key given here.

For example:

Casio *fx-82super*	Others (e.g. only)
1. [shift]	[inv] ; [2nd F]

2. [mode] [.] SD
           [0] COMP
           [5] RAD etc.
	[mode] [?] different numbers for the different modes.

3. To access:	Location of:

(a) x^2 : [shift] [$\sqrt{\ }$]
  (b) x^y : [shift] [x]
  (c) $\frac{1}{x}$: [shift] [Min]
  (d) $\sqrt{\ }$: [$\sqrt{\ }$]
	x^2 : [x^2] on a key
	x^y : above [x^2] ; [a^x]
	$\frac{1}{x}$: [$^1/x$] (on a key) ; above [x^2]
	$\sqrt{\ }$: above [x^2] ; above [ ]

4. [$a\frac{b}{c}$] fraction key	Many brands do not have this feature.

EXAMPLE C.3: Using the [shift] key

Find   **(a)** $(8.2)^2$ **(b)** $(3.5)^{1.2}$ **(c)** $\frac{1}{3.82}$ **(d)** $(-2.5)^3$

       (e) $(5)^{-3}$ **(f)** $\frac{1}{5^3}$

Solution

(a) 8.2 [shift] [$\sqrt{\ }$]   67.24

(b) 3.5 [shift] [x] 1.2 [=]   4.49657305

(c) 3.82 [shift] [Min] 0.261780104

(d) 2.5 [$^+/_-$] [shift] [x] 3 [=]   –15.625

(e) 5 [shift] [x] 3 [$^+/_-$] [=]   0.008

(f) 5 [shift] [x] 3 [=] [shift] [Min]   0.008

Note:

Any calculation involving the use of x^y function **must** be followed by the $\boxed{=}$ key (see (b), (d), (e) and (f) above).

EXAMPLE C.4 Using the $\boxed{a^b_c}$ fraction key

(a) With your calculator, find

(i) $\dfrac{1}{2} + \dfrac{3}{4}$

(ii) $1\dfrac{2}{3} \times 5$

(iii) $\dfrac{5}{8} \div \dfrac{1}{2}$.

Leave each answer in fraction form.

(b) With your calculator, find

(i) $\dfrac{2}{3} + \dfrac{1}{8}$

(ii) $2 \times 1\dfrac{7}{8}$

(iii) $3\dfrac{3}{4} - 1\dfrac{1}{2}$.

Express each answer as a decimal.

Solution

(a) (i) $1\;\boxed{a^b_c}\;2\;\boxed{+}\;3\;\boxed{a^b_c}\;4\;\boxed{=}\;\boxed{1\lrcorner1\lrcorner4.}$

(ii) $1\;\boxed{a^b_c}\;2\;\boxed{a^b_c}\;3\;\boxed{\times}\;5\;\boxed{=}\;\boxed{8\lrcorner1\lrcorner3.}$

(iii) $5\;\boxed{a^b_c}\;8\;\boxed{\div}\;1\;\boxed{a^b_c}\;2\;\boxed{=}\;\boxed{1\lrcorner1\lrcorner4.}$

(b) (i) $2\;\boxed{a^b_c}\;3\;\boxed{+}\;1\;\boxed{a^b_c}\;8$
$\boxed{=}\;\boxed{a^b_c}\;\boxed{0.791666666}$

(ii) $2\;\boxed{\times}\;1\;\boxed{a^b_c}\;7\;\boxed{a^b_c}\;8\;\boxed{=}\;\boxed{a^b_c}\;\boxed{3.75}$

(iii) $3\;\boxed{a^b_c}\;3\;\boxed{a^b_c}\;4\;\boxed{-}\;1\;\boxed{a^b_c}\;1\;\boxed{a^b_c}\;2$
$\boxed{=}\;\boxed{a^b_c}\;\boxed{2.25}$

Note: In (b) above, pressing $\boxed{a^b_c}$ after the $\boxed{=}$ key converts the *fraction* answer on the display to its equivalent *decimal* form.

Specialised operator/function keys

The following keys are used only when

(a) logarithms, or

(b) trigonometric or inverse trigonometric functions are involved.

10^x e^x $\sin^{-1}$ $\cos^{-1}$ $\tan^{-1}$ ← brown functions above actual keys

$\boxed{\log}$ $\boxed{\ln}$ $\boxed{\sin}$ $\boxed{\cos}$ $\boxed{\tan}$

Note: These keys are specific to our **Casio fx-82super** calculator, with those functions written in brown *above* a key being accessed using the $\boxed{\text{shift}}$ key first. Locate the equivalent operator/function keys on your own calculator and be sure that you know how to access them correctly.

EXAMPLE C.5

Evaluate **(a)** $\dfrac{\log 64}{\log 4}$

(b) $\log 45 \times \ln 15$

(c) $10^{0.4} + 5e^{-3}$

(d) $2\sin 45° \times \cos 65°$

(e) $\tan^{-1} 0.6104$ (in ° ' ")

Solution

(a) $64\;\boxed{\log}\;\boxed{\div}\;4\;\boxed{\log}\;\boxed{=}\;\boxed{3.}$

(b) $45\;\boxed{\log}\;\boxed{\times}\;15\;\boxed{\ln}\;\boxed{=}\;\boxed{4.47698248}$

(c) $.4\;\boxed{\text{shift}}\;\boxed{\log}\;\boxed{+}\;5\;\boxed{\times}\;3\;\boxed{+/-}\;\boxed{\text{shift}}\;\boxed{\ln}$
$\boxed{=}\;\boxed{2.760821773}$

(d) (Check that your calculator is in 'DEG' (degree) mode. If not, press $\boxed{\text{Mode}}\;\boxed{4}$ on our **Casio fx-82super.**)
Then,
$2\;\boxed{\times}\;45\;\boxed{\sin}\;\boxed{\times}\;65\;\boxed{\cos}\;\boxed{=}$
$\boxed{0.597672477}$

(e) (in 'DEG' mode)
$.6104\;\boxed{\text{shift}}\;\boxed{\tan}\;\boxed{\text{shift}}\;\boxed{° ' "}$
$\boxed{31° 23° 59.61}$

Note: Pressing $\boxed{\text{shift}}\;\boxed{° ' "}$ in (e) converts the decimal answer (31.39989118) into degrees, minutes and seconds (31° 23' 59.61").

Important comment

Although we have mentioned frequently that different brands and models of scientific calculators will locate and label keys and access functions in several alternative ways, we have concentrated here on presenting the actual sequence of keys to be pressed to evaluate an expression, as they are located and accessed on our **Casio** *fx-82super.*

However, in the body of the text itself, we will overcome the problems associated with different calculators by refraining from indicating the use of the *shift* key. We will simply show the required function on a calculator key and it will be up to the individual reader to locate and access that function on his/her own calculator.

For example:

Expression	Keys to be pressed	Display output
(i) 6^2	6 $\boxed{x^2}$	$\boxed{36}$
(ii) $3^{-1.2}$	3 $\boxed{x^y}$ 1.2 $\boxed{+/-}$ $\boxed{=}$	$\boxed{0.26758052}$
(iii) $e^{0.8}$	.8 $\boxed{e^x}$	$\boxed{2.225540928}$
(iv) $\dfrac{1}{3^4}$	3 $\boxed{x^y}$ 4 $\boxed{=}$ $\boxed{1/x}$	$\boxed{0.012345679}$
(v) $\sin^{-1} 0.3$	.3 $\boxed{\sin^{-1}}$	$\boxed{17.45760312}$
(vi) $10^{0.5}$	.5 $\boxed{10^x}$	$\boxed{3.16227766}$

Finally, not all the calculator keys and associated functions have been mentioned here. Some have been left until the particular chapter where they will be used (e.g. the statistical functions), while others will remain 'untouched' as they are beyond the scope of this book.